Communications
in Computer and Information Science   1929

Rationale

The CCIS series is devoted to the publication of proceedings of computer science conferences. Its aim is to efficiently disseminate original research results in informatics in printed and electronic form. While the focus is on publication of peer-reviewed full papers presenting mature work, inclusion of reviewed short papers reporting on work in progress is welcome, too. Besides globally relevant meetings with internationally representative program committees guaranteeing a strict peer-reviewing and paper selection process, conferences run by societies or of high regional or national relevance are also considered for publication.

Topics

The topical scope of CCIS spans the entire spectrum of informatics ranging from foundational topics in the theory of computing to information and communications science and technology and a broad variety of interdisciplinary application fields.

Information for Volume Editors and Authors

Publication in CCIS is free of charge. No royalties are paid, however, we offer registered conference participants temporary free access to the online version of the conference proceedings on SpringerLink (http://link.springer.com) by means of an http referrer from the conference website and/or a number of complimentary printed copies, as specified in the official acceptance email of the event.

CCIS proceedings can be published in time for distribution at conferences or as postproceedings, and delivered in the form of printed books and/or electronically as USBs and/or e-content licenses for accessing proceedings at SpringerLink. Furthermore, CCIS proceedings are included in the CCIS electronic book series hosted in the SpringerLink digital library at http://link.springer.com/bookseries/7899. Conferences publishing in CCIS are allowed to use Online Conference Service (OCS) for managing the whole proceedings lifecycle (from submission and reviewing to preparing for publication) free of charge.

Publication process

The language of publication is exclusively English. Authors publishing in CCIS have to sign the Springer CCIS copyright transfer form, however, they are free to use their material published in CCIS for substantially changed, more elaborate subsequent publications elsewhere. For the preparation of the camera-ready papers/files, authors have to strictly adhere to the Springer CCIS Authors' Instructions and are strongly encouraged to use the CCIS LaTeX style files or templates.

Abstracting/Indexing

CCIS is abstracted/indexed in DBLP, Google Scholar, EI-Compendex, Mathematical Reviews, SCImago, Scopus. CCIS volumes are also submitted for the inclusion in ISI Proceedings.

How to start

To start the evaluation of your proposal for inclusion in the CCIS series, please send an e-mail to ccis@springer.com.

Rama Krishna Challa · Gagangeet Singh Aujla ·
Lini Mathew · Amod Kumar · Mala Kalra ·
S. L. Shimi · Garima Saini · Kanika Sharma
Editors

Artificial Intelligence of Things

First International Conference, ICAIoT 2023
Chandigarh, India, March 30–31, 2023
Revised Selected Papers, Part I

 Springer

Editors
Rama Krishna Challa ⓘ
National Institute of Technical Teachers
Training and Research
Chandigarh, India

Lini Mathew ⓘ
National Institute of Technical Teachers
Training and Research
Chandigarh, India

Mala Kalra ⓘ
National Institute of Technical Teachers
Training and Research
Chandigarh, India

Garima Saini ⓘ
National Institute of Technical Teachers
Training and Research
Chandigarh, India

Gagangeet Singh Aujla ⓘ
Durham University
Durham, UK

Amod Kumar ⓘ
National Institute of Technical Teachers
Training and Research
Chandigarh, India

S. L. Shimi ⓘ
National Institute of Technical Teachers
Training and Research
Chandigarh, India

Punjab Engineering College (Deemed to be
University)
Chandigarh, India

Kanika Sharma ⓘ
National Institute of Technical Teachers
Training and Research
Chandigarh, India

ISSN 1865-0929 ISSN 1865-0937 (electronic)
Communications in Computer and Information Science
ISBN 978-3-031-48773-6 ISBN 978-3-031-48774-3 (eBook)
https://doi.org/10.1007/978-3-031-48774-3

This Springer imprint is published by the registered company Springer Nature Switzerland AG
The registered company address is: Gewerbestrasse 11, 6330 Cham, Switzerland

Paper in this product is recyclable.

Preface

It is a great privilege for us to present the proceedings of ICAIoT-2023 which contain the papers submitted by researchers, practitioners, and educators to the International Conference on Artificial Intelligence of Things, ICAIoT 2023, held during 30th and 31st March, 2023 at National Institute of Technical Teachers Training and Research, Chandigarh, India. We hope that you will find this book educative and inspiring.

The conference attracted papers from international researchers and scholars in the field of Artificial Intelligence (AI) applications in Internet of Things (IoT). The ICAIoT 2023 conference received 401 papers from around the world out of which 60 were accepted for oral presentation. The total number of papers presented at ICAIoT 2023 was 57. All submitted papers were subjected to strict Single Blind peer review by at least three national and international reviewers who are experts in the area of the particular paper. The acceptance rate of the conference was 15%.

The aim of ICAIoT 2023 was to provide a professional platform for discussing research issues, opportunities and challenges of AI and IoT applications. ICAIoT 2023 received unexpected support and enthusiasm from the delegates. The papers presented at this premier international gathering of leading AI researchers and practitioners from all over the world are collected in two volumes which will connect advances in engineering and technology with the use of smart techniques including Artificial Intelligence (AI), Machine Learning and Internet of Things (IoT).

The book presents the most recent innovations, trends and concerns as well as practical challenges encountered and solutions adopted in the fields of AI algorithms implementation in IoT Systems. The book is divided into two volumes, covering the following topics:

Volume I:

- AI and IoT Enabling Technologies
- AI and IoT for Smart Healthcare

Volume II:

- AI and IoT for Electrical, Electronics and Communication Engineering
- AI and IoT for other Engineering Applications

This book will be especially useful for graduate students, academic researchers, scientists and professionals in the fields of Computer Science and Engineering, Electrical Engineering, Electronics and Communication Engineering and allied disciplines.

Organizing a prestigious conference such as ICAIoT 2023 with the Springer CCIS series as publication partner is a substantial endeavour. We would like to extend our sincere thanks to the organizing committee members for all their support with the conference organization. We would like to thank the authors of all submitted papers for sending high-quality contributions to ICAIoT 2023. We would like to express our sincere gratitude to our sponsors, the advisory committee members, technical program committee members and Ph.D./M.E. scholars for all their support throughout the conference. Our special thanks are due to all the external reviewers who contributed their expertise, insights, and judgment during the review process. Last but not least, we extend our heartfelt thanks to the staff of Springer for their extensive support in the publication of this book under the Springer CCIS Series.

March 2023   Rama Krishna Challa
                                          Gagangeet Singh Aujla
                                                 Lini Mathew
                                                 Amod Kumar
                                                 Mala Kalra
                                                 S. L. Shimi
                                                 Garima Saini
                                                 Kanika Sharma

Message from the Patron

I am happy to learn that the Department of Computer Science and Engineering, Department of Electronics and Communication Engineering and Department of Electrical Engineering of National Institute of Technical Teachers Training & Research, Chandigarh has taken a timely initiative to address the emerging issues on Artificial Intelligence and Internet of Things by organizing the International Conference on Artificial Intelligence of Things (ICAIoT 2023) from March 30–31, 2023.

The organizing Committee has invited many speakers of international and national repute to make the two days' deliberations more meaningful and academically rich. I extend a warm welcome to all the dignitaries, keynote speakers, paper presenters and delegates to the conference. I am sure the conference will provide a platform to researchers, professionals, educators and students to share innovative ideas, issues, recent trends and future directions in the fields of AI, IoT and Industry 4.0 to address industrial and societal needs.

I wish the conference a grand success and all the participants a pleasant stay and great learning.

Bhola Ram Gurjar

Message from the Conference Chairs

We were delighted to welcome delegates from India and abroad to the International Conference on Artificial Intelligence of Things (ICAIoT 2023), organized by Department of Computer Science and Engineering, Department of Electronics and Communication Engineering and Department of Electrical Engineering of the National Institute of Technical Teachers Training & Research (NITTTR), Chandigarh on March 30–31, 2023.

Undoubtedly, AI and IoT technologies have transformed our society in recent times and the pace of change can only be described as revolutionary. The technology is progressing fast and new horizons are being explored. The conference aimed to provide an opportunity to researchers, engineers, academicians as well as industrial professionals from all over the world to present their research work and related development activities. This conference also provided a platform for the delegates to exchange new ideas and application experiences face to face, to establish research relations and to find global partners for future collaboration.

It is rightly said "Alone we can do so little, together we can do so much". We are thankful to all the contributors of this conference who have worked hard in planning and organizing both the academic activities and necessary social arrangements. In particular, we take this opportunity to express our gratitude to the Conference Patron and members of the Advisory Committee for their wise advice and brilliant suggestions in organizing the Conference. Also, we would like to thank the Technical Committee and Reviewers for their thorough and timely reviews of the research papers and our Ph.D./M.E. scholars for their support. We would also like to thank all the sponsors who have supported us in organizing this conference.

We hope all attendees enjoyed this event.

Rama Krishna Challa
Gagangeet Singh Aujla
Lini Mathew
Amod Kumar

Message from the Conference Co-chairs

We were pleased to extend our most sincere welcome to all the delegates to the International Conference on Artificial Intelligence of Things (ICAIoT 2023), being organized by three departments, namely, Department of Computer Science and Engineering, Department of Electronics and Communication Engineering and Department of Electrical Engineering of the National Institute of Technical Teachers Training & Research (NITTTR), Chandigarh during March 30–31, 2023 in association with DSIR, Govt. of India.

Artificial Intelligence plays an enormous role in promoting knowledge and technology which is essential for the educators, researchers, industrial and commercial houses in the present digital age. Being a core part of this conference from the beginning, we feel very enthusiastic about the event and hope that we all will benefit academically through mutual collaboration. This International Conference will provide exposure to recent advancements and innovations in the field of Internet of Things (IoT), Artificial Intelligence (AI) and Industry 4.0. It is expected to be an intellectual platform to share ideas and present the latest findings and experiences in the mentioned areas.

The successful organization of ICAIoT 2023 required the talent, dedication and time of many volunteers and strong support from sponsors. We express our sincere thanks to the paper reviewers, keynote speakers, invited speakers and authors. A special mention about our Ph.D./M.E. scholars who worked hard to make this International Conference a success, would be in order.

We hope all attendees found the event to be a grand success.

Mala Kalra
Garima Saini
S. L. Shimi
Kanika Sharma

Organization

Conference Patron

Gurjar, Bhola Ram — National Institute of Technical Teachers Training and Research, Chandigarh, India

Conference Chairs

Challa, Rama Krishna — National Institute of Technical Teachers Training and Research, Chandigarh, India

Aujla, Gagangeet Singh — Durham University, UK

Mathew, Lini — National Institute of Technical Teachers Training and Research, Chandigarh, India

Kumar, Amod — National Institute of Technical Teachers Training and Research, Chandigarh, India

Conference Co-chairs

Kalra, Mala — National Institute of Technical Teachers Training and Research, Chandigarh, India

Shimi, S. L. — National Institute of Technical Teachers Training and Research, Chandigarh, India and Punjab Engineering College (Deemed to be University), Chandigarh, India

Saini, Garima — National Institute of Technical Teachers Training and Research, Chandigarh, India

Sharma, Kanika — National Institute of Technical Teachers Training and Research, Chandigarh, India

Advisory Committee

Ahuja, Rajeev — Indian Institute of Technology (IIT), Ropar, India

Awasthi, Lalit K. — National Institute of Technology (NIT), Uttarakhand, India

Babu, K. M. — Administrative Management College Institutions, Bengaluru, India

Bollen, Math	Luleå University of Technology, Sweden
Chakrabarti, Saswat	Indian Institute of Technology (IIT), Kharagpur, India
Chatterjee, Amitava	Jadavpur University, India
Das, Debabrata	International Institute of Information Technology (IIIT), Bengaluru, India
Datta, Debasish	Indian Institute of Technology (IIT), Kharagpur, India
Fjeldly, Tor A.	Norwegian University of Science and Technology, Norway
Gaur, Singh Manoj	Indian Institute of Technology (IIT), Jammu, India
Girdhar, Anup	Sedulity Solutions & Technologies, India
Govil, M. C.	National Institute of Technology (NIT), Sikkim, India
Gupta, Savita	Panjab University, India
Hyoung, Kim Tae	Nanyang Technological University, Singapore
Jat, Dharam Singh	Namibia University of Science and Technology, Namibia
Kakde, O. G.	Indian Institute of Information Technology (IIIT), Nagpur, India
Kapur, Avichal	University Grants Commission, India
Kuruvilla, Abey	University of Wisconsin, USA
Limiti, Ernesto	University of Rome Tor Vergata, Italy
Maddara, Ramesh	Cerium Systems Pvt. Ltd., India
Madhukar, Mani	IBM, India
Maurya, U. N.	LM Healthcare, India
Mekhilef, Saad	Swinburne University of Technology, Australia
Mishra, S.	Indian Institute of Technology (IIT), Delhi, India
Naidu, K. Rama	Jawaharlal Nehru Technological University, India
Nassa, Anil Kumar	National Board of Accreditation, India
Pati, Bibudhendu	Rama Devi Women's University, India
Pillai, G. N.	Indian Institute of Technology (IIT), Roorkee, India
Poonia, M. P.	All India Council for Technical Education, India
Popov, Yu. I.	St. Petersburg National Research University of Information Technologies, Russia
Reddy, B. V. Ramana	National Institute of Technology (NIT), Kurukshetra, India
Reddy, Siva G.	Satliva.com, India
Samuel, Paulson	Motilal Nehru National Institute of Technology (MNNIT), Allahabad, India
Sarwat, Arif I.	Florida International University, USA

Sehgal, Rakesh	National Institute of Technology (NIT), Srinagar, India
Sharma, Ajay	National Institute of Technology (NIT), Delhi, India
Shukla, Anupam	Sardar Vallabhbhai National Institute of Technology (SVNIT), Surat, India
Singh, Bhim	Indian Institute of Technology (IIT), Delhi, India
Somayajulu, D. V. L. N.	Indian Institute of Information Technology, Design and Manufacturing, Kurnool, India
Subudhi, Bidyadhar	Indian Institute of Technology (IIT), Goa, India
Tariq, Mohd	Florida International University, USA
Vig, Renu	Panjab University, India
Zin, Thi Thi	University of Miyazaki, Japan

Technical Programme Committee

Agarwal, Ravinder	Thapar Institute of Engineering and Technology, India
Agarwal, Suneeta	Motilal Nehru National Institute of Technology, Allahabad, India
Aggarwal, Naveen	Panjab University, India
Agnihotri, Prashant	Indian Institute of Technology (IIT), Bhilai, India
Aujla, Gagangeet Singh	Durham University, UK
Bakhsh, Farhad IIahi	National Institute of Technology (NIT), Srinagar, India
Bali, S. Rasmeet	Chandigarh University, India
Bali, Vikram	IMS Engineering College, Ghaziabad, India
Bansod, B. S.	CSIR-CSIO, India
Barati, Masoud	Carleton University, Canada
Bashir, Ali Kashif	Manchester Metropolitan University, UK
Behal, Sunny	Shaheed Bhagat Singh State University, India
Bhardwaj, Aditya	Bennett University, India
Bhatia, Rajesh	Punjab Engineering College (Deemed to be University), India
Bhatnagar, Vishal	Netaji Subhash University of Technology, India
Bindhu, Shoba	Jawaharlal Nehru Technological University Anantapur, India
Bindra, Naveen	Postgraduate Institute of Medical Education and Research, Chandigarh, India
Chakraborty, Chinmay	Birla Institute of Technology, Mesra, India
Chawla, Anu	Post Graduate Government College for Girls, Sector 42, Chandigarh, India

Chawla, Naveen K.	University of Oklahoma, USA
Chhabra, Indu	Panjab University, India
Chhabra, Jitender Kumar	National Institute of Technology (NIT), Kurukshetra, India
Chhabra, Rishu	Chitkara University, India
Chinthala, Ashok Praveen	Tata Consultancy Services Limited, India
Dave, Mayank	National Institute of Technology (NIT), Kurukshetra, India
Demirbaga, Umit	University of Cambridge, UK
Fjeldly, Tor A.	Norwegian University of Science and Technology, Norway
Gangadharappa, M.	Netaji Subhash University of Technology, India
Garg, Sahil	Resilient Machine Learning Institute, Canada
Gaur, Manoj	Netaji Subhash University of Technology, India
Gogna, Monika	Post Graduate Government College for Girls, Sector 42, Chandigarh, India
Gupta, Varun	Chandigarh College of Engineering and Technology, India
Gupta, Vipin	U-Net Solutions, India
Handa, Rohit	Citi Canada Technology Services, Canada
Himanshu	Punjabi University, India
Hyoung, Kim Tae	Nanyang Technological University, Singapore
Ilamparithi, T.	University of Victoria, Canada
Jat, Dharam Singh	Namibia University of Science and Technology, Namibia
Jha, Devki Nandan	University of Oxford, UK
Jindal, Anish	Durham University, UK
Karar, Vinod	CSIR-CRRI, India
Kassey, Philip B.	Petrasys Global Pvt. Ltd., India
Kaur, Amandeep	Central University of Punjab, India
Kaur, Gurpreet	SGGS College, India
Kaur, Kuljeet	École de Technologie Supérieure, Montreal, Canada
Kaur, Lakhwinder	Punjabi University, India
Krishna, P. Murali	Broadcom Inc., India
Kumar, Anil	Chandigarh College of Engineering and Technology, India
Kumar, Arvind	National Institute of Technology (NIT), Kurukshetra, India
Kumar, Gaurav	Magma Research and Consultancy Services, India
Kumar, Harish	Panjab University, India
Kumar, Krishan	Panjab University, India
Kumar, Naresh	Panjab University, India

Kumar, Nishant	Indian Institute of Technology (IIT), Jodpur, India
Kumar, Prabhat	LUT University, Sweden
Kumar, Preetam	Indian Institute of Technology (IIT), Patna, India
Kumar, Rakesh	Central University of Haryana, India
Kumar, Ritesh	CSIR-CSIO, India
Kumar, Satish	CSIR-CSIO, India
Kumar, Sunil	Meerut Institute of Engineering and Technology, India
Kuruvilla, Abey	University of Wisconsin, USA
Limiti, Ernesto	University of Rome Tor Vergata, Italy
Maheswari, Ritu	eMANTHAN Inspiring Innovation, India
Maini, Raman	Punjabi University, India
Maswood, Ali I.	Nanyang Technological University, Singapore
Mathew, Jimson	Indian Institute of Technology (IIT), Patna, India
Mehan, Vineet	Maharaja Agrasen University, India
Mekhilef, Saad	Swinburne University of Technology, Australia
Mittal, Ajay	Panjab University, India
Mittal, Meenakshi	Central University of Punjab, India
Mohapatra, Rajarshi	Indian Institute of Information Technology (IIIT), Raipur, India
Murthy, Ch. A. S.	Centre for Development of Advanced Computing, India
Nanayakkara, Samudaya	University of Moratuwa, Sri Lanka
Noor, Ayman	Taibah University, Saudi Arabia
Pal, Sujata	Indian Institute of Technology (IIT), Ropar, India
Panda, Surya Narayan	Chitkara University, India
Panigrahi, Chhabi Rani	Rama Devi Women's University, India
Patel, R. B.	Chandigarh College of Engineering and Technology, India
Patil, Nilesh Vishwasrao	Government Polytechnic, Ahmednagar, India
Pilli, E. S.	Malaviya National Institute of Technology (MNIT), Jaipur, India
Qureshi, Saalim	Quarbz Info Systems, India
Rajpurohit, Bharat Singh	Indian Institute of Technology (IIT), Mandi, India
Ramesh, K.	National Institute of Technology (NIT), Warangal, India
Ravindran, Vineeta	Scania, Sweden
Reaz, Md. Mamun Bin Ibne	Universiti Kebangsaan Malaysia, Malaysia
Reddy, S. R. N.	Indira Gandhi Delhi Technical University for Women, India
Sahu, Benudhar	Institute of Technical Education and Research, India
Saini, Surender Singh	CSIR-CSIO, India

Sangal, A. L.	Dr. B. R. Ambedkar National Institute of Technology, Jalandhar, India
Sarwat, Arif	Florida International University, USA
Sehgal, Navneet Kaur	Chandigarh University, India
Seshadrinath, Jeevanand	Indian Institute of Technology (IIT), Roorkee, India
Shanmuganantham, T.	Pondicherry University, India
Sharma, Shweta	Punjab Engineering College (Deemed to be University), India
Sharmeela, C.	Anna University, India
Singh, A. K.	National Institute of Technology (NIT), Kurukshetra, India
Singh, Amandeep	National Institute of Technology (NIT), Srinagar, India
Singh, Amardeep	Punjabi University, India
Singh, Baljit	Central Scientific Instruments Organization, India
Singh, Balwinder	Centre for Development of Advanced Computing, India
Singh, Brahmjit	National Institute of Technology (NIT), Kurukshetra, India
Singh, Dheerendra	Chandigarh College of Engineering and Technology, India
Singh, Jeetendra	National Institute of Technology (NIT), Sikkim, India
Singh, Paramjeet	Giani Zail Singh Campus College of Engineering & Technology, Maharaja Ranjit Singh Punjab Technical University, India
Singh, Rajvir	Chitkara University, India
Singh, Satwinder	Central University of Punjab, India
Singh, Shailendra	Punjab Engineering College (Deemed to be University), India
Singh, Sukhwinder	Panjab University, India
Singh, Sunil K.	Chandigarh College of Engineering and Technology, India
Sofat, Sanjeev	Punjab Engineering College (Deemed to be University), India
Srinivasa, K. G.	Indian Institute of Information Technology (IIIT), Raipur, India
Sunkaria, Ramesh Kumar	National Institute of Technology (NIT), Jalandhar, India
Tariq, Mohd	Florida International University, USA
Tiwari, Anil Kumar	Indian Institute of Technology (IIT), Jodhpur, India

Toreini, Ehsan	University of Surrey, UK
V. S., Ananthanarayana	National Institute of Technology (NIT), Surathkal, India
Vasantham, Thiru	Durham University, UK
Verma, Harsh K.	Dr. B. R. Ambedkar National Institute of Technology, Jalandhar, India
Verma, Yajvender Pal	Panjab University, India
Vuppala, Anil	Indian Institute of Information Technology (IIIT), Hyderabad
Yadav, Yashveer	Datum Analysis, Jordan
Zin, Thi Thi	University of Miyazaki, Japan

Organizing Committee

Dhaliwal, Balwinder S.	National Institute of Technical Teachers Training and Research, Chandigarh, India
Doegar, Amit	National Institute of Technical Teachers Training and Research, Chandigarh, India
Solanki, Shano	National Institute of Technical Teachers Training and Research, Chandigarh, India
Thakur, Ritula	National Institute of Technical Teachers Training and Research, Chandigarh, India
Sharan, Amrendra	National Institute of Technical Teachers Training and Research, Chandigarh, India

Student Volunteers

Kaur, Amandeep	National Institute of Technical Teachers Training and Research, Chandigarh, India
Agrawal, Deepika Vikas	National Institute of Technical Teachers Training and Research, Chandigarh, India
Agrawal, Lucky Kumar Dwarkadas	National Institute of Technical Teachers Training and Research, Chandigarh, India
Bala, Anju	National Institute of Technical Teachers Training and Research, Chandigarh, India
Bhardwaj, Neha	National Institute of Technical Teachers Training and Research, Chandigarh, India
N., Raj Chithra	National Institute of Technical Teachers Training and Research, Chandigarh, India
Gupta, Mahendra	National Institute of Technical Teachers Training and Research, Chandigarh, India

Gupta, Vinita	National Institute of Technical Teachers Training and Research, Chandigarh, India
Krishna, Banoth	National Institute of Technical Teachers Training and Research, Chandigarh, India
Kulkarni, Atul M.	National Institute of Technical Teachers Training and Research, Chandigarh, India
M. Soujanya	National Institute of Technical Teachers Training and Research, Chandigarh, India
Mahajan, Palvi	National Institute of Technical Teachers Training and Research, Chandigarh, India
M. K., Shajila Beegam	National Institute of Technical Teachers Training and Research, Chandigarh, India
Patel, Anwesha	National Institute of Technical Teachers Training and Research, Chandigarh, India
Pushparaj	National Institute of Technical Teachers Training and Research, Chandigarh, India
Rani, Puja	National Institute of Technical Teachers Training and Research, Chandigarh, India
Shukla, Praveen	National Institute of Technical Teachers Training and Research, Chandigarh, India
Siddiqui, Mohd. Ahsan	National Institute of Technical Teachers Training and Research, Chandigarh, India

Sponsors

 Department of Scientific and Industrial Research, Ministry of Science and Technology, Government of India, New Delhi, India

 Syngient Technologies, India

 IETE Chandigarh Centre, India

Alakh Infotech, India

DesignTech

Design Tech Systems, India

GIGABYTE

Gigabyte Networks, India

FORE

Fore Solutions, India

HITECH

Hitech Solutions, India

Pinnacle Enterprises, India

Luxmi Enterprises, India

Contents – Part I

AI and IoT for Smart Healthcare

Contents – Part II

AI and IoT for Other Engineering Applications

AI and IoT Enabling Technologies

Securing IoT Using Supervised Machine Learning

Sania Iqbal$^{(\boxtimes)}$ (iD) and Shaima Qureshi (iD)

Department of Computer Science and Engineering, National Institute of Technology, Srinagar, J&K, India

{sania_05phd18,shaima}@nitsri.ac.in

Abstract. IoT today is ubiquitous. Its growth has reached a stage where the number of connected devices is growing exponentially daily, raising various security issues of great importance to all stakeholders. Implementing security schemes on IoT devices poses a considerable challenge because of their heterogeneous and restricted existence. Machine learning is already an intricate part of several IoT applications, and it eliminates human errors and enables IoT infrastructure to generate real-time insights to reach its full potential. To protect IoT services from being targeted and raise security awareness throughout all aspects of the network, machine learning algorithms are now explored to address the critical security issues of IoT. Deploying a learned model-based security approach is dynamic and holistically effective as a solution. Supervised machine learning methods have been proposed as a promising approach for IoT security due to their ability to detect and classify malicious data. This paper provides a comprehensive overview of supervised machine learning methods used for IoT security, including various classifiers and data engineering techniques. We have further demonstrated how effectively the performance can be increased by deploying various data engineering methods to improve the overall model performance in terms of accuracy and time consumed. Both these play a crucial role in providing dynamic and real-time services. Overall, this paper aims to raise awareness of the potential of supervised machine learning for IoT security and guide researchers and practitioners in this field.

Keywords: IoT · Machine Learning · Supervised Learning · IoT Security · IoT Applications · Data Engineering

1 Introduction

IoT integrates various processes, such as identifying, sensing, networking, and computation, to develop a smarter environment that saves time, energy, and money. IoT consists of a set of connected devices ranging from a simple sensor to a complex workstation that transfer and exchange data to optimise their performance without any human intervention to provide services that offer personalised interaction with the 'things'. However, for the IoT infrastructure to act intelligently to provide better services with the desired QoS or

© The Author(s), under exclusive license to Springer Nature Switzerland AG 2024
R. K. Challa et al. (Eds.): ICAIoT 2023, CCIS 1929, pp. 3–17, 2024.
https://doi.org/10.1007/978-3-031-48774-3_1

improve its performance, the data must be looked at to find trends, patterns, hidden correlations, and actionable insights. It is where data science comes into play. Data science includes techniques from various fields, including machine learning and data analysis. Data science plays a big part in developing and succeeding IoT applications and services. It draws meaningful conclusions from IoT data. It enables the IoT to provide personalised services, automation, and intelligent decisions [1]. This massive amount of valuable IoT data is analysed using data science to provide better services and applications to its users. IoT environments also generate enormous amounts of traffic, using various data science techniques, it is reviewed intelligently to provide sophisticated security solutions. Thus, when combined with IoT, data science offers better performance to external users and can be effectively used to strengthen the IoT infrastructure by addressing core issues like security. Main causes of concern in IoT that eventually lead to security related issues include factors related to its deployment and execution environment.

Deployment Factors. Incorrect access control due to lack of proper authentication and authorisation of IoT devices resulting in the establishment of a single privilege level throughout the network. Other factors include application vulnerabilities, lack of standard security measures like data encryption, devices with outdated software and with no means to update and further lack of user involvement in deployment and configuration of their devices securely.

Environmental Factors. IoT is a ubiquitous service provider with heterogeneous nodes and services, this provides an overly large and vulnerable attack surface. The inability of the network to report or mitigate any problem, gives rise to an intrusion ignorant environment. Also, no vendor support in case of both device and software development cycle results in the lack of trusted execution environment. Insufficient privacy protection measures in terms of storage and distribution of sensitive information further lead to severe privacy violation.

Several recent studies have emphasized the importance of an intelligent security solution for IoT and have explored the use of machine learning methods. For example, some of the most relevant papers that showcase the potential of machine learning in addressing security challenges in IoT applications are as follows:

[1] provides a comprehensive overview of various machine learning techniques used for IoT security. The paper discusses the potential of machine learning methods to address security challenges and proposes an IDS using multiple ML Classifier techniques for identifying the multi-class intrusion attacks. Similarly, [2] investigates the energy consumption of on-device machine learning models for IoT intrusion detection. The authors compare cloud computing-based, edge computing-based, and IoT device-based approaches for training ML models. They find that the Decision Tree algorithm deployed on-device provides better results in terms of training time, inference time, and power consumption. [3] focused on the challenges and requirements of IoT-based systems in terms of security. It also analyses limitations of the existing ML based security solutions for IoT security. The paper offers a thorough investigation of ML-based security solutions and a critical evaluation of their drawbacks for IoT devices, particularly those with limited resources. [4] presents a systematic literature review on recent research trends in IoT security and machine learning. The study identifies key research gaps that need to be addressed to secure IoT devices from launching large-scale attacks on critical

infrastructures and websites. The paper highlights the importance of integrating big data and machine learning to detect IoT attacks in real or near real-time. [5] survey discusses the security and privacy challenges in the implementation of IoT. The paper provides a systematic study of ML and its integration with IoT to make the system more secure. The survey concludes by outlining some research challenges that need to be addressed, including scalability, system throughput, and secure computation, among others. [6] discusses the security and privacy challenges in IoT networks and the limitations of traditional cryptographic approaches. The authors argue that ML can provide embedded intelligence in IoT devices and networks to address these security issues. They review existing security solutions and highlight the gaps that call for ML and DL approaches. [7] discusses the challenges and security concerns surrounding the growing use of IoT devices and the need for efficient solutions to address them. The authors emphasize the importance of ML in securing IoT devices and present an overview of existing ML techniques used for detecting anomalies and attacks. In conclusion, the use of machine learning methods in IoT security is an active area of research. The papers reviewed highlight the importance of an intelligent security solution for IoT devices, which have limited memory and processing power. Machine learning techniques have been proposed to address security challenges, such as intrusion detection and anomaly detection, and to enhance the security of IoT networks. However, there are still challenges to overcome, such as scalability, system throughput, and secure computation.

This paper first presents an in-depth analysis of how various machine learning paradigms are used to achieve multiple security goals in IoT. This paper discusses how machine learning has become an integral part of IoT security and how we can efficiently and effectively use various supervised learning models in IoT intrusion detection. This paper aims to stepwise chalk out the integration of various data engineering methods with the existing supervised learning models and, using comparative analysis, build a model with performance parameters better suited for emerging IoT networks. The following sections include the investigation details conducted to re-introduce the use of supervised machine learning models in IoT security to ensure better performance in terms of the discussed parameters. It is followed by the results and their interpretations and subsequent discussions.

2 IoT and Machine Learning

IoT has seen overwhelming growth in terms of its diverse applications and the number of connected devices, significantly impacting people's lives. Such pervasiveness results in tremendous amounts of data exchange between diverse devices throughout the globe. This magnitude of connectivity also calls for equally pervasive security measures that secure the IoT network from edge to cloud. Earlier research on IoT security and privacy solutions mainly focused on adapting the security solutions aimed at wireless sensor networks and the internet to secure the IoT. However, the inherent differences in the features of the IoT from its constituent elements render these efforts insufficient [8].

Machine learning as a computational paradigm allows systems to learn and improve from experience without requiring human intervention. Here, systems are trained using algorithms or statistical models over sample data, usually characterised by features and output values known as labels. The system finds the correlation between features and labels during the training phase, and this information is then used to analyse new data. based on the learning style, we have three major machine learning algorithms based on the learning style: supervised, unsupervised, and reinforcement learning. Besides these traditional machine learning approaches, we have advanced machine learning techniques: Deep learning, Incremental Learning, and Transfer Learning. Of these advanced approaches, Deep Learning is more often used in IoT because of its ability to model the complexities of diverse datasets. A machine learning algorithm is deployed as an integrated part of an IoT end-device or an embedded sensor system for the IoT application. [9–11] present the summary of IoT-based applications that use the above-mentioned machine learning algorithms.

The following sub-sections present the various machine learning methods used in IoT and their potential security solutions for which they can be implemented [6, 12–15].

2.1 Supervised Machine Learning

Supervised learning is used in IoT for prediction and classification. It produces an inferred function and needs external assistance in the form of a labelled training dataset. Multivariate Linear Regression (MLR), Logistic Regression (LR), Support Vector Machines (SVM), Naïve Bayes (NB), k-Nearest Neighbours (kNN), Decision Tree (DT) and Random Forests (RF) are commonly used supervised learning methods used in IoT as shown in Table 1 below. Supervised Learning, however, requires a massive, labelled training dataset, and it is costly to generate labelled data. On the other hand, it is suitable for problems dealing with security where each data point necessarily belongs to one of the labels- attack or non-attack.

Table 1. Use of Supervised Learning methods for addressing security issues in IoT

Potential Security solutions	Supervised Learning Methods				
	SVM	NB	k-NN	DT	RF
Intrusion Detection	✓	✓	✓	✓	✓
DDoS attacks	✓	x	✓	x	✓
Anomaly or Malware Detection	✓	x	✓	x	✓
Detection of suspicious traffic	x	x	x	✓	x
Network Access Control	x	✓	x	x	x
Authentication	✓	x	x	x	x
Detection of unauthorised nodes	x	x	x	x	✓

2.2 Unsupervised Machine Learning

Unsupervised learning depicts a concealed structure by learning a few features from the previous data to identify new data. It is used to find hidden structures in the unlabelled datasets with zero human intervention. This makes unsupervised learning suitable for large and complex datasets that have no labels. However, due to lack of prior knowledge of expected output, results tend to be more unstable than supervised learning. Commonly used unsupervised learning methods in IoT are Association Rule Algorithm, Ensemble Learning, k-means clustering, and Principal Component Analysis (PCA) as shown in Table 2 below.

Table 2. Use of Unsupervised Machine Learning methods for addressing security issues in IoT

Potential Security solutions	Unsupervised Learning Methods			
	Association Rule Algorithm	Ensemble Learning	k-means Clustering	PCA
Intrusion Detection	✓	✓	✓	x
Anomaly or malware detection	x	✓	x	✓
Real-time Detection	x	x	x	✓
Data Anonymisation	x	x	✓	x
Sybil Detection	x	x	✓	x

2.3 Deep Learning

This type of machine learning utilises a hierarchical learning process where several layers learn multiple levels of representation. It performs feature extraction using neural network. This learning model is more configurable and flexible than classical machine learning and achieves much better accuracy. However, compared to traditional supervised modelling its training is comparatively slow. Also, it requires a large amount of labelled data, huge computing resources like high-end processors. Deep learning-based models used in IoT security include Recurrent Neural Network (RNN), Convolution Neural Networks (CNN), Auto-Encoders (AE), Restricted Boltzmann machines (RBM), Deep Belief Networks (DBN) and Generative Adversarial Networks (GAN). Their use cases are enlisted in Table 3 below.

Table 3. Use of Deep Learning methods for addressing security issues in IoT

Potential Security solutions	Deep Learning Methods					
	RNN	CNN	AE	RBM	DBN	GAN
Intrusion Detection	✓	✓	✓	x	✓	x
Anomaly or Malware Detection	✓	✓	✓	✓	x	✓
Network traffic classification	✓	x	x	x	x	x
Effective against time-series-based threats	✓	x	x	x	x	x
Securing the cyberspace of IoT systems	x	x	x	x	x	✓

2.4 Reinforcement Learning

Reinforcement learning model learns by interacting with the environment to achieve a goal using feedback in terms of credit. The model has no prior idea of the underlying learning environment, hence no knowledge of which action to take until it has a situation. This type of learning is suitable for problems having continuous interaction between the model and the environment. However, it suffers from credit assignment problems while distributing credits to decisions taken. Commonly using reinforcement learning methods in IoT security are Q-Learning and Deep Reinforcement Learning (DRL) as described in Table 4 below.

Table 4. Use of Reinforcement Learning methods for addressing security issues in IoT

Potential Security solutions	Reinforcement Learning Methods	
	Q-Learning	DRL
Prevent Eavesdropping	✓	x
Malware detection	x	✓
Access Control and Anti-jamming	✓	✓
DDoS Detection	✓	x
Authentication	✓	✓
Eavesdropping prevention	✓	x

Machine learning is a promising approach for addressing IoT security challenges. One significant advantage of using machine learning for IoT security is its ability to analyse large amounts of data and identify patterns that may be difficult for humans to detect. Machine learning algorithms can learn from past data and adapt to new threats, making them more effective in detecting new attacks. Additionally, machine learning can help reduce false positives, making security systems more efficient and effective. Furthermore, machine learning can provide real-time insights, which is essential for IoT applications that require timely responses. Overall, machine learning can enhance the security of IoT devices and networks by providing a dynamic and holistic approach to security.

Supervised machine learning involves the training of models on labelled data, which is readily available for security-related tasks, hence more favourable for use in IoT security. With supervised machine learning, it is possible to train models to detect anomalies or classify data as benign or malicious, making it an effective tool for intrusion detection, network monitoring, and threat detection. Moreover, supervised machine learning models can be fine-tuned and updated as new security threats emerge, providing a more dynamic and adaptable security solution for IoT devices. Additionally, supervised machine learning models can be optimized for performance, which is crucial for IoT devices, which often have limited computing resources. Overall, supervised machine learning provides a powerful and effective approach for IoT security, and its ability to learn from labelled data makes it a natural fit for many security-related tasks.

3 Methods

In this section, we discuss the various methods used in this experiment, the dataset used and training and analysis of machine learning models. The simulations are conducted in a level-wise manner, each next level simulation incorporating the results of previous one. This investigation aims to train a supervised model reinforced to detect anomalous behaviour in an IoT network with maximum accuracy and a reasonable timeframe.

3.1 UNSW-NB15 Dataset

The UNSW-NB15 [16, 17] is created by the Australian Centre for Cyber Security's cybersecurity research group. It has 49 features along with the class label. The IXIA PerfectStorm program generated a combination of genuine modern normal activities and synthetic current attack behaviours. The dataset contains network traffic data captured in a realistic and controlled environment, simulating various attack scenarios on a typical enterprise network. It includes nine different types of attacks, such as Fuzzers, Analysis, Backdoors, DoS, Exploits, Generic, Reconnaissance, Shellcode, and Worms. The dataset is widely used in research to develop and evaluate machine learning-based intrusion detection systems for network security. It has become a standard benchmark dataset in the field of cybersecurity and machine learning. The selected dataset has two main issues that need to be handled: imbalanced data distribution and redundant data. Both these issues are handled, and further improvisation of the dataset is performed during the experiment.

3.2 Experimental Setup

Each machine learning classifier consists of three basic steps: data pre-processing, classifier training, and data prediction (Fig. 1).

In the first step the data is pre-processed in substages to ensure data quality. Data cleaning handles incorrect, incomplete, irrelevant, duplicated, or improperly formatted data. Normalisation of the feature values is done to condense then to a same relative scale

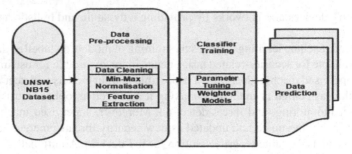

Fig. 1. Machine Learning Classifier Setup

and that none of the value dominates the model. Feature extraction ensures that essential and the most relevant features are used in model training. In the second step the model is trained, and its performance is evaluated using the performance measures described below. In this study, we evaluate five supervised learning algorithms, including logistic regression (LR), linear support vector classification (LSVC), decision tree (DT), random forest (RF), gradient boosted decision tree (XGB), and an ensemble-based voting classifier (V). It is in the second step of data pre-processing, where various data engineering methods are deployed to obtain a model with balanced performance in terms of time and accuracy. The third and last step is a functional supervised model that makes predictions, and its performance is evaluated based on the performance measures discussed in the next sub-section.

3.3 Performance Measures

Various accuracy measures used to evaluate the trained models are defined in terms of the confusion matrix. As shown in Fig. 2, confusion matrix is a tabular performance measuring matrix for machine learning models. It includes combinations of both predicted and actual values.

		Actual Values	
		Positive (1)	Negative (0)
Predicted Values	Positive (1)	*True Positives*	*False Positives*
	Negative (0)	*False Negatives*	*True Negatives*

Fig. 2. Confusion Matrix

The above confusion matrix defines True positives (*TP*) as correct positive predictions, False positives (*FP*) as incorrect positive predictions, True negatives (*TN*) as correct negative predictions and False negatives (*FN*) as incorrect negative predictions. Based on the confusion matrix defined, the accuracy measures are as described below:

AUC (Area Under the ROC Curve). A ROC curve (receiver operating characteristic curve) is a graph that shows how well a classification model performs at different classification thresholds by plotting True Positive Rate vs False Positive Rate,

$$true\ positive\ rate = TP/(FN + TP) \tag{1}$$

$$false\ positive\ rate = FP/(TN + FN) \tag{2}$$

AUC presents a combined measure of performance across various classification thresholds. AUC value ranges from 0 to 1, and a model with all correct predictions scores an AUC of 1.0. AUC is a preferred performance measure as:

- AUC is scale-invariant. It assesses the ranking of predictions and is unaffected by their absolute values.
- AUC is classification-threshold-invariant and evaluates the accuracy of the model's predictions regardless of the classification threshold.

F1 Score. It considers the precision (number of correct positive predictions out of total positive predictions) and recall (number of correct positive predictions out of total positive cases in the dataset) at a particular threshold value. The F1 score is the harmonic mean of the two and tries to strike a balance between them. It lies between [0,1] and is high only when both precision and recall are high.

$$F1\ score = 2 * (precison * recall)/(precision + recall)$$
$$such\ that, \quad precison = TP/(TP + FP) \tag{3}$$
$$recall = TP/(TP + FN)$$

The advantages of using F1 score:

- It combines various matrices into one, thus capturing many aspects of the model.
- It showcases the model's accuracy as it tells how precise and robust the model is, especially when the dataset is imbalanced.
- The F1 score is a mean value sensitive to the lowest one.

FAR (False Acceptance Rate). It is used to measure the average number of false acceptances, and it is the percentage of the time an invalid data point is declared valid.

$$FAR = FP/(FP + TN) \tag{4}$$

FAR is considered a significant measure of security errors as it gives unauthorised nodes access to systems with explicit security mechanisms to avoid such nodes.

Run Time. To make the comparison of models more exhaustive, we include the time component as one of the performance measures. The run time of the model is expressed in terms of training time of the model (in seconds) over the training dataset and the testing time of the model (in seconds) over the test dataset. It gives us an idea of the time and subsequently the resource consumption of the model.

4   Results and Inferences

In this study, we have evaluated the performance of the seven supervised learning models mentioned above, under various conditions of data processing. It also gives us an overview of how various data engineering levels have an impact on the model's performance.

4.1   Using Hyperparameter Tuned Models

Hyperparameter tuning is the process of selecting the ideal values for these parameters, which are the values that define the learning model architecture. Various works [18–20] have discussed the importance of hyperparameter tuning to improve model performance. Fundamental grid-search-based hyperparameter tuning is applied to a model's key hyperparameter set and performs an exhaustive sampling of hyperparameter space. Although expensive in terms of computation, it ensures a set of hyperparameters from the specified space that are optimal (Table 5).

Table 5. Performance of hyperparameter tuned supervised ML models

Classifier	AUC	F1	FAR	Run Time (seconds)	
				Training	Testing
LR	0.9772	0.9891	0.0228	2.1966	1.0123
LSVC	0.9905	0.9885	0.0095	0.1489	0.0604
DT	0.9877	0.9905	0.0123	0.3411	0.1747
RF	0.9860	0.9943	0.0140	10.0986	5.0060
XGB	0.9917	0.9965	0.0083	4.1298	2.2584
Voting (XGB+DT)	0.9912	0.9961	0.0088	4.0462	2.7309
Voting (XGB+LR)	0.9921	0.9949	0.0079	3.8904	2.4269

We have trained two voting classifiers based on the results of the first four models, with a combination of two classifiers in each. Since the XGB outperforms both in terms of AUC and F1 scores, we have combined it with the next best classifiers, DT and LR, in terms of run time and for which probability estimates exist.

4.2   Using Feature Selection on Parameter-Tuned Models

Feature selection reduces the model's computational cost and improves the model's performance. In the previous section, we examined the performance of various supervised models. We use a filter method on the outperforming supervised model to retrieve the essential features. We use the *feature_imporatances_* property on the trained XGBoost model, which filters the features based on their gain (depicting the contribution of each

Table 6. Performance of hyperparameter tuned Supervised ML models with Important Feature Selection

Classifier	AUC	F1	FAR	Run Time (seconds)	
				Training	Testing
LR	0.9879	0.9889	0.0121	0.1440	0.0609
LSVC	0.9756	0.9887	0.0244	0.1465	0.0650
DT	0.9877	0.9905	0.0123	1.4234	0.6009
RF	0.9870	0.9948	0.0130	13.7093	5.5812
XGB	0.9922	0.9967	0.0078	4.9958	1.9662
Voting (XGB+DT)	0.9916	0.9963	0.0084	19.1731	9.1446
Voting (XGB+LR)	0.9926	0.9953	0.0074	12.0534	6.0838

feature for every tree in improving the model's accuracy). It reduced the feature size by almost one-fourth, from 197 features in the sparse matrix to only 55 features (Table 6).

Based on the above results, the XGB model outperforms all other models in terms of AUC, F1 and FAR. In terms of run time, it gives an average performance with almost real-time testing. Thus, we can see that the XGB model gives a balanced performance regarding the accuracy and run time. The second-best model is DT in terms of balanced performance, with AUC and FAR less than XGB but better in terms of runtime.

4.3 Using Class Weights on Parameter-Tuned Models

The UNSW- NB15 dataset is class imbalanced [21]. The number of negative data points outnumbers the positive data points by more than 75 per cent. It can result in the model being biased towards the negative class. To mitigate the effects of the skewed dataset, we assign class weights. It changes the cost function of the model so that misclassifying an observation from the minority class carries a heavier penalty than misclassifying an observation from the majority class. Using class weights increases the model's accuracy by rebalancing the class distribution (Table 7).

Using only the class weight balancing for the same set of models, we notice an overall improvement in model performance and significantly decreased FAR. Also, a slight increase (by 1–2 s) in training time is noticed. In this setup, XGB outperforms all other models in balanced performance, followed by DT, which gives us a comparable performance in lesser runtime.

4.4 Using Class Weights with Important Features on Parameter-Tuned Models

In the final model, we combine the top three approaches and deploy the models with hyper-parameters assigned by tuning, training it for important features only and on classes represented equally.

In the above table, the AUC and F1 scores have shown considerable improvement compared to previous tables, and the FAR value has decreased sufficiently by more than

Table 7. Performance of hyperparameter tuned Supervised ML models with Class Weight Balancing

Classifier	AUC	F1	FAR	Run Time (seconds)	
				Training	Testing
LR	0.9904	0.9883	0.0096	2.0153	0.8607
LSVC	0.9929	0.9880	0.0071	0.1480	0.0580
DT	0.9932	0.9888	0.0068	0.3526	0.1628
RF	0.9932	0.9916	0.0068	12.3393	5.2906
XGB	0.9951	0.9961	0.0049	5.3205	2.3786
Voting (XGB+DT)	0.9951	0.9932	0.0049	5.7287	2.3744
Voting (XGB+LR)	0.9940	0.9920	0.0060	5.4244	2.3782

Table 8. Performance of hyperparameter tuned Supervised ML models with Important Feature Selection and Class Weight Balancing

Classifier	AUC	F1	FAR	Run Time (seconds)	
				Training	Testing
LR	0.9911	0.9875	0.0089	0.1527	0.0674
LSVC	0.9928	0.9879	0.0072	0.1407	0.0608
DT	0.9932	0.9888	0.0068	1.3736	0.5940
RF	0.9926	0.9930	0.0074	13.1711	5.5996
XGB	0.9951	0.9961	0.0049	5.7474	2.2005
Voting (XGB+DT)	0.9950	0.9932	0.0050	7.3054	2.6404
Voting (XGB+LR)	0.9943	0.9926	0.0057	5.3685	2.2309

one-tenth. However, the training time has increased considerably, but changes in testing time are negligible and still give a real-time performance. From the above results, we can conclude that XGB followed by DT gives us better and balanced performance in terms of accuracy and execution time.

5   Observations

Based on the inferences of the final experiment (Table 8), it is observed that out of the seven supervised learning models XGB and XGB followed by DT show drastic improvement in overall performance by applying vigorous data engineering methods before training. The voting models also show performance comparable to XGB in terms of AUC, F1 and FAR but have very high runtime values, which however, cannot be treated as real-time responses. The important observations of this study can be summarised as:

- XGB model is a top performer as it scores more than 99.5% in AUC and F1. Besides, it has the least FAR, the most crucial parameter of a security module, as it prevents unauthorised access or denial to the IoT network.
- DT model is the second-best performer where little lesser values of performance measures AUC, F1 and FAR than XGB is compromised by very small runtime for DT.
- Data Engineering methods were applied keeping in view the dataset characteristics - the unbalanced nature of IoT traffic data and the presence of irrelevant and redundant data.
- The results obtained can be used to successfully deploy a particular supervised learning model based on its performance matrix and IoT network requirements. For example, if the only requirement of a network is real-time security, then LSVC (Table 8) with the least test time is the best choice.
- The above results deliberate the need to develop methods to increase the F1 score and decrease the run time.

Even though for all the models an overall improvement in the AUC and FAR values was observed, the F1 score remained almost unchanged. The only trade-off was an increase in the runtime of the models, especially the training time, while the testing time with a slight increase was still real-time in nature. One of the reasons for this could be that the models were trained and evaluated on a regular computer (Intel(R) Core (TM) i7-8750H CPU @ 2.20 GHz, 32 GB RAM). Hence, one of the methods to improve execution time is to use high-end computational resources for model training.

6   Conclusion

The quality and quantity of the data generated by IoT make machine learning approaches more appropriate for it and suggest that they need more than just traditional processing methods, be it for providing end-user services or securing the IoT infrastructure itself. While incorporating a machine learning algorithm in IoT infrastructure, we must ensure that the model utilises the data efficiently and in real-time. This study re-evaluates the use of supervised machine learning for IoT security by exploiting the dataset characteristics. The discussed approach of considering dataset characteristics while deploying a machine learning model imparts intelligent security to these IoT networks that is also efficient and real-time, making them more independent and reliable than traditional networks. The experiment shows the AUC, F1, FAR, and runtime performance of seven supervised models. The stated results demonstrate that XGB and DT are the models with the most balanced overall performances.

References

1. Saran, N., Kesswani, N.: A comparative study of supervised machine learning classifiers for intrusion detection in internet of things. Proc. Comput. Sci. **218**, 2049–2057 (2023). https://doi.org/10.1016/j.procs.2023.01.181

2. Tekin, N., Acar, A., Aris, A., Uluagac, A.S., Gungor, V.C.: Energy consumption of on-device machine learning models for IoT intrusion detection. Internet Things **21**, 100670 (2023). https://doi.org/10.1016/j.iot.2022.100670
3. Farooq, U., Tariq, N., Asim, M., Baker, T., Al-Shamma'a, A.: Machine learning and the Internet of Things security: solutions and open challenges. J. Parallel Distrib. Comput. **62**, 89–104 (2022). https://doi.org/10.1016/j.jpdc.2022.01.015
4. Ahmad, R., Alsmadi, I.: Machine learning approaches to IoT security: a systematic literature review. Internet Things **14**, 100365 (2021). https://doi.org/10.1016/j.iot.2021.100365
5. Mohanta, B.K., Jena, D., Satapathy, U., Patnaik, S.: Survey on IoT security: challenges and solution using machine learning, artificial intelligence and blockchain technology. Internet of Things. **11**, 100227 (2020). https://doi.org/10.1016/j.iot.2020.100227
6. Hussain, F., Hussain, R., Hassan, S.A., Hossain, E.: Machine Learning in IoT Security: Current Solutions and Future Challenges. arXiv:1904.05735 [cs, stat]. (2019)
7. Moh, M., Raju, R.: Machine learning techniques for security of Internet of Things (IoT) and fog computing systems. In: 2018 International Conference on High Performance Computing Simulation (HPCS), pp. 709–715 (2018). https://doi.org/10.1109/HPCS.2018.00116
8. Zantalis, F., Koulouras, G., Karabetsos, S., Kandris, D.: A review of machine learning and IoT in smart transportation. Future Internet **11**, 94 (2019). https://doi.org/10.3390/fi11040094
9. Shanthamallu, U.S., Spanias, A., Tepedelenlioglu, C., Stanley, M.: A brief survey of machine learning methods and their sensor and IoT applications. In: 2017 8th International Conference on Information, Intelligence, Systems & Applications (IISA), pp. 1–8 (2017). https://doi.org/10.1109/IISA.2017.8316459
10. Duarte, D., Ståhl, N.: Machine learning: a concise overview. In: Said, A., Torra, V. (eds.) Data Science in Practice. SBD, vol. 46, pp. 27–58. Springer, Cham (2019). https://doi.org/10.1007/978-3-319-97556-6_3
11. Sharma, S.K., Wang, X.: Toward massive machine type communications in ultra-dense cellular IoT networks: current issues and machine learning-assisted solutions. IEEE Commun. Surv. Tutor. **22**, 426–471 (2020). https://doi.org/10.1109/COMST.2019.2916177
12. Xiao, L., Wan, X., Lu, X., Zhang, Y., Wu, D.: IoT security techniques based on machine learning: how do IoT devices use AI to enhance security? IEEE Sig. Process. Mag. **35**, 41–49 (2018). https://doi.org/10.1109/MSP.2018.2825478
13. Kotstein, S., Decker, C.: Reinforcement learning for IoT interoperability. In: 2019 IEEE International Conference on Software Architecture Companion (ICSA-C), pp. 11–18 (2019). https://doi.org/10.1109/ICSA-C.2019.00010
14. Al-Garadi, M.A., Mohamed, A., Al-Ali, A.K., Du, X., Ali, I., Guizani, M.: A survey of machine and deep learning methods for internet of things (IoT) security. IEEE Commun. Surv. Tutor. **22**, 1646–1685 (2020). https://doi.org/10.1109/COMST.2020.2988293
15. Iqbal, S., Qureshi, S.: A top-down survey on securing IoT with machine learning: goals, recent advances and challenges. IJWMC **22**, 38 (2022). https://doi.org/10.1504/IJWMC.2022.122484
16. Moustafa, N.: UNSW_NB15 dataset (2019). https://ieee-dataport.org/documents/unswnb15-dataset
17. Moustafa, N., Slay, J.: UNSW-NB15: a comprehensive data set for network intrusion detection systems (UNSW-NB15 network data set). In: 2015 Military Communications and Information Systems Conference (MilCIS), pp. 1–6 (2015). https://doi.org/10.1109/MilCIS.2015.7348942
18. Yang, L., Shami, A.: On hyperparameter optimization of machine learning algorithms: theory and practice. Neurocomputing **415**, 295–316 (2020). https://doi.org/10.1016/j.neucom.2020.07.061

19. Muhajir, D., Akbar, M., Bagaskara, A., Vinarti, R.: Improving classification algorithm on education dataset using hyperparameter tuning. Proc. Comput. Sci. **197**, 538–544 (2022). https://doi.org/10.1016/j.procs.2021.12.171
20. Weerts, H.J.P., Mueller, A.C., Vanschoren, J.: Importance of Tuning Hyperparameters of Machine Learning Algorithms. http://arxiv.org/abs/2007.07588 (2020)
21. Patel, H., Singh Rajput, D., Thippa Reddy, G., Iwendi, C., Kashif Bashir, A., Jo, O.: A review on classification of imbalanced data for wireless sensor networks. Int. J. Distrib. Sens. Netw. **16**, 1550147720916404 (2020). https://doi.org/10.1177/1550147720916404

DDoS Attack Detection in IoT Environment Using Crystal Optimized Deep Neural Network

C. Karpagavalli[1]([✉]) [ID] and R. Suganya[2] [ID]

[1] Department of Artificial Intelligence and Data Science, Ramco Institute of Technology, Rajapalayam, India
karpagavalli@ritrjpm.ac.in
[2] Department of Information Technology, Thiagarajar College of Engineering, Madurai, India

Abstract. Internet of Things (IoT) is the interconnection of many devices through the internet for different real-time applications. One of the major issues of IoT is Distributed Denial of Service attack(DDoS) and many research works have been carried out to circumvent the DDoS attack; however, they failed to attain the accurate classification of DDoS and normal traffic. In context with this, we propose a novel Deep Neural Network (DNN) based Crystal Search algorithm (CSA) (DNN-CSA). The modified Multilayer Preceptor (MLP) based DNN enhances the classification outcomes. Prior to classification the data are collected by sniffer tool and preprocessed using the min-max approach. Experimental analyses are carried out to analyze the performance of our proposed approach and the results are compared with other state-of-art works. The proposed methodology offers better detection accuracy, precision, recall, and F1-score for DDoS attack detection and also effective results in terms of throughput, energy consumption, and memory utilization.

Keywords: Denial of service attack · IoT nodes · DNN · CSA · Traffic analyzer · traffic · entropy metrics

1 Introduction

The internet of things refers to a collection of items that are embedded with software and sensors and that connect to and transfer data with other machines over a network [1, 2]. Light, motion, the temperature can be calculated by using sensors. It gathers information through sensors and automatically detects and analyzes the data through IoT. In our day-to-day life, the applications of the Internet of Things are very important. It generally consists of so many applications that are used to detect and avoid errors. Smart homes pertain to the Internet of Things. It is usually connected with a set-top box, locking system, electricity, etc. Next, the self-driven cars were tested by multiple organizations with the help of this technology [3]. The sensors are connected by the use of IoT. The major application used here is named 'smart grid' that applies in Information Technology and it reduces cost and wastage caused by electricity. It is also used in the healthcare system and the security is also improved. Moreover, it can access information from

R. K. Challa et al. (Eds.): ICAIoT 2023, CCIS 1929, pp. 18–36, 2024.
https://doi.org/10.1007/978-3-031-48774-3_2

everywhere. Safety is the main concern, it senses and warns the users are the advantages of IoT. One of the drawbacks of IoT is smart things, which may increase the inflation. If there is an error in the network it will spread all over the system [4].

The IoT devices consume more power and are necessary to design an IoT network with low consumption of power. Moreover, the transmission of multimedia data also requires more power, and in concern with these issues, some researchers utilize the neural network for the detection and mitigation of intrusion. However, the detection of DDoS attack is not precise and some works failed to reduce the DDoS attack in IoT networks. Hence a novel DNN based CSA method is proposed which effectively detects the attack from the IoT network and also mitigates it. The main contributions of the proposed work are listed below,

- To gather the traffic details from the subjected IoT networks, the proposed work utilize sniffer tools.
- The gathered data sets are preprocessed by using the min-max approach.
- The features are separated with the aid of a filtering approach known as Correlation-Based Feature (CBF) selection
- Then, the DNN based CSA approach is used to detect the DDoS attacks and also classifies the collected details as normal and intruder.
- Then, the DDoS attackers are mitigated with the exploitation of some of the entropy metrics such as throughput, allocation of bandwidth, the flow of traffic, generalized information divergence (GID), deviation of bandwidth, projected entropy, and generalized entropy.

The rest of the work is organized as; the relevant works are revised in Sect. 2. Section 3 explains the system model, entropy metrics in a detailed manner. The proposed DNN-CSA approach is elucidated in Sect. 4. The work is summarized in Sect. 5.

2 Literature Survey

Zaminkar et al. [8] proposed a robust hybrid method to determine the loss in power and network through IoT devices. The sinkhole attack has been detected by the Detection of sinkholes- Low-Power and Lossy Networks (RPL) (DSH-RPL) method. In IoT networks, the sinkhole layer poses a threat. A node that attacks the network is referred to as launching a selective forwarding attack. The transmitted data in IoT devices are secured in the proposed algorithm.

Latif et al. [9] have designed a random neural network (RNN) algorithm to detect cyber security attacks. The proposed method has focused on the level that has to be increased. The complexity is reduced by implementing IoT devices. However, the experiments are evaluated in real-time applications.

Silva et al. [10] developed software-defined networking (SDN) prototype to overcome the restrictions of an attack that employs a single controller strategy. It provides greater reliability in the attack. A single controller using an attack causes security and scalability issues. However, high-potential technology is recommended.

Pathak et al. [11] have proposed a novel low powered wide area network that acts as transmission technology in IoT applications. The data is transmitted across a distance of

10 to 20 km, and the nodes are used in IoT applications such as billing, smart homes, etc. The technique was primarily concerned with security. The SDN is used in the proposed method for security measures. The advantages are that it is flexible, durable, and low-cost.

Mandal et al. [12] demonstrated that a SDN can identify Mediam Access Control spoofing attacks using a multiplicative increase and additive reduction method. The accuracy of this method is high and reduces the false-positive rate. When the traffic is high it minimizes the threshold. The method attained a higher rate and security level. Traffic analysis, on the other hand, should reduce the amount of time spent.

Low-power wide-area networks, according to Torres et al. [13], can detect attacks in IoT applications. To verify the results Narrow Band –Internet of Things (NB-IoT) and Long Range(LoRa) are used. The bandwidth, cost, and efficiency are higher when compared with the previous method. However, limited databases are used.

Wazirali et al. [14] have proposed a hyperparameter based on a k-nearest neighbor (KNN) algorithm that employs an Intrusion detection system. It determines the sturdy nature of the algorithm. The algorithm identifies the different types of attacks. The distance is calculated from each data. From every point, the data are identified. The hyperparameter identifies the attack and rectifies it. It performs the accuracy of distance functions, parameters, weight. However, the performance is improved by alternating the size of the records.

Ravi et al. [15] have analyzed a novel learning-driven detection mitigation (LDDM) algorithm in DDoS attacks. The algorithm is used to detect the efficiency of the two critical parameters used. It increases the throughput solutions up to 21%. The method guards against DDoS attacks conducted on IoT applications. However, the changes must be implemented in one-stop security solutions for all violations.

Li et al. [18] have described the Real-Time Edge Detection Scheme for Sybil on the Internet of Vehicles (IoV). In real-time, the messages are enclosed as packages. To track vehicles, the user messages are collected from the extricated messages. The two features are identified clearly in the attack. The number of vehicles can be recognized by entropy-based to detect suitable id. The detected technique is perspective and fast. However, difficult to analyze the large scale dataset.

To detect and mitigate the attacks Yin et al. [19] have proposed a software-defined Internet of things (SD-IoT). The data layer of IoT are routers, base stations, two-layer, etc. It analyses and controls directly the IoT devices from the terminals. The similarities in the vectors and threshold are based on the frame of this method. It improves the exposure in the IoT. Furthermore, an efficient algorithm can be used for the investigation purpose.

Baig et al. [22] have presented a dependence estimator-based scheme to detect DDoS attacks. Here, volume, features are techniques followed to analyze the traffic. The traffic volume progressed and it makes large changes to attack. The sample of the data classifier can be classified as a data classifier in features. The other dataset can be represented as a tuple. The classification can be done by averaging the dependent technique. The scalability and density of traffic are the cons of this technique. Thus, various mining techniques can be improved by this method.

Lawal et al. [23] have demonstrated an Intrusion Detection System (IDS) to secure a peculiar network in IoT. The classification problems can be solved by the intrusion method. The concept is based on network or host level. The application is hosted in a node or system. The network traffic can be identified by this method. The clusters are grouped to monitor the reports from nodes. The performance of this method is the low positive rate and high accuracy. There are no steps to reduce the attack.

Alamri et al. [26] have proposed an Extreme Gradient Boosting (XGBoost) Algorithm. To reduce the packets bandwidth profile is used to detect the attack. The traffic attack is determined by the threshold value of the SDN. It controls the internet protocol from the average parameters. The counters are calculated and are stored as tables. The threshold values are evaluated. The performance is improved in a software-defined context. Further, the SDN is approached in a multi-controller.

3 System Model

The most challenging issue in IoT nodes is Distibuter Denial of Service (DDoS) and several work have been carried out to detect as well as to enhance the performance [28]. However, there still exist some gaps and our proposed approach is conducted to detect the DDoS in IoT nodes with the adoption of DNN based CSA approach. The DNN exploits only little resources for the detection of DDoS attacks. The steps involved in this process are enlisted below,

a. Estimation of network traffic intensity in the IoT nodes
b. Collection of similar parameters for the detection of DDoS with the aid of unsupervised DNN.
c. Categorizing the traffic flow as common traffic and incoming network traffic.

The DDoS can affect the IoT networks by forwarding affected data or flooding it with a maximum number of network packets. The DDoS is of two kinds: Direct and Reflected. If the targeted node is flooded with numerous data requests by the host represents, the direct attack and in the latter one, the host manages another set of hosts known as reflectors. With the hidden IP address, the host sends more multimedia data. The IoT nodes can be easily affected by both types of attacks.

3.1 Performance Improvement by Feature Extraction

The feature selection is based on three types: (i) Wrapper method, in which the feature subsets are selected based on the predefined values by the machine learning approach. This estimation criterion depends on the classification performances. (ii) Embeddded method which is based on the selection of features by providing training to machine learning method and known as an embedded method. (iii) This methos is based on filtering the features based on the common attributes. With the statistical approach, the ranking of the features is made and then selects the features with the highest ranking.

Besides, the network flow can be estimated by gathering the sequence of the data between the source and destination IoT nodes. The network flow can be detected with the parameters such as Port address of source and destination, IP address of source and destination, and protocol. To find the attack, the transmitted packets are classified and investigated. This ensures an optimal solution to the network. The steps involved in Correlation Based Feature (CBF) selection in DNN are listed below,

a. To find the subset of redundant features, the Pearson coefficient is evaluated for all features from the dataset.

$$PC(i, j) = \frac{Cov(A, B)}{\sqrt{SD^2(A)SD^2(B)}} \tag{1}$$

The covariance of respective features A and B are denoted as Cov (Covariance) and the evaluation of standard deviation is made for the selected features and is represented by Standard Diviation (SD). The redundant features are obtained.

b. Estimating the correlation for many features and choosing similar features to form a new dataset. The high relevant features are designed using the outcomes acquired by the above process.

3.2 Traffic Analyzer Module Based DDoS Detection

The malicious traffic is merged along with the metrics such as throughput, generalized entropy, bandwidth, traffic flow, projected entropy, deviation and generalized divergence information in this module [29]. Figure 1 illustrates the semi-centralized architecture of the traffic analyzer system. This is to maintain the involvement of locally generated Integrated Passive Device Systems (IPDS) [23] for the local routers. The detection of spoof-based attacks when the flooding of collaborates can be performed using the proposed approach. The components used in the traffic analyzer to mitigate the attacks when there is a collaboration to involve four components.

To analyze the traffic abnormalities, the threshold values are forwarded to each IoT node in the network by the bandwidth monitor. On the other hand, the abnormalities are monitored with the aid of a timer, admission controller, and bandwidth monitor by the collaborated reduction manager.

The profile such as IP address, timer value, Mac address, location address and details of allotted bandwidth of the IoT nodes are maintained in the global IPDS. Meanwhile, local profiles are sustained at local IPDS. The data transmission between each IoT node is also stored in it along with the timer value, port number, and network flow. The proposed work focused to consider the following entropy metrics such as throughput, allocation of bandwidth, the flow of traffic, generalized information divergence (GID), deviation of bandwidth, projected entropy, and generalized entropy.

Throughput. All the IoT nodes in the network system ensure a minimum throughput of δ within the system. It can be formulated as,

$$\sum_{m \in M} g_m \leq \delta \tag{2}$$

However, the throughput also depends on the allotted bandwidth for each IoT node. Thus, the allocation of bandwidth by the bandwidth allocation protocol is also deemed as the pivotal factor in the accomplishment of required throughput. Here, g_m is the fraction of the allotted bandwidth for each IoT node in the network.

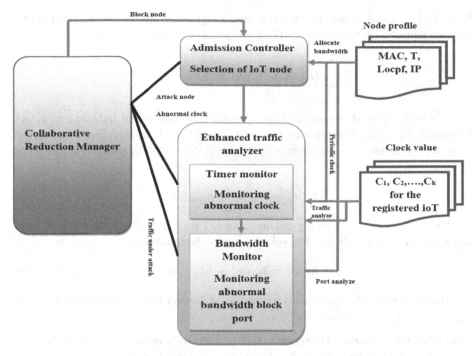

Fig. 1. Framework of the traffic analyzer

Allocation of Bandwidth. The total bandwidth allotted by the bandwidth allocation protocol in the network system can be given as G and the bandwidth allotted for each newly joined IoT node in the network can be estimated as,

$$G_r = G - G_{mb} \qquad (3)$$

Here, G represents the total bandwidth of the IoT network, the bandwidth that is given for the global and local IPDS is indicated as G_{mb}. Then, the limited bandwidth can be expressed as,

$$g_m \leq G_r/M \qquad (4)$$

Flow of Traffic. The overall communication that has been performed in the entire IoT system can be defined by this metric. However, the consolidated flow of traffic in global IPDS can be expressed as,

$$h(F_n) = \sum_{n=1}^{i} h_{out}(D_n) \qquad (5)$$

The sum of D_n provides the entire traffic flow in the confined IPDS. To transmit data from one IoT node to other local IPDS has been exploited as mentioned earlier. Then, the overall traffic flow in the local IPDS can be determined as,

$$h(D_n) = \sum_{m \in M} h_{in}(m) + \sum_{m \in M} h_{out}(m) \qquad (6)$$

Here, $h_{in}(m)$ and h_{out} are the incoming and outgoing traffic of the user IoT nodes correspondingly. Then, the total traffic of the user can be represented as,

$$h(m) = \sum_{e=1}^{i} h_e(m) + \sum_{k=1}^{i} h_k(m) \tag{7}$$

The traffic flow of the user can be controlled, measured and indicated as $h_e(m)$. Moreover, the traffic flow from individual IoT node and the user node can be denoted as $h_k(m)$. Meanwhile, the measure of the traffic control over the user IoT node can be evaluated as,

$$h_e(m) = h_{ein}(m) + h_{eout}(m) \tag{8}$$

The control flow of incoming and outgoing IoT users can be represented as $h_{ein}(m)$ and $h_{eout}(m)$ correspondingly. The velocity can be determined as,

$$h_k(m) = h_{kin}(m) + h_{kout}(m) \tag{9}$$

The incoming and outgoing data traffic flow at the user IoT node can be indicated as $h_{kin}(m)$ and $h_{kout}(m)$ respectively.

Generalized Information Divergence (GID). The GID can be formulated by deeming two various probabilities such as $S = (s_1, s_2, \ldots, s_n)$ and $T = (t_1, t_2, \ldots, t_n)$ which is given below,

$$L_\eta(S\|T) = \frac{1}{1-\eta} \log_2\left(\sum_{i=1}^{M} s_i^\eta t_i^{1-\eta}\right), \quad here \ \eta \geq 0 \tag{10}$$

Deviation of Bandwidth. It can be defined as,

$$Dv(g_m, g_{m'}) \leq W \tag{11}$$

The IoT nodes in the network system must use the allocated bandwidth g_m only. If it fails, then the deviated bandwidth is denoted as $g_{m'}$. If the deviation exceeds the value of W, i.e., 0.1, then the network will automatically deny the IoT nodes.

Projected Entropy. Based on the stochastic processes, the entropy of two random processes are similar and can be denoted as $E(a)$. It can be formulated as,

$$E(a) = \lim_{m \to \infty} \frac{1}{m} E(a_1, a_2, \ldots, a_m) \tag{12}$$

The flow can be considered an attack flow when the threshold value is higher or the same i.e., $E(a) \leq Th$ as the entropy value. Then, the proposed system will discard the IoT node. Th denotes the threshold value.

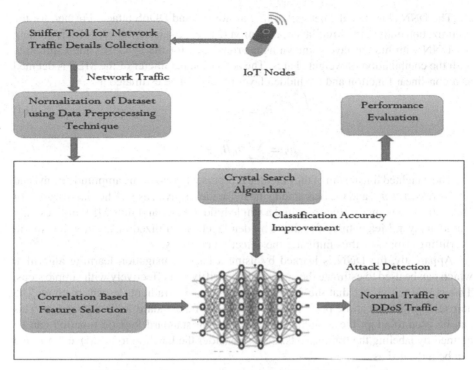

Fig. 2. Overall architecture of proposed DNN-CSA approach

3.3 Crystal Optimized Deep Neural Network

The steps involved in the detection of DDoS in the IoT nodes using the DNN are (i) gathering the network traffic, (ii) pre-processing stage, (iii) detection of the DDoS module. Overall architecture of the proposed work is illustrated in Fig. 2.

Gathering Network Traffic. Sniffer tools are utilized to gather the network traffic details in the IoT node in which the attack is made by the attackers. The tool is installed at the edge of the victim network. This usually gathers the incoming network packets by the routers of the attacked IoT node.

Pre-processing Stage. The collected features of DDoS attacks are preprocessed for normalization purposes. For the pre-processing, the min-max approach is utilized which falls under the range of 0 to 1.

$$z_m = \frac{z - \min(z)}{\max(z) - \min(z)} \tag{13}$$

The relevant features are determined as z_m the value that is relevant to the normalization in the time window is determined as z. For the detection of DDoS, the precise parameters are selected by the modules.

Detection of DDoS. This step usually categorizes the incoming network traffics of the IoT node that is attacked by the attackers. It employs three processes which are listed below.

The DNN classifies the network traffic as normal and DDoS attack. Further, for the optimization output, the Multilayer perceptron (MLP) [24] is exploited. MLP is nothing but DNN with hidden layers known as neurons used for the synapses and estimation with the combination of weighted arcs. The activation parameter of the MLP is deemed as a non-linear function and the hidden layer inputs can be estimated as,

$$B_i = h(\omega_i) \tag{14}$$

$$\omega_i = \sum \vartheta_{ij} B_i + \gamma_i \tag{15}$$

The weighted linear sum of the outcome is represented as ω_i, the amplitude from i and j is expressed as ϑ_{ij}, and the bias among the IoT nodes is given as γ_i. The classification of DDoS from the other traffics required a single hidden layer and thus MLP utilizes only a hidden layer. Meanwhile, if multiple hidden layers are utilized means it will result in overfitting errors and thus mitigates the detection accuracy.

Apparently, the DNN is learned by using a back propagation learning algorithm which can be used to estimate the computation swiftly and effectively with simple steps. This is due to the fact that the utilization of stochastic gradient descent method. The activation of MLP can be performed with the aid of a standard logistic function. This can be used to adopt the K number of classes. The standard logistic function can be defined by labeling the training sets. Let us consider the labeling to be a(i) and b(i) and can be indicated as

$$F\varphi(a) = \frac{1}{1 + e^{(-\varphi^T a)}} \tag{16}$$

Here φ is used to reduce the cost function. Most probably, the standard logic function [27] incorporated with cross-entropy mitigates the cost and achieves a better cost function for the DDoS classification in IoT networks. Further, to improve the classification and mitigation of DDoS in the IoT network, CSA is adopted. This effectively chooses the features from the traffic flow datasets and classifies them accordingly without falling for local optima. The following section explains the CSA in a wider manner.

Crystal Search Algorithm (CSA)

To enhance the classification with the circumvention of overfitting and cost issues, the CSA algorithm is adopted [29]. This relies on the framework of solid crystalline that contains atoms, molecules, or ions. The formation of crystal is made with the repeated organized above said components. Meanwhile, the diverse and isotropic characteristics of the crystal are at maximum. The lattice denotes the predetermined spaces and does not mean for the identification of the position of the atom i.e., detection of DDoS attacked IoT node. However, for the identification of position a concept called basis is used incorporated with lattice point. Hence, the crystals are formulated with two components such as lattice and basis.

The configuration of the atoms is represented by basis and the framework of various geometrical structures is determined as Lattice. According to the Bravais model [26], the lattice can be numerically determined. Prior to this, the periodic crystal structure is determined as vector form as shown below,

$$V = \sum r_i s_i \tag{17}$$

The principal crystallographic directions can be determined as the shortest vector s_i. The integer is denoted as r_i; i is the total number of corners in the crystals like the neighboring IoT nodes in the entire network system.

Numerical Expression. To analyze the optimized detection of DDoS, the solutions are considered in the lattice space of the single crystal. The total number of crystals is estimated for the purpose of iterations.

$$C = \begin{bmatrix} C_1 \\ C_2 \\ \vdots \\ C_3 \\ \vdots \\ C_n \end{bmatrix} = \begin{bmatrix} m_1^1 & m_1^2 & \dots & m_1^j & \dots & m_1^d \\ m_2^1 & m_2^2 & \dots & m_2^1 & \dots & m_2^d \\ \vdots & \vdots & \vdots & \vdots & \vdots & \vdots \\ m_i^1 & m_i^2 & \dots & m_i^j & \dots & m_i^d \\ \vdots & \vdots & \vdots & \vdots & \vdots & \vdots \\ m_n^1 & m_n^2 & \dots & m_n^j & \dots & m_n^d \end{bmatrix}, \tag{18}$$

$$\begin{cases} i = 1, 2, \dots, n \\ j = 1, 2, \dots, d \end{cases}$$

The number of IoT nodes can be indicated as n i.e., a number of crystals. The dimensionality of the issue is denoted as d. For simplicity, the IoT nodes are distributed randomly in the network that can be represented as,

$$m_i^j(0) = m_{i,\min}^j + \psi \left(m_{i,\max}^j - m_{i,\min}^j \right), \quad \begin{cases} i = 1, 2, \dots, n \\ j = 1, 2, \dots, d \end{cases} \tag{19}$$

The initial position of IoT nodes in the network is shown as $m_i^j(0)$. The minimum and maximized values are defined as $m_{i,\min}^j$ and $m_{i,\max}^j$ respectively. This is the value obtained for the ith candidate solution at j^{th} decision variable. ψ is the arbitrary value which comes under the range of 0 to 1.

The nodes that are located at the corner are considered as the base nodes C_B. This relies on randomly generated initial crystals. The detection of IoT node in the network can be analyzed and updated by using the lattice principle and are illustrated in four steps,

Simple Cubicle

$$C_{New} = C_{old} + rC_B \tag{20}$$

Cubicle Along with the Best Nodes

$$C_{New} = C_{old} + r1C_B + r2C_O \tag{21}$$

Cubicle with Respect to the Mean Nodes

$$C_{New} = C_{old} + r1C_B + r2M_C \tag{22}$$

Cubicle with Respect to the Best and Mean Nodes

$$C_{New} = C_{old} + r1C_B + r2M_C + r3C_O \tag{23}$$

C_O is the node with the best configuration mean value of the randomly selected IoT nodes is indicated as M_C. The updated position is represented as C_{New}. The location before the upgrading is denoted as C_{old}. . The random values are r1, r2 and r3. For the exploration and exploitation estimation, we have used equations from (20) to (23). The algorithm used for the enhancement of DDoS detection in IoT nodes is illustrated in algorithm 1.

Algorithm 1: DNN based Crystal Search Algorithm
Set the initial location of m_i^j
Evaluate the fitness value for each IoT node
When (t<terminating condition)
For i=I: number of initial IoT nodes
Produce C_B
Create a new location using eqn. (21)
Create C_O
Create a new location using eqn. (22)
Create M_C
Create a new location using eqn. (23)
Create a new location using eqn. (24)
If any IoT nodes violate the constraints scenarios then control the boundaries of the location to avert it
End if
Estimate the fitness values for the generated new located nodes
The DDoS is detected after analyzing the nodes
End for
$t = t + 1$
End when
Return the result
End

The proposed DNN based CSA approach to detect the DDoS attack in the IoT environment is explained in this section. The adopted CSA will enhance the classification accuracy of the MLP based DNN method. This method effectively classifies the normal and DDoS traffic and thereby helps to mitigate the DDoS traffic in the IoT network.

4 Experimental Results and Analysis

The efficiency of the proposed technique is evaluated in an IoT testbed and UNB-ISCX. The IoT testbed dataset is formed using the following configurations. The IoT devices are controlled using a Contiki open-source operating system and the cooja simulator is the one that is implemented in the Contiki OS. The Edimax EW-7416Apn is the access point used for the testbed and different IoT sensors such as Raspberry Pi are also used with a raspberry stretch OS. The proposed methodology is implemented in Matlab and the simulations are conducted on an Intel® Core™ i7-1185G7 Processor with vPro. The optimized parameter settings of the proposed model are presented in Table 1.

Table 1. Parameter settings of the proposed DNN-CSA approach

Parameter	Description
Balance parameters $(\theta_1 - \theta_4)$	0.7,0.5,0.01, and 0.8
Number of hidden layers	5
Number of hidden neurons	175
Learning rate	0.02
Number of epochs	500

4.1 Dataset Description

UNB-ISCX dataset is the benchmark dataset which mainly consists of synthetically recorded packet details acquired from 7 days [38, 39]. These details mainly imitate the real-time network traffic and it is a labeled attack dataset. The total number of tuples present in this dataset is $225, 745$. The two classes of this dataset are normal and which is used to test the efficiency of the proposed classifier with others. The dataset was partitioned into two where 70% is used for training and the remaining 30% is used for testing. The efficiency of the proposed classifier is mainly evaluated by comparing it with different classifiers such as RNN [9], DSH-RL, LDDM [15] and KNN algorithm [14].

4.2 Performance Metrics

The DDoS attack detection accuracy is measured via the following metrics:

Network Accuracy. It is mainly associated with the capability of the model to classify the normal and abnormal attacks correctly.

$$Network\_accuracy = \left(\frac{AA' + JJ'}{AA' + JJ' + JA' + AJ'} \right) \times 100 \tag{24}$$

Precision. It represents the ability of the classifier to differentiate the normal samples from the abnormal samples correctly.

$$Pr\,ecision = \left(\frac{AA'}{AA' + AJ'} \right) \times 100 \tag{25}$$

Recall. It determines the model's capacity to differentiate the abnormal attacks from the normal ones.

$$Recall = \left(\frac{AA'}{AA' + JA'}\right) \times 100 \tag{26}$$

F-measure. It mainly detects the model's ability to predict the DDoS attack by taking into account both the precision and recall.

$$F - Score = \frac{2}{\frac{1}{Pr\,ecision} \times \frac{1}{Recall}} \tag{27}$$

In the above equations, AA' represents the true positive, JJ' represents true negative, JA' represents false negative, and AJ' represents false positive.

4.3 Experimental Results

The results of the proposed methodology when compared to the existing techniques such as RNN, KNN, and LDDM are presented in Tables 2, 3, 4 and 5 in terms of network accuracy, precision, recall, and F-score respectively. The comparison is done with the number of true positives to estimate the sensitivity of the classifier. The true positive rate is mainly used to evaluate the DoS detection performance of the classifiers. The number of attacks is increased to test the accuracy of the systems. The higher performance offered by the proposed techniques is mainly due to the usage of the CRYSTAL algorithm for parameter tuning.

Table 2. Test results for accuracy

Tp	Accuracy			
	Proposed	LDDM [15]	RNN [9]	KNN [14]
50	93.64845	93.40259	92.63341	91.44459
60	95.52087	94.7461	93.72315	93.23712
70	96.46863	95.3407	93.67522	92.2164
80	97.6153	96.46534	94.71338	94.3849
90	98.63136	97.44616	96.59809	95.48255

The detection response time is the time taken by the proposed model to detect the attack and respond to it. This metric is evaluated using the testbed dataset via a background process. The background process mainly had a malicious IoT node generating an increased number of UDP packets. The results demonstrate the minimal time taken by the proposed approach to detect and respond to the attack with a varied UDP packet count.

Table 3. Test results for precision

Tp	Precision			
	Proposed	LDDM [15]	RNN [9]	KNN [14]
50	92.65845	91.41544	90.62744	89.28123
60	93.77847	93.08487	91.87761	90.42005
70	94.36319	93.46274	92.77468	91.49029
80	95.21276	94.15813	92.54428	91.31023
90	96.62384	95.75007	94.78902	93.82861

Table 4. Test results for Recall

Tp	Recall			
	Proposed	LDDM [15]	RNN [9]	KNN [14]
50	87.36456	86.71381	85.61267	84.40199
60	88.376	87.25288	85.64009	84.05751
70	89.09367	88.41491	88.19004	87.52577
80	90.8092	89.95438	88.84768	87.86042
90	91.82212	90.23145	89.74011	89.64176

Table 5. Test results for F-measure

Tp	F-measure			
	Proposed	LDDM [15]	RNN [9]	KNN [14]
50	89.93367	89.00258	88.04871	86.77307
60	90.99712	90.07457	88.64927	87.12277
70	91.65275	90.86878	90.42428	89.46413
80	92.95886	92.00826	90.65831	89.55211
90	94.1618	92.90888	92.19549	91.68741

4.4 Throughput

The throughput metric mainly identifies how much normal data packets reached the IoT server in a certain time period. The throughput performance is analyzed under attack using both the CRYSTAL algorithm optimized DNN and standard DNN. The results obtained are shown in Table 6. The throughput is low for the DNN technique when compared to the DNN-CSA technique. The throughput loss is mainly due to the malicious IoT node which floods the IoT server with UDP packets which affect the throughput of the normal packet in reaching the server.

Table 6. Throughput results

Time (seconds)	Throughput (pps)	
	DNN	Proposed DNN-CSA
0	0	110
20	10	125
40	15	135
60	20	148
80	23	152
100	27	165
120	32	180

4.5   Average Throughput

It mainly measures how many normal packets have traveled the IoT server safely during the DDoS attack. The Fig. 3 shows the results of the average throughput obtained when compared to the existing techniques. Based on the results, it is confirm that the proposed methodology offers a 28% improvement in terms of throughput when compared to the existing techniques.

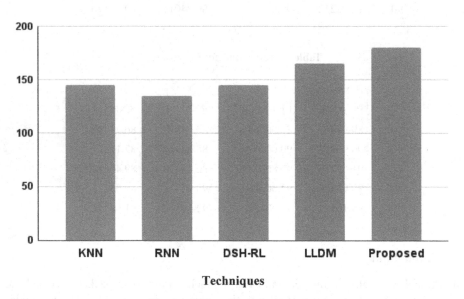

Fig. 3. Average throughput results

4.6   Energy Consumption

The average energy consumed by each IoT node is computed in Joules at the time of simulation and the results are provided in Table 7. The time consumption is graphically represented in Fig. 4 by comparing the proposed approach with different existing techniques such as RNN [9], DSH-RL [8], and KNN algorithm [14]. Based on the results, it is observed that the proposed methodology shows low energy consumption when compared to the existing techniques. The energy consumption of the proposed DNN-CSA algorithm is lower than the KNN, DSH-RL, and RNN algorithm.

Table 7. Comparative analysis using energy consumption

Time (Seconds)	RNN [9]	DSH-RL [8]	KNN ALGORITHM [14]	Proposed DNN-CSA
200	1	0.95	0.9	0.105
300	1.15	1.02	0.98	0.79
400	1.54	1.39	1.32	0.98
500	1.87	1.45	1.47	1

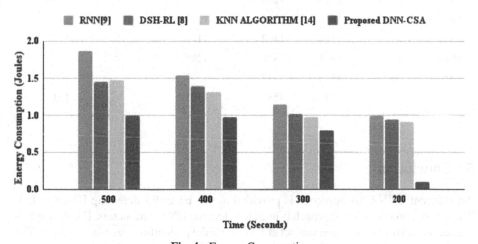

Fig. 4. Energy Consumption

4.7   Memory Utilization

Memory utilization is the memory utilized by the proposed methodology to detect DDoS attacks and it is usually measured on the size of the data traffic generated. The memory utilization is statistically represented as follows:

$$Memory\_utilization = Data\_traffic \times Memory\_required \tag{28}$$

The memory consumption is mainly computed in terms of Megabyte in the proposed work. The memory utilization efficiency of the proposed technique is compared with the existing techniques such as RNN, DSH-RL and KNN algorithms. The results obtained are shown in Table 8. The data traffic size is varied from 100–1000 Mbps. The results show that the proposed methodology utilizes less memory in detecting DDoS attacks when compared to the existing techniques.

Table 8. Memory utilization results

Data traffic size (Mbps)	Memory Utilization (MB)			
	RNN [9]	DSH-RL [8]	KNN ALGORITHM [14]	Proposed
100	525	856	854	631
250	550	901	878	654
300	690	956	895	698
350	721	987	904	700
400	839	995	926	758
450	954	1012	936	800
500	1014	1245	968	950
750	1230	1366	1014	995
800	1324	1402	1139	1121
850	1358	1456	1205	1187
950	1456	1524	1301	1201
1000	1526	1589	1401	1310

5 Conclusion

An efficient DNN-CSA approach is provided in this paper for detecting DDoS in IoT. The Crystal optimization approach is used to select the DNN parameters. For the experiments, the maximum iteration was set at 100, while the population size was set at ten. The features are extracted using the wrapper method, embedded method, and filtering using common attributes. Furthermore, by aggregating the sequence of data between the source and destination IoT nodes, the network flow may be computed. The bandwidth monitor sends the threshold values to each IoT node in the network to assess traffic abnormalities. The abnormalities, on the other hand, are monitored by the collaborated reduction manager using a timer, admission controller, and bandwidth monitor. Using the UNB-ISCX and IoT testbed dataset, the efficiency of the proposed methodology is evaluated with other techniques. The efficiency of the technique will be measured using several performance indicators such as accuracy, F1-Score, recall, precision, energy consumption, throughput, etc. For DDoS attack detection, the proposed DNN-CSA algorithm has a detection accuracy, precision, recall, and F1-score of 98.6%, 96.6%, 91.8%, and 94.1%, respectively. The results show that the proposed methodology is efficient in attaining higher throughput, energy consumption, and memory utilization.

References

1. Lee, I., Lee, K.: The Internet of Things (IoT): applications, investments, and challenges for enterprises. Bus. Horiz. **58**(4), 431–440 (2015)
2. Gubbi, J., Buyya, R., Marusic, S., Palaniswami, M.: Internet of Things (IoT): a vision, architectural elements, and future directions. Futur. Gener. Comput. Syst. **29**(7), 1645–1660 (2013)
3. Soumyalatha, S.G.H.: Study of IoT: understanding IoT architecture, applications, issues and challenges. In 1st International Conference on Innovations in Computing & Net-working (ICICN16), CSE, RRCE. International Journal of Advanced Networking & Applications, no. 478 (2016)
4. Su, X., Wang, Z., Liu, X., Choi, C., Choi, D.: Study to improve security for IoT smart device controller: drawbacks and countermeasures. Secur. Commun. Netw. **2018**, 1–14 (2018)
5. Liang, L., Zheng, K., Sheng, Q., Huang, X.: A denial of service attack method for an iot system. In: 2016 8th International Conference on Information Technology in Medicine and Education (ITME), pp. 360–364. IEEE (2016)
6. Džaferović, E., Sokol, A., Almisreb, A.A., Norzeli, S.M.: DDoS and vulnerability of IoT: a review. Sustain. Eng. Innov. **1**(1), 43–48 (2019)
7. Ge, M., Fu, X., Syed, N., Baig, Z., Teo, G., Robles-Kelly, A.: Deep learning-based intrusion detection for IoT networks. In: 2019 IEEE 24th Pacific Rim International Symposium on Dependable Computing (PRDC), pp. 256–25609. IEEE (2019)
8. Zaminkar, M., Sarkohaki, F., Fotohi, R.: A method based on encryption and node rating for securing the RPL protocol communications in the IoT ecosystem. Int. J. Commun Syst **34**(3), e4693 (2021)
9. Latif, S., Zou, Z., Idrees, Z., Ahmad, J.: A novel attack detection scheme for the industrial internet of things using a lightweight random neural network. IEEE Access **8**, 89337–89350 (2020)
10. Dantas Silva, F.S., Silva, E., Neto, E.P., Lemos, M., Venancio, A.J., Esposito, F.: A taxonomy of attack mitigation approaches featured by SDN technologies in IoT scenarios. Sensors **20**(11), 3078 (2020)
11. Pathak, G., Gutierrez, J., Rehman, S.U.: Security in low powered wide area networks: opportunities for software defined network-supported solutions. Electronics **9**(8), 1195 (2020)
12. Mandal, S., Khan, D.A., Jain, S.: Cloud-based zero trust access control policy: an approach to support work-from-home driven by COVID-19 pandemic. N. Gener. Comput. **39**(3), 599–622 (2021)
13. Torres, N., Pinto, P., Lopes, S.I.: Security vulnerabilities in LPWANs—an attack vector analysis for the IoT ecosystem. Appl. Sci. **11**(7), 3176 (2021)
14. Wazirali, R.: An improved intrusion detection system based on KNN hyperparameter tuning and cross-validation. Arab. J. Sci. Eng. **45**(12), 10859–10873 (2020). https://doi.org/10.1007/s13369-020-04907-7
15. Ravi, N., Shalinie, S.M.: Learning-driven detection and mitigation of attack in IoT via SDN-cloud architecture. IEEE Internet Things J. **7**(4), 3559–3570 (2020)
16. Ghahramani, M., Javidan, R., Shojafar, M., Taheri, R., Alazab, M., Tafazolli, R.: RSS: an energy-efficient approach for securing IoT service protocols against the DDoS attack. IEEE Internet Things J. **8**(5), 3619–3635 (2020)

17. Li, J., Liu, M., Xue, Z., Fan, X., He, X.: Rtvd: a real-time volumetric detection scheme in the internet of things. IEEE Access **8**, 36191–36201 (2020)
18. Li, J., Xue, Z., Li, C., Liu, M.: RTED-SD: a real-time edge detection scheme for sybil in the internet of vehicles. IEEE Access **9**, 11296–11305 (2021)
19. Yin, D., Zhang, L., Yang, K.: A attack detection and mitigation with software-defined Internet of Things framework. IEEE Access **6**, 24694–24705 (2018)
20. Mubarakali, A., Srinivasan, K., Mukhalid, R., Jaganathan, S.C., Marina, N.: Security challenges in internet of things: distributed denial of service attack detection using support vector machine-based expert systems. Comput. Intell. **36**(4), 1580–1592 (2020)
21. DDoShi, R., Apthorpe, N., Feamster, N.: Machine learning detection for consumer internet of things devices. In: 2018 IEEE Security and Privacy Workshops (SPW), pp. 29–35. IEEE.(2018)
22. Baig, Z.A., Sanguanpong, S., Firdous, S.N., Nguyen, T.G., So-In, C.: Averaged dependence estimators for DDoS attack detection in IoT networks. Futur. Gener. Comput. Syst. **102**, 198–209 (2020)
23. Lawal, M.A., Shaikh, R.A., Hassan, S.R.: Security analysis of network anomalies mitigation schemes in IoT networks. IEEE Access **8**, 43355–43374 (2020)
24. Aljuhani, A.: Machine learning approaches for combating distributed denial of service attacks in modern networking environments. IEEE Access **9**, 42236–42264 (2021)
25. Perez-Diaz, J.A., Valdovinos, I.A., Choo, K.K.R., Zhu, D.: A flexible SDN-based architecture for identifying and mitigating low-rate attacks using machine learning. IEEE Access **8**, 155859–155872 (2021)
26. Alamri, H.A., Thayananthan, V.: Bandwidth control mechanism and extreme gradient boosting algorithm for protecting software-defined networks against attacks. IEEE Access **8**, 194269–194288 (2020)
27. Fuentes-García, M., Camacho, J., Maciá-Fernández, G.: Present and future of network security monitoring. IEEE Access **9**, 112744–112760 (2021)
28. Zhou, J., Cao, Z., Dong, X., Vasilakos, A.V.: Security and privacy for cloud-based IoT: challenges. IEEE Commun. Mag. **55**(1), 26–33 (2017)
29. Sahoo, K.S., Puthal, D., Tiwary, M., Rodrigues, J.J., Sahoo, B., Dash, R.: An early detection of low rate attack to SDN based data center networks using information distance metrics. Futur. Gener. Comput. Syst. **89**, 685–697 (2018)

Prediction Based Load Balancing in Cloud Computing Using Conservative Q-Learning Algorithm

K. Valarmathi$^{(\boxtimes)}$ ⓘ, M. Hema ⓘ, A. Mahalakshmi, S. Jacqulin Elizabeth ⓘ, and E. Arthi ⓘ

Department of Information Technology, Easwari Engineering College, Ramapuram, Chennai, India
Valar.me964@gmail.com

Abstract. Cloud computing is a striking expertise trend that provides Computing assets as a service experiencing a revolution for the IT industry and academics researchers. The potential of emergent cloud computing technology is efficiently hooked by the primary requisite such as resource management. The furthermost vital aspect of asset controlling techniques in Cloud environment depends on scheduling and Load Balancing techniques. Load Balancing strategy is attracting and generating considerable interest in order to utilize resources, thereby increasing the enactment of the cloud datacenter. The energy consumption in the datacenter is customarily due to improper utilization of resources in terms of overloading the Servers or sometimes due to idle Servers. Load is flourishing in the recent era; the Internet based Computing proposals shared resources such as hardware, software, and information on the demand basis. Cloud Computing carry amendments, and the revolution of the Information Technology industries emerged with its popularization and applications. Balancing is one of the best solutions for efficient utilization of resources and is extensively considered to be the most important method to decrease energy utilization. The prime aim of the exploration work is to scrutinize various Load Balancing techniques and propose an energy-aware Load Balancing strategy for ideal utilization of the assets in cloud Computing environment. The proposed Conservative Q-learning algorithm (CQA) for maintaining efficient equilibrium between the work load among virtual machines and optimally lessens the energy ingesting through the Load Balancing algorithms proposed by us.

Keywords: Cloud Computing · Load Balancing · CQA · Energy Efficiency · Cost

1 Introduction

Cloud Computing is a up-to-the-minute, leading-edge figuring model connected over the Internet. It is flourishing in the recent era; the This proposals sharing the resources such as hardware, software, and information on the basis of demand. Cloud Computing carry amendments, and the revolution of the Information Technology industries emerged with its popularization and applications. The service vitality, suppleness, sturdiness and bounciness managed by this scalable technology make cloud computing an essential part of giant business handling environments [1]. SaaS, IaaS and PaaS are provided to the businesses through the Internet by Cloud Computing. Operative treatment of the cloud environment is fundamental to get greatest guides out of it. Figure 1 illustrates the typical cloud computing paradigm. Cloud Computing is a virtualized image-based paradigm that becomes prevalent for providing a lively infrastructure and remote distribution for many applications. It is the Internet based Computing, that is fit for restructuring IT processes and commercial centers.

1.1 Characteristics of Cloud Computing

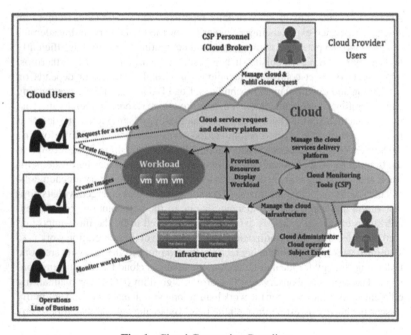

Fig. 1. Cloud Computing Paradigm

The cloud features can be labeled as fundamental and as the common characteristics of Cloud Computing, which differentiates from other Computing advancements.

1.2 Cloud Service Models

Cloud Computing provides various kinds of Services such as Software as a Service, Platform as a Service and Infrastructure as a Service [2]. All three Service models are used for abstracting the physical resources and offering these resources as services for the cloud users.

1.3 Load Balancing

It is the most common way of designating the load across the different virtual machines in the data center. The work-load has to be allocated to the resources, so that each one of them ought to have similar measure of load anytime. In the event that the load is imbalanced, the performance of the framework will be definitely reduced. The system should follow an efficient Load Balancing techniques to enable the promotion of the availability of the assets and to upsurge the performance [3].

This technique in a cloud environment is a crucial issue for ensuring optimal utilization of resources and fast processing time. The capability of each VM in the datacenter is determined by the summation of expected Reckoning Time of autonomous jobs allocated. Load Balancing works with resource usage which furnishes throughput with least Response Time by sharing jobs. The foremost aim of this procedure is to further develop execution by adjusting jobs among virtual machines.

Likewise, it accomplishes ideal asset use, expands reaction time, amplifies throughput and stays away from over-burden. Different Load Balancing calculations are utilized for various frameworks. The Load Balancing algorithms are specified in Fig. 2 based on the following strategies:

Data policy. It indicates the data based on which workload has been assigned when it is to be assigned, and where to be assigned.

Triggering policy. This policy determines the proper time period in which the balancing operations can be done.

Resource policy. It categorizes the cloud resources according to availability.

Location policy. It finds an appropriate partner for a Server based on the resource type policy.

Choice policy. It isolates a precise task, which can be migrated from over- loaded VM to the idlest VM.

1.4 Classification of Load Balancing Techniques

There are two basic techniques in load balancing.

Static Method. It uses prior information of the features of the task, Computing assets, and Networks. This technique does not rely upon the system's current state. The balancing decisions are made at compile time. A Static Load Balancing algorithm designates load among the virtual machines before the execution of an algorithm. Dynamic changes of attributes as well as workload are unable to be overseen in a Static technique [4].

Dynamic Method. The Dynamic Load Balancing algorithm doesn't record the previous state behavior of the resources; it keeps track of only the present status behavior. It makes changes to the workload among virtual machines during runtime. The Load Balancing decision is made on the basis of current load information. In Dynamic Load Balancing techniques, the Task Migration can happen assuming the task has been allocated to over loaded VM [5].

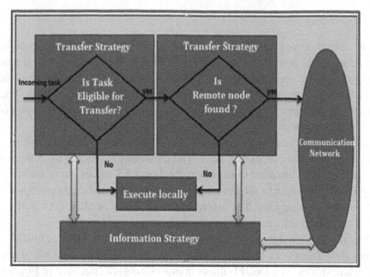

Fig. 2. Load Balancing Policies

1.5 Load Balancing Metrics

Load Balancing approach is the distribution of tasks assigned equal among the available VMs. The qualitative metrics in a cloud environment are discussed as follows:

Scalability. Scalability is the ability to perform Load Balancing of the system with an ample quantity of nodes or resources.

Response time. The minimum time taken for the response of the Load Balancing algorithm. To increase the performance of the system, the response time should be minimized.

Throughput. Through put is defined as the total number of tasks that has completed execution. Higher throughput is required for better performance.

Fault Tolerance. The ability of a system which continues processing though there is a failure on a specific processing unit in the system.

Migration time. The amount of time required for transferring the task from one VM to the other. Performance will be better when migration time is lesser.

Resource utilization. The degree wherein the cloud resources are used, for better performance, for which the maximum use of resources is required.

Overhead: The amount of overhead involved while executing the Load Balancing algorithm. Less overhead means the algorithm is more efficient.

Performance. Performance is defined as the efficiency of a system. Performance has to be more.

Make span. The maximum completion time of tasks that are allocated to the resources is called Make span.

1.6 Motivation

Cloud Computing can go about as an unavoidable and get through major advantage in every task of resource concentrated applications like service provisioning, collaborative strategies, operating design models and end-user high quality services.

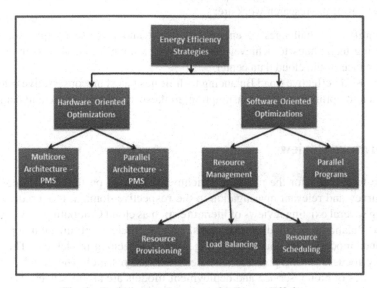

Fig. 3. Optimization Measures for Energy Efficiency

The main motivation behind Cloud Services is to offer a fast and simple way of access to the virtual machines and a vast effective circulative application. Due to a large number of tasks submitted to the virtual machines in public cloud, Load Balancing is the tasks among VM is a very important criterion for increasing performance and reducing the power and cost consumption.

Figure 3 shows the different energy efficient strategies in cloud computing. The performance is limited, when the virtual machines are overloaded and result in downtime and outages, with a consequent drop in the system utilization in public cloud. Approximately 80% of the energy consumption is due to the inefficient usage of computing resources as well as the existence of idle servers in the datacenter that consumes up to 60% of the peak power. High energy consumption leads to increase in TCO and decrease in Return on Investment.

1.7 Problem Statement

To investigate on available Load Balancing techniques and propose energy efficient Load Balancing algorithms for optimal utilization of virtual machines thereby optimizing energy consumption, response time, make span and datacenter cost.

1.8 Research Objectives

The prime goal of this research work is to enable an energy-aware Load Balancing scheme for utilizing the cloud resources optimally and increases the performance of the system. This work aims to identify the research gaps and deduce efficient Load Balancing techniques based on nature-inspired optimization approach to solve the issues identified. The objectives of the research work are:

- To identify the challenges of energy-efficiency and propose energy-aware Load Balancing techniques to achieve balanced load among VMs, also to increase the performance of the cloud data center.
- To propose an effective Load Balancing techniques based on conservative q-learning approach to optimize energy consumption, makespan, response time and data center cost.

2 Literature Review

The basic foundation for the research enrichment is based on the comprehensive literature survey and relevant investigation in the respective domain. It is predominantly discussing several existing reviews of literature such as cloud Computing and virtualization, Load Balancing in cloud environments, energy efficient optimization approaches, performance modeling in cloud Computing based on queuing model, etc. This survey principally focuses on the problem statement and the research The cloud definition, characteristics of cloud, service and deployment models are all objectives. The current research methodologies and their respective strengths are identified with their limitations. The NIST has provided a structure for cloud Computing [6]. Shafiq DA et al. has made an elegant presentation of the cloud Computing architecture, that has a datacenter that stocks an enormous network of computers [7]. All kind of web administrations is provided by the Central Repository known as the datacenter in the cloud. The organization of computers are housed in the datacenter to support the requirements in SLA are presented.

Alqahtani B et al. has represented a comparison of grid Computing and the cloud Computing in diverse viewpoint with appropriate technologies [9]. The IaaS techniques and tools used for provisioning in virtualized mode for the end users are presented. Sohani M et al. have explained cloud Computing in a business perspective. The weakness, opportunities, and strengths of the cloud are briefed, and stakeholders related issues with recommended practitioners for operating and managing the services has focused on market-situated resource oriented procedures that embrace the risk managing techniques and service management approaches based on the client to help the SLA-based resource allocation have elegantly presented the benefits and the opportunities of inter-cloud for consumers. The scalable applications are provided and operated in different datacenters on various geographical location also developed a tool called Cloud Sim tool for simulating the performance measures in a cloud environment [10].

3   Methodology

Conservative Q-learning Algorithm (CQA) is a meta-heuristic approach, inspired by the metallurgical process, Annealing. The reenacted annealing process begins with an initial and an upgraded solution. The procedure's solution will be generated if the value of the Fitness f(S*) is smaller than f(S).

$$\hat{Q}^{\pi}_{CQL} \leftarrow arg \min_{Q} \max_{\mu(a|s)} \underbrace{\left(\mathbb{E}_{s\sim data, a\sim\mu}[Q(s, a)] - \mathbb{E}_{s\sim data}[Q(s, a)]\right)}_{CQL\ regularizer}$$

$$+\frac{1}{2\alpha} \underbrace{\mathbb{E}_{s,a,s'\sim data}\left[r(s, a) + \gamma\mathbb{E}_{\pi}[\overline{Q}(s',\ a')] - Q(s, a))^2\right]}_{} \tag{1}$$

Equation (1) represents the formula of Conservative Q- Learning Algorithm, where,

- 's' stands for State
- 'a' stands for Action
- 'Q(s, a)' stands for the Q-function under
- (state, action) pairs
- 'E' stands for Estimate Function

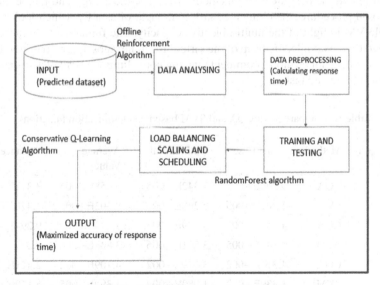

Fig. 4. Proposed System Architecture

The greater fitness value of S* is considered with the defined probability. This specific approach empowers the looking through cycle to keep away from the trap in neighborhood optima. Here f(S*) represents the fitness function of the neighbor solution, and f(S) represents the fitness function of the current solution. Temperature Tm defines the control parameter. The balance state is accomplished in light of the progression of moves and in view of the cooling rate; the temperature control not entirely set in stone.

Figure 4 shows the overall system architecture of proposed system. The control parameter Tm influences the performances of the worldwide surfing. In the event that the temperature gets more initial value, then the calculated annealing process gets a greater chance. After the progression of a reduction in temperature, the CQA technique will be ended, on a chance there are no upgrades. The chance of finding a solution for worldwide is additionally restricted, assuming the initial temperature is low, and the computation time will be limited [11].

4 Experiment Results

In this section, the different experiments are represented. The comparison of CQA with Hybrid SVM is represented first. The experiment use four notable bench-mark optimization functions to compare original CQA with SVM. We correlate the optimization of correctness and the concurrence speed using tight iterations and community size. We used fly size as 40 and the number of iterations as 90 in our experiment. The experimental results are depicted in Table 1. The order of magnitude is better for CQA analyzed with SVM for all the four functions. From Table 2, it can be seen that the nearer to the calculated value and the circulation frequency of CQA is lower than SVM. The SA approach is utilized to find the best ideal arrangement in view of the energy and temperature. In our proposed procedure, the eliminated task is thought of as a fly, which looks for the reasonable VM in light of the multi-objective capacity. The fundamental requirements are followed, for example, the heap of the virtual machine, subsequent to doling out the undertaking ought not be more prominent than the upper limit worth to pick a reasonable VM for the eliminated task.

Table 1. Comparison of CQA and SVM based on Optimization Functions

Optimization Function	Method	Worst Case	Optimal Value	Average Value	Variance
f(x)	CQA	9.294E−005	7.342E−005	8.258E−005	3.311E−001
	SVM	1.201E−005	7.268E−006	8.402E−006	1.47E−005
f2(x)	CQA	4.319E−005	3.589E−005	9.294E−00	9.294E−002
	SVM	4.343E−005	3.551E−006	3.848E−006	9.294E−005
f3(x)	CQA	4.311E−002	3.779E−002	4.109E−002	2.421E−005
	SVM	1.339E−002	1.089E−002	1.399E−005	5.179E−007
f4(x)	CQA	1.852E−002	1.389E−002	1.613E−002	1.191E−006
	SVM	3.12E−003	1.275E−003	1.29E−003	7.512E−008

On the off chance that there is a more noteworthy number of the VM is accessible, cutoff time limitation is thought of. The cutoff time of the assignment is basic to move the undertaking from weighty stacked VM to low stacked VM. In the event that the cutoff time of the eliminated task is high, the VM having at least higher cutoff time task is chosen. On the off chance that the cutoff time of the undertaking is medium, the VM having fewer higher and medium cutoff time task is chosen. The VM gathering depends on the ongoing burden LVM(t) of the virtual machine. We take two kinds of groups like overloaded VM group and underloaded VM group. The task that's overloaded is removed from the overloaded group and allocated to the underloaded group on the basis of objective function. The process of removing the task from overloaded set is continued, till the underloaded is φ. This work centers around Load Balancing as well as distillates on saving the energy consumed in the datacenter to lessen cost. The extreme course of energy preservation depends on making the virtual machines to ON and OFF state, which isn't being used [12]. It distinguishes the Material virtual machines in the datacenter, which is underutilized and significantly having an impact on the state from dynamic to rest.

Table 2. Comparison of Energy Consumption of Various Load Balancing Algorithms

No of Tasks	CQA	SVM	PSO
100	1.13	1.17	1.78
200	1.20	1.28	3.12
300	2.01	2.21	4.34
400	3.08	3.20	7.21

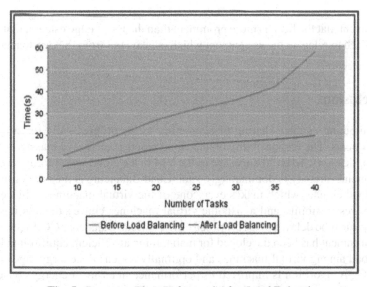

Fig. 5. Response Time Before and After Load Balancing

Figure 5 shows the response time of proposed techniques. Figure 6 shows the energy consumption of proposed technique. The proposed approach centers around Load Balancing issues as well as lessens the energy and datacenter cost is additionally stressed. In view of the edge esteem, the responsibilities doled out to each VM are adjusted in the datacenter alongside the dozing system to make the VM in a rest state, on the off chance that there is no heap relegated to the virtual machine.

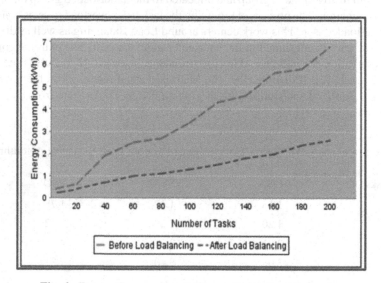

Fig. 6. Energy Consumption Before and After Load Balancing

In the event that the heap is more prominent than the lower edge esteem, that specific VM is utilized to relegate the assignment which is taken out from the overloaded virtual machine.

5   Conclusion

Cloud Computing enables sharing of computing assets progressively to an extensive variety of clients. The responsibility on the cloud assets is increased massively towards the new applications. To build the usage of the VM in the cloud server farm, an effective Load Balancing strategy is dominatingly vital. Load Balancing is the core of the server farm in cloud climate which makes every one of the virtual machines achieve similar measure of responsibility and assists the virtual machines convey the administrations with negligible time delay. So this proposed efficient Load Balancing CQA techniques in cloud environment has been developed for maintaining an efficient equilibrium between the workload among virtual machines and optimally lessens the energy ingesting. Our proposed CQA algorithm is improved its performance in terms of energy consumption by 37% than the existing PSO algorithm and 21% by SVM algorithm. As part of our future work, it might be improved with other QOS factors, for example, versatility and organization traffic data in a cloud environment. The organization traffic in the cloud likewise consumes a non-trivial measure of energy which builds the datacenter cost and it should carry out the energy effective Load Balancing approaches in Fog Computing.

References

1. Souravlas, S., Anastasiadou, S.D., Tantalaki, N., Katsavounis, S.: A fair, dynamic load balanced task distribution strategy for heterogeneous cloud platforms based on Markov process modeling. IEEE Access **10**, 26149–26162 (2022)
2. Kishor, A., Niyogi, R., Chronopoulos, A., Zomaya, A.: Latency and energy-aware load balancing in cloud data centers: a bargaining game based approach. IEEE Trans. Cloud Comput. (2021)
3. Saxena, D., Singh, A.K., Buyya, R.: OP-MLB: an online VM prediction-based multi-objective load balancing framework for resource management at cloud data center. IEEE Trans. Cloud Comput. **10**(4), 2804–2816 (2021)
4. Hung, L.H., Wu, C.H., Tsai, C.H., Huang, H.C.: Migration-based load balance of virtual machine servers in cloud computing by load prediction using genetic-based methods. IEEE Access **9**, 49760–49773 (2021)
5. Alqahtani, J., Hamdaoui, B., Langar, R.: Ernie: scalable load-balanced multicast source routing for cloud data centers. IEEE Access **9**, 168816–168830 (2021)
6. Dong, Y., Xu, G., Zhang, M., Meng, X.: A high-efficient joint 'cloud-edge' aware strategy for task deployment and load balancing. IEEE Access **9**, 12791–12802 (2021)
7. Nezami, Z., Zamanifar, K., Djemame, K., Pournaras, E.: Decentralized edge-to-cloud load balancing: Service placement for the Internet of Things in IEEE. Access **9**, 64983–65000 (2021)
8. Shafiq, D.A., Jhanjhi, N.Z., Abdullah, A., Alzain, M.A.: A load balancing algorithm for the data centres to optimize cloud computing applications. IEEE Access **9**, 41731–41744 (2021)
9. Wu, X., You, L., Wu, R., Zhang, Q., Liang, K.: Management and control of load clusters for ancillary services using internet of electric loads based on cloud–edge–end distributed computing. IEEE Internet Things J. **9**(19), 18267–18279 (2022)
10. Jasim, M.A., Siasi, N., Rahouti, M., Ghani, N.: SFC provisioning with load balancing method in multi-tier fog networks. IEEE Netw. Lett. **4**(2), 82–86 (2022)
11. Valarmathi, K., Kanaga Suba Raja, S.: Resource utilization prediction technique in cloud using knowledge based ensemble random forest with LSTM model. Concurr. Eng. **29**(4), 396–404 (2021)
12. Kanagasubaraja, S., Hema, M., Valarmathi, K., Kumar, N., Kumar, B.P., Balaji, N.: Energy optimization algorithm to reduce power consumption in cloud data center. In: 2022 International Conference on Advances in Computing, Communication and Applied Informatics (ACCAI), pp. 1–8. IEEE (2022)

An Attribute Selection Using Propagation-Based Neural Networks with an Improved Cuckoo-Search Algorithm

Priyanka$^{(\boxtimes)}$ and Kirti Walia

University Institute of Computing, Chandigarh University, Gharuan, Mohali, Punjab, India
priyankatuli1986@gmail.com

Abstract. Sentiment Analysis is getting an area of implication of the researchers in the business as well as in the research. To know about the opinions of human beings through artificial machines have always been interesting thing to note and that has been getting updated from all over the world as the time has passed by. The research article presents an attribute selection mechanism by using enhanced Cuckoo Search algorithm which is known as meta-heuristic categorized algorithm. A novel fitness function which has been designed and Neural Networks have been used for the training and validation to get the proposed solution. The proposed algorithm has also been compared to state art of the art techniques based on quantitative parameters. Accuracy has been considered as a main objective. The detailed result and analysis have shown that integration of neural network have demonstrated distinguishing results in terms of performance parameters, namely, precision, recall, f-measure, and accuracy of polarity classification.

Keywords: Artificial Neural Network · Cuckoo Search · Sentiment Analysis · Opinion Mining

1 Introduction

Messaging, social media connections, blogging, and tweeting are now the most popular online activities. Twitter is one of the most famous microblogging platforms and might be regarded as one of the biggest user-generated data sites with a vast quantity of organised and unstructured information [1]. According to the interests of the viewers, the uploaded tweets might indicate their thoughts on various issues and their polarity regarding these subjects.

Social media has evolved into a legitimate means of communication for individuals to express their opinions and points of view on any topic [2]. Numerous studies must investigate these consumer viewpoints. In addition, they depend on the comments offered by diverse Internet users. This may greatly alter the product's purchasing behavior. Consequently, studying the views or feelings of the user arises as a crucial area of research. Sentiment analysis (SA) is the study of recognizing and classifying the public's views, sentiments, emotions, and attitudes about any subject, person, or event. Additionally, the view is classified as good, negative, or neutral.

R. K. Challa et al. (Eds.): ICAIoT 2023, CCIS 1929, pp. 48–59, 2024.
https://doi.org/10.1007/978-3-031-48774-3_4

Sentiments and opinions are both very indispensable in almost every human behavior and it is very major to provide effect on the manner of human. Human values and explanation of facts and their actions depend on how someone explains and assess the surrounding at some level of significance. While the individual needs to make a conclusion, they always depend to a considerable level on other opinions for a person or an organization. The fast growth in this field conflict with the social platform on the internet, such as micro forums, forum discussions, reviews, Facebook, Twitter and social networks, due to the extensive amount of personalized data recorded in digital form. So, mining will be examined for the data and point out the consumer emotions, therefore one of the prime tasks that has drawn the research circle attention from the last decade. Judgement mining is to take out the information from different variety of text and mark it as constructive, obstructive, or unbiased on the basis of its sentiment or polarity [1].

Opinion Mining (OM) is well known for vast in sequence on spiritual insertion. SA is the study of the views of elements, conclusions, and subjectivity. The element represents a person, occasion, or point. The basic work in information mining can be arranged in two corresponding ways: grouping, and the cluster. Sentiment mining is dealt about investigate a person's sensation, feelings, and way of thinking from a given bit of text. "Mining is required to the use of NLP, computer linguistics and content analysis to classify and get subjective knowledge in source materials". Sentiment mining deals with investigate and forecast the hidden details stored in the text [2]. Some samples of sentiment mining are given below:

a) Subjective: This camera by Sony is "remarkable". This explained sentence has a sentiment and so, wind up that it is subjective.
b) Objective: I purchase this camera twelve months before. This sentence is indicating the reality of past, and hence it is objective. The subjective text can be further classified in below provided three separate categorizations based on the sentiments conveyed in the text.
   i. Constructive: I love to use this camera.
   ii. Obstructive: The picture quality of the camera is bad.
   iii. Unbiased: I usually take pictures in the noon.

1.1 Corpus-Based Approach

This method is implemented to address the challenges of context-specific orientations to locate the opinion mining words present in the textual sample. This method works based on the grammatical patterns that are a part of typical opinion words usually present in the large textual samples [3].

The designed framework for the sentiment analysis is separated into two parts. The first division is aimed to establish a proposed model and the second part comprises of testing the performance of the designed model. The training of the corpus model has been done by collecting data from various social sites such as Facebook, Instagram, Twitter and getting characterized by them on the basis of positive or negative opinions. The textual information is used as the sample is not always normalized. Therefore, it requires a process of pre- processing in order to combine different forms and remove unnecessary words from the main textual information (Fig. 1).

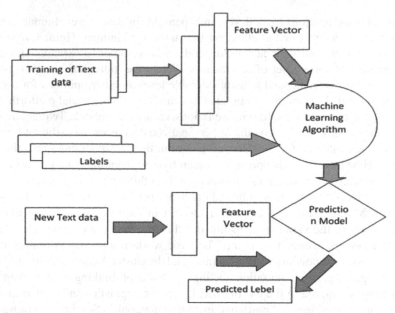

Fig. 1. Adopted Corpus-based approach.

1.2 Machine Learning Utilization and Swarm Intelligence Approach

Imparting knowledge to a machine in order to get some tasks done is categorized under machine learning (ML) and Swarm Intelligence (SI) is a branch of study under meta-heuristic architecture system. SI has been observed mostly for the feature selection approach in case of sentiment analysis or pattern analysis research works [4, 5]. In order to enhance the prediction or the classification accuracy, the proposed work contributes in the following manner.

The proposed work extends the usage of machine learning by applying novel feature selection architecture inspired by Cuckoo-Search algorithm. In addition to this, natural computing inspired enhanced GA has been used for the attribute set assortment mechanism. The paper is organized in the following manner. Section 2 describes the related work whereas Sect. 3 describes the methodology. Section 4 describes the result and the evaluated parameters whereas the paper is concluded in Sect. 5.

2 Related Work

The work is dedicated to presenting the detailed literature analysis of the past research work pertaining to the field of the text mining leading to the analysis of the opinions and sentiments. The leading research work revolving around polarity classification based on the improvement at different stages such as the text mining, feature extraction and selection using optimization approaches, and various methods employed for the improved training and classification presented by the researchers are summarized below (Table 1).

Table 1. Related Work

Author	Techniques	Dataset	Results
Pak, A et al. (2010) [6]	SVM	Twitter	The accuracy using bi-gram is approximately equal to 68.18% and is better as compared to unigram and trigram approaches
Glorot, X. et al. (2011) [7]	Deep learning and SVM	Amazon	It was analyzed that Stacked De-noising Auto encoder provides better results than SVM
Patil et al. (2015) [9]	SVM & ANN	Facebook	Support Vector Machine (SVM) gave better results with text classification than the artificial neural network (ANN)
Sosa and P. M (2017) [10]	CNN	Twitter	The average accuracy obtained by using LSTM-CNN is high about 75.2%
Wehrmann et al. (2017) [11]	CNN	Twitter	The experiment was carried out in 4 languages, including English, German, Portuguese, and Spanish
Xia et al. (2017) [1]	–	'Twitter	The performance as per the accuracy of classification and computation efficiency has been measured

(continued)

Table 1. (*continued*)

Author	Techniques	Dataset	Results
Pandey et al. (2017) [5]	K-mean algorithm along with Cuckoo search technique	Twitter	The comparison has been performed with various optimization techniques such as PSO, DE, CS and two n-gram schemes
Alarifi, A et al. (2018) [17]	Swarm Intelligence	Amazon	The result shows the maximum accuracy of 96.89%
Ali et al. (2019) [2]	Deep Learning	Twitter	Proposed algorithm has been compared with other classification algorithms namely, SVM and Naïve Bayes that were utilized previously on English datasets
Chowdhury et al. (2019) [20]	Support Vector Machine	Movie reviews	Using Support Vector Machine algorithm, this model achieves 88.90% accuracy on the test set and by using Long Short Term Memory network, the model manages to achieve 82.42% accuracy. Furthermore, a comparison with some other machine learning approaches is presented here
Jagdale et al. (2019) [21]	Support Vector Machine (SVM) & NB	Amazon	In this work authors have concluded that machine learning techniques provide best results to classify the products reviews. In case of camera reviews, the obtained accuracy by utilizing Naïve Bayes and Support Vector Machine (SVM) was correspondingly 98.17% and 93.54%

(*continued*)

Table 1. (*continued*)

Author	Techniques	Dataset	Results
Kermani et al. (2020) [22]	SVM and NB	Twitter	The experimental results described that the proposed work exhibited higher accuracy as compared to existing approaches
Aljameel et al. (2021) [23]	SVM, NB and KNN	COVID-19	The experimental analysis using SVM classifier outperformed with a high accuracy of 85%

3 Methodology

The proposed work is divided in two parts. The first part illustrates the training pattern analysis about the data to be passed for the training and the second part explains the training and the classification mechanism of the proposed work. The proposed work utilized two algorithm enhancements namely from the natural computing block and from the meta-heuristic block. The proposed algorithm views the training as a problem to train the system with two things namely the attribute and the attribute set. The record of each category with all the selected features are called the attribute set whereas one attribute is termed as one feature of the data. The proposed algorithm uses Genetic Algorithm (GA) for the selection of the attribute set whereas meta-heuristic inspired cuckoo search (CS) algorithm for the attribute selection. The proposed algorithm enhances the Cuckoo-Search by introducing a random placement scenario of the cuckoo egg to a random nest. There are multiple datasets available for the processing of the proposed algorithm. This research article uses two different datasets from the Kaggle repository entitled "Twitter Sentiment Analysis". The algorithmic description can be given by following pseudo code.

Pseudo Code: Process – Train and Classify
Inputs: Raw Data, Output: Classified data // The inputs to the system will be the raw data that is downloaded from the Kaggle repository. The output to this algorithm would be the classified data //

Extract data labels//The dataset comes along with two information a) The dataset itself and b) the ground truth values or the labels of the category. Ground truths are extracted here//
GTclasses = Arrange data according to ground truth values//Arrange the data as per their ground truth values//
//process of attribute selection starts here//
Totaleggs =Raw Data.colos.count //Counting total number of attributes, each attribute is considered as one egg//
Cuckoofitnessresults = [ ] // Initialize Cuckoo fitness results to be empty//
For i = 1: TotalEggs //For each egg in egg list
    Cuckoonest = 3 // Taking 3 cuckoo nests for the placement
    Cuckooresult = [ ]// initialize an empty array for the cuckoo judgement//
    For j =1: CuckooNests
        Eggnests.othereggs = Choose 3 random eggs// Generate a 3 egg vector, the purpose of the selection of the other eggs is to check the placement result of the egg to egg   nest and to check the possibility of egg to produce higher precision//
        Eggnest [1] = Totaleggsi
        Eggnest.Append (othereggs)
        Eggsentiment = Extract sentiments of Eggnest.Egg from GTclasses // Extracting the sentiments of each egg, obviously the sentiment of each egg would be same and hence these features will be labelled as nest 1 and they will check against 3 other nest values that belong to other GT class//
        f1 = choose Random. Next Emotion (GT.values$\neq$Eggsentiment) //Extract a sentiment that is not same as that of the current sentiment//
        choose 3 nest eggs similar to that of current egg
        Egg nest2 = f1.egg.values //Extract the values of the eggs from the dataset//
        Dataset temp = [Egg nest; Egg nest2] // Create a temp ground truth//
        GroundTruth temp =[Egg sentiment; f1] // Create temp ground truth //
        Train SVM (Dataset_temp, GroundTruth_temp)
        Classify-semi.supervised // perfrorm semi-supervised classification //
        Evaluate Classification Accuracy, Append to Cuckoo result
    End for
    Cuckoo fitness result.Append (Cuckoo result)
End for 2

$$\beta = \int_{j=1}^{totaleggs} \frac{\sum_{i=1}^{n} Cuckoo_{result}}{n} \, dt$$ where β is the fitness threshold of the enhanced cuckoo search algorithm and is computed by taking average of every computed cuckoo result in the given cuckoo bird egg list

Cuckoo result selected =[] // Create an empty list of selected attributes to be replaced by 0 and 1, 1 for the selected and 0 for non selected
For each entry in the Cuckoo fitness result
    0 if Cuckoo fitness result.entry < β
    1 Otherwise
End for

The selected features have been passed to Genetic Algorithm containing the following information architecture shown in Table 2.

Table 2. Parameters

Mutation Type	Linear
Crossover	Intermediate
Selection Function	Selection Mu
Distribution Order	Random
Fitness	Binary
Selection type	Polynomial
Selection Order	Sigmoid

The GA is an additive tool in the MALAB simulation toolbox under optimizations system architecture and hence the mentioned architecture got implemented with the toolbox itself. Furthermore, the selected attributes along with the selected attribute set of the respective classes have been passed to the training algorithm which is modelled by the Neural Networks (NN). In order to train the system, the pseudo code illustrates the considered measure of Feed forward Back Propagation Neural Networks (FFBPNN).

1. Initialize FFBPNN
2. Training. Data. Dataset $=$ *Selected. Data (Cuckoo $+$ GA)* // Pass data to Neural Network with the selected data from Cuckoo Search and Genetic Algorithm//
3. Trainingmodel. name $=$ Levenberg // Use Levenberg training model //
4. Trainingmodel. Maximum epochs $=$ 500; // Total number of maximum simulation epochs is set to be 500 //
5. Trainingmodel. validation $= \{t, \gamma, \vartheta, \mu\}$ where μ is the gradient of Levenberg, ϑ is the total epoch validation i.e. 500, t is the time and γ, is total number of neurons.
   //For the proposed scenario, the total number of neurons is 20 //
6. Train. Neural (Training. *Data. Dataset*, GT. Selected) // Select the Ground Truths of the selected data and train the system //
7. Validation. ratio $= \{70, 30\}$ // 70% data has been considered for the training, and the 30% of this data is for the classification purpose//
8. Evaluate Classification Accuracy and other quantitative parameters
9. Return Classification Accuracy

The presentation of the projected algorithm is calculated on the basis of classification required such as accuracy, precision, recall and F-measure illustrated in the next section.

4 Results

The evaluation of the results has been done on the basis of the overall number of records which have been used for the evaluation in the proposed structure against each evaluated parameter. The illustrations have been provided in Table 3.

Table 3. Evaluation of Accuracy

Total Records	Proposed Accuracy	Accuracy [13]	Accuracy [16]	Accuracy [6]
200	97.3	92.11	92.651	91.887
500	97.3483693	92.1346737	92.7407692	91.9103755
1000	97.3792138	92.2102873	92.7499183	91.9399702
1200	97.381985	92.3017298	92.7740321	92.0006722
1500	97.4211679	92.3543487	92.7876649	92.0976585
1800	97.4584491	92.4337151	92.8736132	92.1379138
2000	97.514173	92.4907977	92.8843864	92.2108231
2300	97.568172	92.5260146	92.9654821	92.2851587
2500	97.6254371	92.5945519	93.0406489	92.3309656
2800	97.6925013	92.6853649	93.0625922	92.4107006
3000	97.7915751	92.7325452	93.097465	92.4596121
3500	97.8391496	92.7704718	93.1520847	92.4782373
4000	97.9038639	92.7872444	93.198813	92.5089543

Table 4. Evaluation of Precision

Total Records	Precision Proposed	Precision [13]	Precision [16]	Precision [6]
200	0.94212	0.925543	0.910023	0.902543
500	0.94303604	0.92628178	0.91058893	0.90346253
1000	0.94379396	0.92661981	0.91085775	0.90374232
1200	0.94384444	0.92673073	0.91117721	0.9039654
1500	0.94394423	0.92745656	0.91179825	0.90423615
1800	0.94449152	0.92833078	0.9121959	0.90488613
2000	0.94515285	0.92917118	0.91309	0.90584655
2300	0.94553129	0.92957045	0.91374497	0.90681526
2500	0.94604124	0.93016567	0.91441843	0.90691823
2800	0.94690475	0.93115537	0.91444391	0.90716737
3000	0.94762644	0.93207384	0.91445929	0.90789732
3500	0.94772726	0.93220691	0.91455104	0.90834037
4000	0.94840402	0.93301069	0.91553529	0.9086549

The evaluation of the proposed algorithm architecture depicts that it surpassed the rest of the techniques provided in the related work section by a significant margin. The effectiveness of the proposed algorithm can be viewed from Table 3 where the proposed

Table 5. Evaluation of Recall

Total Records	Recall Proposed	Recall [13]	Recall [16]	Recall [6]
200	0.90334	0.90221	0.901467	0.90116
500	0.90430164	0.90290688	0.9022109	0.90181453
1000	0.90512993	0.90331453	0.90288987	0.90230943
1200	0.90611587	0.90422294	0.90346477	0.90252001
1500	0.90642841	0.90511261	0.90377749	0.90323761
1800	0.90715851	0.90563452	0.90475407	0.9037842
2000	0.90782391	0.90607414	0.90569003	0.90447854
2300	0.90816287	0.90614861	0.90637663	0.90462823
2500	0.90836067	0.90623392	0.90692768	0.90552006
2800	0.90934613	0.90665449	0.90727421	0.90606435
3000	0.90936748	0.90708652	0.90794843	0.90657618
3500	0.91036499	0.90773054	0.90846621	0.90715674
4000	0.91066157	0.9079608	0.90930244	0.9079792

accuracy lies between the 97% bar whereas the accuracies of other algorithms are behind by 5–7%. In order to be more precise on the evaluated parameters, the proposed algorithm is also compared on other quantitative parameters mentioned in Table 4 and 5. Two set of parameters namely the precision and recall have been evaluated for a maximum of 4000 records and the f-measure has been calculated by utilizing precision and recall. The f-measure is represented as in Fig. 2.

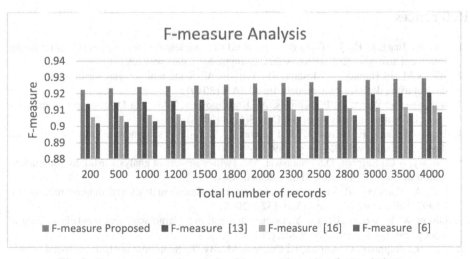

Fig. 2. F-measure of proposed algorithm vs. other state of art techniques.

The maximum F-measure is noted to be 0.9278 for the proposed algorithm at 3500 records and the trend of observations has been found to be same for each algorithm. The maximum accuracy attained by the proposed algorithm is 97.88% whereas the other algorithms have been in the range of 90–94% only.

5 Conclusion

This paper illustrates a new meta-heuristic behavior of CS algorithm for the assortment of the attributes required from the given dataset. The proposed algorithm introduced a new fitness behavior and nest placement architecture compared to the traditional cuckoo search algorithm. The selected attribute set has been passed to Genetic Algorithm for the attribute set selection. The selected attributes are further passed to Levenberg based Feed Forward Back Propagation Training Algorithm. The training has been kept semi-supervised and hence 30% data has been kept as the data and the rest of that data has been kept as the classification data. The maximum classification accuracy of the proposed algorithm has been evaluated as 97.88% for 3500 data records. The proposed algorithm has opened wide gates for the future researchers. The hybridization of the meta-heuristics could be an alternative to replace the current solution. Furthermore, future studies will investigate the possibility of improving accuracy by incorporating a feature selection approach and using other optimization technique variations. In addition, there is room for improvement in dealing with sarcastic and ironic tweets. In addition, domain-specific ontologies and contextual data at the keyword and post levels may be utilized for classifying tweets. Future research may involve including "neutral" tweets in the suggested approach by modifying the feature extraction and classification procedures to effectively identify these tweets. A multi-objective categorization model is paired with an efficient feature selection strategy to improve the precision of Twitter sentiment assessment.

References

1. Xia, R., Jiang, J., He, H.: Distantly supervised lifelong learning for large-scale social media sentiment analysis. IEEE Trans. Affect. Comput. **8**(4), 480–491 (2017)
2. Ali, N.M., El Hamid, A., Mostafa, M., Youssif, A.: Sentiment analysis for movies reviews dataset using deep learning models. In: ALIAA (2019)
3. Sohangir, S., Wang, D., Pomeranets, A., Khoshgoftaar, T.M.: Big Data: deep Learning for financial sentiment analysis. J. Big Data **5**(1), 3 (2018)
4. Bravo-Marquez, F., Mendoza, M., Poblete, B.: Meta-level sentiment models for big social data analysis. Knowl.-Based Syst. **69**, 86–99 (2014)
5. Pandey, A.C., Rajpoot, D.S., Saraswat, M.: Twitter sentiment analysis using hybrid cuckoo search method. Inf. Process. Manag. **53**(4), 764–779 (2017)
6. Pak, A., Paroubek, P.: Twitter as a corpus for sentiment analysis and opinion mining. In: LREC, vol. 10, no. 2010, pp. 1320–1326 (2010)
7. Glorot, X., Bordes, A., Bengio, Y.: Domain adaptation for large-scale sentiment classification: a deep learning approach. In: ICML (2011)
8. Neri, F., Aliprandi, C., Capeci, F., Cuadros, M., By, T.: Sentiment analysis on social media. In: 2012 IEEE/ACM International Conference on Advances in Social Networks Analysis and Mining, pp. 919–926 (2012)

9. Patil, P.K., Adhiya, K.P.: Automatic sentiment analysis of Twitter messages using lexicon based approach and Naive Bayes classifier with interpretation of sentiment variation. Int. J. Innov. Res. Sci. Eng. Technol. 9025–9034 (2015)
10. Sosa, P.M.: Twitter sentiment analysis using combined LSTM-CNN models, pp. 1–9. Eprint Arxiv (2017)
11. Wehrmann, J., Becker, W., Cagnini, H.E., Barros, R.C.: A character-based convolutional neural network for language-agnostic Twitter sentiment analysis. In: 2017 International Joint Conference on Neural Networks (IJCNN), pp. 2384–2391 (2017)
12. Rosenthal, S., Farra, N., Nakov, P.: SemEval-2017 task 4: sentiment analysis in Twitter. In: Proceedings of the 11th International Workshop on Semantic Evaluation (SemEval-2017), pp. 502–518 (2017)
13. Zafra, S.M.J., Valdivia, M.T.M., Camara, E.M., Lopez, L.A.U.: Studying the scope of negation for Spanish sentiment analysis on Twitter. IEEE Trans. Affect. Comput. 10(1), 129–141 (2017)
14. Daniel, M., Neves, R.F., Horta, N.: Company event popularity for financial markets using Twitter and sentiment analysis. Expert Syst. Appl. 71, 111–124 (2017)
15. Yoo, S., Song, J., Jeong, O.: Social media contents based sentiment analysis and prediction system. Expert Syst. Appl. 105, 102–111 (2018)
16. El Alaoui, I., Gahi, Y., Messoussi, R., Chaabi, Y., Todoskoff, A., Kobi, A.: A novel adaptable approach for sentiment analysis on big social data. J. Big Data 5(1), 1–18 (2018)
17. Alarifi, A., Tolba, A., Al-Makhadmeh, Z., Said, W.: A big data approach to sentiment analysis using greedy feature selection with cat swarm optimization-based long short-term memory neural networks. J. Supercomput. 76(6), 4414–4429 (2020)
18. Sharma, N., Pabreja, R., Yaqub, U., Atluri, V., Chun, S.A., Vaidya, J.: Web-based application for sentiment analysis of live tweets. In: Proceedings of the 19th Annual International Conference on Digital Government Research: Governance in the data Age, pp. 1–2 (2018)
19. Clark, E.M., et al.: A sentiment analysis of breast cancer treatment experiences and healthcare perceptions across Twitter. arXiv preprint arXiv:1805.09959 (2018)
20. Chowdhury, R.R., Hossain, M.S., Hossain, S., Andersson, K.: Analyzing sentiment of movie reviews in Bangla by applying machine learning techniques. In: International Conference on Bangla Speech and Language Processing (2019)
21. Jagdale, R.S., Shirsat, V.S., Deshmukh, S.N.: Sentiment analysis on product reviews using machine learning techniques. In: Cognitive Informatics and Soft Computing, pp. 639–647 (2019)
22. Kermani, F.Z., Sadeghi, F., Eslami, E.: Solving the Twitter sentiment analysis problem based on a machine learning-based approach. Evol. Intell. 13(3), 381–398 (2020)
23. Aljameel, S.S., et al.: A sentiment analysis approach to predict an individual's awareness of the precautionary procedures to prevent COVID-19 outbreaks in Saudi Arabia. Int. J. Environ. Res. Public Health 18(1), 218 (2021)

Payable Outsourced Decryption for Functional Encryption Using BlockChain

R. Sendhil[✉] ⓘ, S. Asifaⓘ, and A. Savitha Sreeⓘ

School of Computer Science and Engineering, Vellore Institute of Technology, Chennai, India
sendhildit6464@gmail.com, {asifa.s2022,
savithasree.a2022}@vitstudent.ac.in

Abstract. In many new implementations that needs data storage and exchange (such as cloud storage services), the notion of Functional Encryption (*FE*) was initiated to label the defects of Public-key Encryption (*PKE*). One of the biggest issues with most FE charts is performance, as they are calculated on very expensive pairs of two lines. The general acquired solution to this issue is to load a massive load on a strong third party and let the client to do the simple calculations. However, it is unrealistic to accept that a mediator provides a free service. According to our perception, *FE* plans using External Decryption Scheme (*FEOD*) will not terminate the payment process between the user and the third party only if both are untrusted. In this article, the proposed method is to design a *FE* with Payable Outsourced Decryption (*FEPOD*) in the treaty. The payment according to FEPOD method is made by a blockchain based cryptocurrency. This allows users to pay a mediator when the allocated decoding is successfully completed. After defining the paradoxical model of the **FEPOD** design, the asymmetric structure of the **FEPOD** design is introduced. It also measures the overall performance of the proposal by executing a specific **FEPOD** scheme on the blockchain policy.

Keywords: Cloud Storage · Blockchain · Outsourced Decryption · Functional Encryption

1 Introduction

In cloud storage scenario where all data is encoded using an encryption apparatus such as Functional Encryption (FE) and saved in an encrypted format to preserve the data (Data reliability and seclusion). Cloud storage permits you to save information and documents in an off-website area that you access either through the public web or a committed private organization association. Information that you move off-site for capacity turns into the duty of an outsider cloud supplier. The supplier has, gets, oversees, and keeps up with the workers and related foundation and guarantees you approach the information at whatever point you need it.

Cloud storage conveys a savvy, versatile option in contrast to putting away documents on-premises hard drives or capacity organizations. PC hard drives can just store a limited measure of information. At the point when clients run out of capacity, they need to move documents to an outer stockpiling gadget. Generally, associations fabricated and kept up with capacity region organizations (SAN) to document information and records. SAN is costly to keep up with, nonetheless, because put away information develops, organizations need to put resources into adding workers and framework to oblige expanded interest.

Cloud storage administrations give versatility, which implies you can scale limit as your information volumes increment or dial down limit if fundamental. By putting away information in a cloud, your association save by paying for capacity innovation and limit as a help, as opposed to putting resources into the capital expenses of building and keeping up with in-house stock piling organizations. You pay for just precisely the limit you use. While your expenses may increment over the long haul to represent higher information volumes, you do not need to overprovision stock piling networks fully expecting expanded information volume.

Blockchain is an arrangement of recording data such that makes it troublesome or difficult to change, hack, or cheat the framework. A Blockchain is basically an advanced record of exchanges that is copied and appropriated across the whole organization of PC frameworks on the blockchain. Each square in the chain contains various exchanges, and each time another exchange happens on the blockchain, a record of that exchange is added to each member's record. The decentralized data set oversaw by numerous members is known as Distributed Ledger Technology (DLT). Blockchain is a kind of DLT wherein exchanges are recorded with an unchanging cryptographic mark called a hash.

For example, let us say Alice, the privileged client of a cloud depository app, is using a tool with limited resources. Alice wants to approach encoded data saved in the cloud, but it is inadequate to execute many calculations (such as double editing) to decrypt it. A direct solution to this problem is Identity-Based Encryption (IBE) using External Encryption or Attribute-Based Encoding (ABE) using allocated Decoding. This, for example, allows users (like Alice) to outsource most of their IT load to powerful mediators for decryption without having to enter delicate data into the initial data. In fact, third parties are reluctant to offer free services and expect others to pay for what they have done. Alice can send a math problem to the car next to Bob. Bob manages the network and promises to "pay a dollar as soon as he does the right calculations". Bob gets the information, computes it, and shares the outcome to Alice. For this reason, there are two other things to consider. The first is a mechanism that allows Alice to see Bob's response before pushing it towards him. Second, a mechanism that alleviates Bobs concerns that Alice is refusing the correct answer to avoid payment.

Inherently, these two issues can be easily resolved by having a believed agent (e.g., a bank) as a negotiator. However, this procedure is not sufficient for users who expect privacy and make all transactions transparent to authorities. The new blockchain technology issues a disseminated, self-executing infusion for honest payments among Bob and Alice. Blockchain-based operating systems can execute smart contracts among Bob and Alice. Alice can publish paid allocating contracts on the blockchain to explain IT outsourcing activities and reward rules. According to the smart contract, anyone who finds the right solution can receive a predetermined reward. Basically, blockchains here traditionally act as believed minor parties. Alice accumulates money on the blockchain, if and only if Bob can give the correct answer, the blockchain returns the money to Bob. In order to participate in account obligations, Bob must deposit "money" on the blockchain, and malicious actions are punished.

These blockchain-based solutions provide an accurate and efficient way to test Bob's solution without compromising the security of the cryptographic system. The traditional outsourcing approach provided as a trapdoor (with password) with Alice (who has the decryption key) can effectively confirm Bob's answer [2, 3] etc. In principle, anyone who can see Bob's answer can recalculate Bob's outsourcing. Unfortunately, it does not work in a blockchain-based allocating surroundings. Public validation, on the other hand, is inefficient because it requires recalculation, but for efficiency the validation cost must be lower than the calculation. However, Alice is unreliable and may reject valid assumptions, so she has no listener role to help decide whether to continue or not. The fair settlement protocol suggested by [4, 5] is different from that obtained in this article. A blockchain-based fair settlement outsourcing service offers the possibility of integrating the FEOD plans into the Bitcoin policy. The honest interchange deal is a fair exchange of cryptocurrency payments for receipts. We provide solutions to your potential. Improves the stability of stable configurations for both users and service providers.

In this paper, FEPOD method where cryptocurrency-based payments are made via a blockchain development (e.g., a specific architecture must be implemented in most FEOD methods). The FEPOD system should ensure the systematic and common viability of the results of external decryption operations. This means that everyone in the FEPOD plan should be able to validate results based solely on the common instructions provided.

The rest of this paper is organized as follows. Section 2 outlines the definitions and preliminaries suitable to this article. Section 3 describes the system architecture and security model for functional encryption using the FEPOD decryption method. Section 4 introduces the specific FEPOD project and its common architecture and shows how to consolidate the FEPOD project into the blockchain. Section 5 simulates a typical FEPOD design on a blockchain board and evaluates its execution. Eventually, this document is summarized in Sect. 6.

1.1 Challenges and Contribution

A major provocation in establishing functional encoders using external decoders (FEPOD) is to achieve efficient public validation of external decoder responses. Modern functional encryption with outsourced decryption systems (FEODs) perform authentication (i.e., decryption) by a decryptable user. FEPOD Scheme is used to achieve consensus among all blockchain nodes, validation should be based only on common information with maximum efficiency.

Challenge 1: The first idea is a verifiable cryptocurrency protocol. This allows the target recipient to prove that the explicit M decrypted ciphertext in the decryption key satisfies the normally defined properties. FEPOD schemes typically generate ciphertexts in the form of ciphertexts initiated by common key cryptography schemes such as the El Gamal cryptography system. Alice can validate or invalidate that Whistle Bob's result can be encrypted with the DK decoded key, which can result in the loss of message M without delicate instruction passing through a verifiable decoded protocol. This method appears to help gain public confidence in the FEPOD program. However, according to our perception, commonly verifiable cryptographic deals work only with a limited number of Public-Key Encryption (PKE) classes that correspond to private and not public properties. Such as, the commonly verifiable decryption technique in assumes a Decisional Composite Residuosity (DCR) based on the PKE technique. Therefore, it does not meet the requirements for creating a public FEPOD plan that can be implemented in most FEOD plans.

Challenge 2: The Zero-Knowledge Contingent Payment (ZKCP) [6] could vend verifiable information in any form for bitcoin payments. This is because ZKCP is implemented as concise mutual dispute (SNARGs) and can be run on SNARGs knowledge edge. SNARG permits a suspicious minor party (such as Bob) to create account records without secrets. This looks like a possible solution. To give Bob the ability to compute a SNARG, Alice's initiates two common keys: an account key along with a verification key. Then Bob calculates the amount of the item (generated by the account key) and Alice uses it based on the account's public key to see the outcome. Anyone can use the common key to verify Bob's exit for a given item. At first glance, this FEPOD map generation method works. Unfortunately, implementing such an approach in a FEPOD plan requires extensive computation, including list generation, which can be tedious if the FEPOD plan is too convoluted. What is the most efficient way to ensure fair settlement between vendor and client?

Possible Solution: We found that FE parameter validations could be generated more efficiently without the use of ZK tests. The idea of validation is very difficult to adapt to some nature of public procurement where validation keys are generated by computing additional random keys. For more information, ask Alice to generate two public keys [7]. One is the *TK (TransportKey)* region account key, and the other is the real-derived *VK* (validation key). The *TK* key is the account decryption key that Alice uses to hide her password. The validation key is linked to a validation key that other users use to validate multiple chapters (i.e., randomness of *VK* validation key generation). The control button is not released by Alice until Bob sends the exchange response. After a certain waiting period, to deal with the situation where Alice does not want to pay Bob and Bob does not release the order button after finding a solution, he sends the money directly to Bob. Also, Bob can malfunction in the blockchain network and must deposit when outsourcing. If an event occurs (e.g., if a participant does not necessarily adhere to the protocol), depending on the protocol and the parties, Bob may mislay the installment and rewards the proceedings on behalf of Alice, or else Alice may be confused.

In summary, following contributions are made in thin paper:

- Offers a functional outsourced decryption (FEPOD) concept where anyone can verify the response to computational work outsourced by an untrusted third party instead of processing payments through Blockchain based cryptocurrency.
- Define the safety model for the FEPOD project, provide the overall architecture, and analyze the security.

Implement a specific FEPOD project for the public service sector through a blockchain platform and assess its suitability and application.

1.2 Related Works

Outsourcing Computation: Account allocating allows users to outsource their extended accounts, including network operations, to third parties [8, 9]. These approaches are often based on cryptographic principles such as digital signature schemes [10, 11], Identity-Based Cryptography (IBE) [12] and Attribute-Based Encryption (ABE) plans for performing heavy computational tasks on resource-constrained devices. Currently, a major security issue with outsourced software is the verification of return results from untrusted third parties, which are considered inaccurate responses. Based on these observations, various account outsourcing schemes with validation functions have been proposed to improve security or performance. Unfortunately, these existing solutions do not work in the low-cost third-party cryptocurrency scenarios described in this paper. These schemes do not allow for effective third-party authentication because they obtain public authentication from a third-party running external account again.

Zero-Knowledge Contingent Payment: ZKCP allows merchants to vend any verifiable details for payment in Bitcoin. However, if you sell services on behalf of information, ZKCP will not pay any appropriate fee. To overcome these limitations, ZKCP has been expanded to include an unconditional payment service (ZKCSP) where sellers can sell certain collateral to buyers. Unfortunately, de facto evidence for ZKCSP differentiation (BISE) is inconclusive [13]. Short Non-Interactive Reasoning (SNARG) and SNARG of knowledge aim to provide a well-ordered way to validate outsourced operations on untrusted entities whose solutions are proposed in different configurations [14–16] The assignee may have other entities, regardless of which private verifiable arithmetic register the test is published. We need to be able to generate vk_x verification keys so we can get feedback from third parties. It also assigns arithmetic functions to third-party x items. Check the answer. There are several publicly verifiable SNARGs [17, 18] etc. to ensure that anyone receiving the results can verify them. Blockchain-based decryption can be used to design payment mechanisms. However, when used in complex FE formulations, its effectiveness is counterproductive due to the ZK antagonist formulation. This article is looking for a solution to make fair payments without using ZK Proof.

2 Preliminaries

This part briefly describes the applicable symbols and denotations which are used in this paper.

2.1 Blockchain and Smart Contracts

Splendid arrangements possibly founded on cryptographic types of cash [19] to engage the denotation and implementation of concurrences on the blockchain, and Ethereum [20] is the initial blockchain that maintains Turing-complete agreements. Quickly talking, a splendid understanding is a bit of program code managing over blockchain. The implementation among program code is set off by events, e.g., the trades that are attached to the blockchain, of which the precision is assured through the arrangement show of the blockchain. In an optimal manner, brilliant agreements can be viewed as being executed by a confided in agreement PC that dependably adheres to every guidance. Consequently, with the cryptographic money ability among blockchain, a keen agreement above blockchain understands the implementation among mind boggling business agreements including financial exchanges.

Ethereum has two kinds of records called remotely claimed records and agreement reports. A remotely claimed report has an equilibrium noted above blockchain and is related with an interesting common and personal key pair where the personal key is utilized via proprietor to sign exchanges their record. An agreement account keeps an equilibrium however is not related with any personal key, and it saves code of an agreement to choose the progression of the ethers in the record.

An exchange in the Ethereum is a guidance developed and endorsed by a proprietor of a remotely claimed report in a cryptographic manner. Every exchange is related with two address fields to indicate the source and the collector. An agreement can be started by submitting an exchange with the collector being a location of another agreement report and information field setting the agreement code. An exchange is possibly utilized as a note to summon a capacity in an agreement too, where the collector's location is the agreement account putting away the agreement code and the capacity to be summoned alongside contentions in the information field of the exchange. The conduct of an agreement is absolutely controlled by the execution of its code. A brief knowledge about Ethereum blockchain smart contract vulnerabilities with day-to-day life examples with preventive methods are identified. The Ethereum blockchain smart contract vulnerabilities have three important reasons and seventeen other reasons groups. Known Security attacks and privacy issues of Blockchain named Double-Spending Attacks will be prevented by making the invalidity of sender's location for the transaction, Majority Occupation Attacks, Privacy Disclose can be prohibited by presenting various techniques like Mixing, Fungibility. To avoid the computational overhead in the decryption process, the outsourcing of resource utilization is avoided by Fair payment between the users is introduced.

Table 1 describes the definitions of symbols and the functions. Policies can be identities, public keys, attributes/access policies.

Table 1. Definitions of symbols

$\lambda \in N$: Security parameter	$F\lambda$: function space
parX: scheme X's public parameter	$M\lambda$: message space
mskX: scheme X's master private key	$P\lambda$: policy space
CTp: ciphertext for policy p	p: policy
Skfid: id's private key for function f	f: function
tkfid: id's transformation key for function f	M: message
dkid: id's decryption key	id: user identity
CTid: transformed ciphertext for id	$\perp$: failure symbol

2.2 Functional Encryption with Outsourced Decryption

Accompanied by documentations portrayed within Table 1, a FEOD conspire FEOD comprises of an arrangement calculation $FEOD. Setup(1^\lambda) \rightarrow (par_F, msk_F)$, a key generation calculation FEOD.

$KeyGen(par_F, msk_F, id, f) \rightarrow sk_{id}^f$, a transformation algorithm FEOD.

$TranKG\left(par_F, id, sk_{id}^f\right) \rightarrow \left(tk_{id}^f, dk_{id}\right)$, an encryption algorithm FEOD.

$Encrypt(par_F, p, M) \rightarrow CT_p$, a transformation algorithm FEOD.

$Transform\left(par_F, id, tk_{if}^f, CT_p\right) \rightarrow CT_{id}/\bot$, and a decryption algorithm FEOD.

$Decrypt(par_F, dk_{id}, CT_{id}) \rightarrow M/\bot$.

An *FEOD* plan *FEOD* is supposed to be accurate, signifies that for all safety parameters $\lambda \in \mathbb{N}$, all functions $f \in \mathcal{F}_\lambda$ suit all plans $p \in \mathcal{P}_\lambda$ and all messages $M \in \mathcal{M}_\lambda$, if $\left(par_F, msk_F\right) \leftarrow FEOD.Setup(1^\lambda)$, $sk_{id}^f \leftarrow FEOD.KeyGen(par_F, msk_F, id, f)$, $(tk_{id}^f, dk_{id}) \leftarrow FEOD.TranKG\left(par_F, id, sk_{id}^f\right), CT_p \leftarrow$ $FEOD.Encrypt(par_F, p, M), CT_{id} \leftarrow FEOD.Transform(par_F, id, tk_{id}^f, CT_p)$, we have $FEOD.Decrypt\left(par_F, dk_{id}, CT_{id}\right) = M$.

3 Security and Framework Definitions

This portion defines the complex architecture and safety related to featured encoding using FEPOD for blockchain networks.

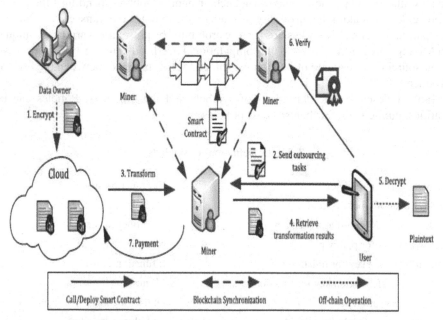

Fig. 1. Architecture of the FEPOD Scheme over a blockchain.

3.1 Overview

As in Fig. 1, FEPOD applications include blockchains, data holders, users (e.g., devices with limited resources), clouds (e.g., powerful servers) and extractors. The data holder (the data holder can also be the user if decryption is enabled) encrypts data items through the Functional Encryption Program (FE) to ensure data security. Users can receive encrypted key text messages from the cloud depending on their access rights. The cloud stores data owner's data items and enables users to perform large-scale calculations (if needed). Miners are responsible for tracking and refining agreements on the blockchain.

Expect that a client, say Alice, who approaches a sum of ciphertexts put away above cloud, plans to get the decoded of a ciphertext CT with the assistance of the cloud above hefty calculation without bargaining the safety and protection of the first information. Alice transfers the re-appropriating calculation function, as a savvy agreement, to the blockchain along with two common keys. Alice likewise stores a certain measure of digital forms of money to this brilliant agreement and indicates that "whoever posts the right outcome can guarantee their ward (e.g., \$1). Anyone can check the blockchain also, discover the reevaluated calculation task. On the off chance that a cloud, say Bob, will hold this errand, he stores some digital currencies to the shrewd agreement. Weave then, at that point executes the calculation to deliver a changed ciphertext CT_A, and presents the changed ciphertext CT_A to the savvy agreement. Alice can recover and decode the changed ciphertext CT_A utilizing her decoding key. From there on, Alice shows whether the given changed ciphertext CT_A is right as "1" or "0" and discharges the verification regarding the rightness of the public keys also, the change result CT_A with the end goal that excavators can decide if Bob ought to be paid for his work. If Alice specifies "1", Bob will be returned his store and get the installment right away. If Alice designates "0", the excavator confirms the accuracy of the changed ciphertext CT_A. On the off chance that the change result CT_A passes the check, the digger will play out the change to yield the changed ciphertext CT'_A. On the off chance that CT'_A equivalents to CT_A, the digger yields $\top$, implying that Bob will be rewarded, and Alice will reward more exchange expenses (than that in an ordinary circumstance). Something else, the digger yields $\bot$, implying that Bob (instead of Alice) will mislay a piece of his store to reward for the exchange charge also, Alice will be remitted her store. On the off chance that the change result CT_A neglects to pass the confirmation, the excavator inspects the legitimacy of the common keys. In the event that the common keys are not shaped well, digger yields "1", implying that Bob will be rewarded furthermore, gave back his store. Something else, the digger yields "0", implying that the exchange comes up short with no installment.

Note that if in a specific time-frame (e.g., 60 min), Alice does not transfer any verification with respect to the accuracy of the outcome CT_A, the installment to Bob will be continued of course.

Generally, the FEPOD plan includes configuration algorithm, *PrivKG* function key generation algorithm, transport key creation algorithm *TranKG*, *VeriKG* authentication key creation algorithm, encryption algorithm, transformation algorithm, decryption algorithm as given in the following:

   i. **Setup**$(1^\lambda) \rightarrow$ **(par, msk)**: Managed by the believed Key Generation Center (KGC), this algorithm uses input security parameters to extract public parameters and MSK master keys.

ii. **$PrivKG(par, msk, id, f)$** → sk_{id}^f: Considering the public boundary standard, the expert personal key msk, a capacity f (i.e., a character, a bunch of properties, etc.) for a client id as the insert, this calculation, execute by the KGC, yields a personal capacity key sk_{id}^f over the capacity f for client id.

iii. **$TranKG(par, id, sk_{id}^f)$** → (tk_{id}^f, dk_{id}): Considering the common boundary standard, a personal capacity key sk_{id}^f for a client id with a capacity f as the info, this calculation, execute via client id himself/herself, yields a common change key tk_{id}^f and an unscrambling key dk_{id} for the client id.

iv. **$VeriKG\left(par, id, tk_{id}^f\right)$** → $\left(vk_{id}^f, wk_{id}\right)$: Considering the common boundary standard, a common change key tk_{id}^f for a client id with a capacity f as the information, this calculation, run by the client id himself/herself, yields a common confirmation key vk_{id}^f then an observer key wk_{id} for the client id.

v. **$Encrypt(par, p, M)$** → CT_p: Taking the common boundary standard, a strategy p (i.e., a character, an entrance strategy, et al.) and a note M as the information, this calculation, execute via information proprietor, yields a ciphertext CT_p related along with approach p.

vi. **$Transform(par, id, tk_{id}^f, vk_{id}^f, CT_p)$** → CT_{id}: Taking the public boundary standard, a change key tk_{id}^f and a confirmation key vk_{id}^f for a client id with a capacity f and a ciphertext CT_p over an approach p as a info, this calculation, execute via cloud, yields a changed ciphertext CT_{id} if the capacity f of change key tk_{id}^f suites the arrangement p of the ciphertext CT_p else a disappointment image $\perp$.

vii. **$Decrypt(par, id, dk_{id}, CT_{id})$** → $M/ \perp|$ **"1/0"**: Considering the common boundary standard par, the unscrambling key dk_{id} for a client id and a changed ciphertext CT_{id} as the information, this calculation, execute via client id himself/herself, yields a plaintext M and remits the decoding result "1" or a disappointment image $\perp$ and remits the decoding result "0".

viii. **$Verify(par, id, tk_{id}^f, vk_{id}^f, wk_{id}, "1/0", CT_{id})$** → $1/0/\perp$: Considering the common boundary standard, the common change key tk_{id}^f, the common check key vk_{id}^f, the observer key wk_{id} for a client id with a capacity f, the decoding result "1/0" and the changed ciphertext CT_{id} as the info, this calculation, run by the excavator, yields 1 if there should be an occurrence of installment affirmation (e.g., the decoding result is "1") or 0 demonstrating a fruitless exchange or else $\perp$ for the occasions where the cloud will reward for the exchange expense or else $\top$ for the occasions where the client id will pay extra exchange charges.

Note that a FEPOD plot is right, implying that for all stable boundaries $\lambda \in \mathbb{N}$, all capacities $f \in \mathcal{F}_\lambda$ coordinating with all arrangements $p \in \mathcal{P}_\lambda$ and all messages $M \in \mathcal{M}_\lambda$, if $(par, msk) \leftarrow Setup(1^\lambda)$, $sk_{id}^f \leftarrow PrivKG(par, msk, id, f)$, $(tk_{id}^f, dk_{id}) \leftarrow TranKG(par, id, sk_{id}^f)$, $(vk_{id}^f, wk_{id}) \leftarrow VeriKG(par, id, tk_{id}^f)$, $CT_p \leftarrow Encrypt(par, p, M)$, $CT_{id} \leftarrow Transform(par, id, tk_{id}^f, vk_{id}^f, CT_p)$. We have $Decrypt(par_f, id, dk_{id}, CT_{id}) = M|"1"$, and $Verify(par, id, tk_{id}^f, vk_{id}^f, wk_{id}, "1", CT_{id}) = 1$.

3.2 Adversarial Model

For functional encryption security by the FEPOD decoder, the following conditions are expected to be met:

- Only super users can know the plaintext associated with the ciphertext.
- One should not make cloud payments without proper accounting for external computing tasks.
- Users cannot withdraw payments after cloud maintenance. In other words, user get the correct calculation results in the cloud.

Additionally, the cloud or its users must be "fined" for malicious actions that cause miners to perform additional actions (i.e., they may be charged more for their transactions).

The adversarial model assumes that the blockchain is based on trust and exposes the entire internal state to the public. In particular, the blockchain can always store, calculate, and use data correctly. In addition, KGC believes in the cloud, which is trusted and untrusted, and can work with other malicious users (together with unauthorized clients) to extract plaintext particulars from passwords goal pass. In other words, everyone should be able to perform any activity performed by the user. Thus, the user has access to the private key, decryption key and control key of the user with the required functions.

4 Description of Protocol

This portion first shows a general architecture for programming applications using *FEPOD* plan, after that it shows a typical *FEPOD* design. The following step shows methods to combine the *FEPOD* design into blockchain.

4.1 Generic Construction

Assume,
$FEOD = (FEOD.Setup, FEOD.KeyGen, FEOD.TranKG, FEOD.Encrpt, FEOD.$
  Transform, *FEOD.Decrypt*, specifies a *FEOD*. In this case, secret feature key is reversible and is normal function programming. Architecture using an External Decoding System (FEPOD), which consists of the below algorithms:

  i. *Setup*: This algorithm implements the *FEOD.Setup* algorithm on a stable parameter λ to create the common parameter *par* and the master personal key *msk*.
  ii. **PrivKG**: This algorithm implements the *FEOD.KeyGen* algorithm on the user *id* of function f (in function space F) uses the prescribed function f to generate the user $id's$ secret function key sk_{id}^f.
 iii. **TranKG**: This algorithm implements the *FEOD.TranKG* algorithm with sk_{id}^f private key user function key, which initiates a tk_{id}^f common transport key and a dk_{id} decryption key for the f function user key.

iv. **VeriKG:** This algorithm randomly deploys a global tk_{id}^f switch and receives a tk_{id}^f redistribute switch. Then run *FEOD.TranKG* algorithm on the tk_{id}^f over switch f for the user id to initiate common verification key vk_{id}^f and the personal witness key wk_{id} with functional f for user id.

Note. Record to decrease the calculation elevates, the client can produce and save numerous sets of change keys and confirmation keys by controlling TranKG and *VeriKG* calculations disconnected. When we have a rethinking function, the client essentially relates there thinking function with a picked set of modification key also, check key.

v. **Encrypt:** This encryption algorithm implements the *FEOD.Encrypt* algorithm on a note M (in the note space $\mathcal{M}_\lambda$) and a strategy p (in $\mathcal{P}_\lambda$) to produce a ciphertext CT_p.

vi. **Transform:** This algorithm implements the *FEOD.Transform* algorithm above change key tk_{id}^f and the check key vk_{id}^f over a capacity f of a client id, separately, and the ciphertext CT_p to create a changed ciphertext $CT_{id} = (CT_{tk}, CT_{vk})$.

vii. **Decrypt:** This algorithm implements the *FEOD.Decrypt* calculation on the decoding key dk_{id} of a client id and changed ciphertext CT_{id} to recuperate the plaintext M what's more, decide the decoding result, "1/0".

viii. **Verify:** On the off chance that the unscrambling result is, "1", this calculation yields "1" straight forwardly. Something else, this calculation runs the *FEOD.Decrypt* calculation on the observer key wk_{id}^f with a capacity f of a client id and the changed ciphertext CT_{id} to confirm the accuracy of the change. If the change is right, this calculation runs the *FEOD.Transform* calculation to create a changed ciphertext $CT_{id}^{'}$. If $CT_{id}^{'} \neq CT_{id}$, this calculation yields $\bot$, and $\top$ something else. If the change is not right, this calculation runs the *FEOD.VeriKG* calculation on the observer key wk_{id}^f with a capacity f of a client id and the change key tk_{id}^f to check the rightness of the public confirmation key. On the off chance that the check key is not very much framed, this calculation yields 1, and 0 in any case.

For the rightness of the above conventional development on a FEPOD plan FEPOD, we necessitate that the output of $FEOD.Decrypt(par_f, id, wk_{id}, FEOD.Transform(par, id, vk_{id}^f, CT_p))$ equivalents to the output of $FEOD.Transform(par, id, tk_{id}^f, CT_p)$. That is, the condition $FEOD.Decrypt\left(par_f, id, wk_{id}, CT_{vk}\right) = CT_{tk}$ holds.

Hypothesis 1: Suppose the basic *FEOD* system *FEOD* is stable, and afterward the given conventional FEPOD system FEPOD is safe.

Evidence: Besides, the Type One foe who needs to get to know the decoded of the ciphertext, there are other two types of opponents in the show *FEPOD*: The Type-2 enemy who seeks to make "wrong change and affirmation keys" to proceed the check and the Type-3 opponent who desires to give "fake changed ciphertexts" to proceed the affirmation. The Type-1 foe possibly thwarted since the secret *FEOD* plan *FEOD* is safe. The Type-2 foe is hindered clearly through the essential yield of every critical confidential including the haphazardness (i.e., the onlooker key) of the check keys if essential. The Type-3 enemy is thwarted by re-attempting the execution when required. Thus, the above shown *FEPOD* is safe.

4.2 Instantiation

The following proposes a FEPOD design derived from the full FEPOD structure based on a concrete programming design (ABE). Let G be the set of initial sequences p. where $g \in G$ is the relevant generator and $\hat{e} : G \times G \to G_1$ is the 2-line map. Define the note space as G_1 and the attribute space as Z_p. It is centred on the Rouselkis-Waters (CP-ABE) ciphertext properties in and the outsourcing method. The design of the ABEPOD is:

- *Setup*: This algorithm accepts safety parameter λ as input. Using the $g \in G$ generator, select a set of G from the first instruction and a 2-line map of $\hat{e} : G \times G \to G_1$. Also, $u, h, w, v \in G, \alpha \in Z_p$, public parameters $par = (g, w, v, u, h, \hat{e}(g \times g)^{\alpha})$ and primary private key $msk = g^{\alpha}$ are chosen randomly.
- **PrivKG:** This algorithm accepts common parameter par, private master key msk and client id with attribute set $A = \{A_1, \ldots, A_k\}$ with input. It randomly selects $r, r_1, \ldots, r_k \in Z_p$ and computes

$$sk_1 = g^{\alpha}.w^r, \quad sk_2 = g^r, \quad sk_{i,3} = g^{r_i}, \quad sk_{i,4} = (u^{A_i}h)^{r_i}.v^{-r} \tag{1}$$

It returns the personal attribute-key $sk_{id}^A = (sk_1, sk_2, \{sk_{i,3}, sk_{i,4}\}_{i \in [1,k]})$ for client id with attributes A.

- **TranKG:** This algorithm considers input as common parameter par, client id along with the personal attributes key sk_{id}^A in the set of attributes A (if it consists of $A_1, \ldots, A_k$). To calculate

$$tk_1 = sk_1^{t_0}, \quad tk_2 = sk_2^{t_0}, \quad tk_{i,3} = sk_{i,3}^{t_0}, \quad tk_{i,4} = sk_{i,4}^{t_0} \tag{2}$$

It returns the common transformation-key $tk_{id}^A = \left(tk_1, tk_2, \{tk_{i,3}, tk_{i,4}\}_{l \in [1,k]}\right)$ and the personal decoded-key $dk_{id} = t_0$ for the client id with attributes A.

- *VeriKG*: This algorithm grabs the common parameter par and a client id with a common transformation-key tk_{id}^A above an attribute set A (presume it contains $A_1, \ldots, A_k$) as the input. It erratically selects $t_1, r', r_1', \ldots r_k' \in Z_p$, and computes

$$vk_1 = \left(tk_1.w^{r'}\right)^{t_1}, \quad vk_2 = \left(tk_2.g^{r'}\right)^{t_1}, \quad vk_{i,3} = \left(tk_{i,3}.g^{r_i'}\right)^{t_1},$$
$$vk_{i,4} = \left(tk_{i,4}.\left(u^{A_i}h\right)^{r_i'}.v^{-r'}\right)^{t_1} \tag{3}$$

It outputs $vk_{id}^A = \left(vk_1, vk_2, \{vk_{i,3}, vk_{i,4}\}_{i \in [1,k]}\right)$ as a common verification-key, and remains $wk_{id} = \{t_1, r', r_1', \ldots r_k'\}$ as a personal verification-key for the client id with an attribute set A.

Note. Record that the requirement on people in general change key tk_{id}^f is vital. Something else, the ABEPOD system, *ABEPOD* would be helpless against an assault as follows. The cloud can produce a changed ciphertext, it proceeds check by registering $C_0' = \hat{e}(g^x, tk_1)$, and $C_1' = \hat{e}(g^x, vk_1)$, where $x \in Z_p$.

- **Encrypt:** This algorithm is used as input common parameter *par*, LSSS approach system $(\mathbb{M}, \rho)$ and message M. Assuming $\mathbb{M}$ is an $l \times n$ matrix, we assign the rows of matrix $\mathbb{M}$ to features and randomly choose the vector $\vec{v} = (\mu, y_2, \ldots, y_n)^{\perp} \in Z_p^n$. The factor used is μ. If $i = 1$ to l then calculate $v_i = \mathbb{M}_i$ where $\mathbb{M}_i$ is the $i - th$ row of $\mathbb{M}$. Also $\mu_1, \ldots, \mu_l \in Z_p$ are chosen and randomly calculated.

$$C_0 = \hat{e}(g, g)^{\alpha\mu}.M, \quad C_1 = g^{\mu}, \quad C_{i,2} = w^{v_i}.v^{\mu_i}, \quad C_{i,3} = \left(\mu^{\rho(i)}h\right)^{-\mu_i} \quad (4)$$

It outputs the ciphertext $CT_{\mathbb{M},\rho} = \left((\mathbb{M}, \rho), C_0, C_1, \{C_{i,2}, C_{i,3}, C_{i,4}\}_{i\in[1,l]}\right)$.

- **Transform:** This algorithm grabs inputs as public parameter *par*, client *id* with tk_{id}^A public key, vk_{id}^A public validation key via function set A and CT_A ciphertext via access structure $(\mathbb{M}, \rho)$. Assume A assures the access structure $(\mathbb{M}, \rho)$. Assume I, described as $I = \{i : \rho(i) \in A$. The notation for $\{w_i \in Z_p\}_{i\in I}$ is a group of constants, and if $\{v_i\}$ are justifiable portions of any secret μ according to $\mathbb{M}$, then the constant $\sum_{i\in I} w_i v_i = \mu$.

1) It inspects CT, and enumerates C_0' with the common transformation-key tk_{id}^A as

$$C_0' = \frac{\hat{e}(C_1, tk_1)}{V_0} = \hat{e}(g, g)^{\alpha\mu t_0} \quad (5)$$

where $V_0 = \Pi_{i\in I}\left(\hat{e}(C_{i,2}, tk_2).\hat{e}(C_{i,3}, tk_{i,3}).\hat{e}(C_{i,4}, tk_{i,4})\right)^{w_i}$
2) It enumerates C_1' with the common verification-key vk_{id}^A as

$$C_1' = \frac{\hat{e}(C_1, vk_1)}{V_1} = \hat{e}(g, g)^{\alpha\mu t_0 t_1} \quad (6)$$

where $V_1 = \Pi_{i\in I}\left(\hat{e}(C_{i,2}, vk_2).\hat{e}(C_{i,3}, vk_{i,3}).\hat{e}(C_{i,4}, vk_{i,4})\right)^{w_i}$
It outputs the transformed ciphertext $CT_{id} = (C_0, C_0', C_1')$.
- **Decrypt:** This algorithm takes input as common parameter *par*, and user *id* with decryption key dk_{id} and transformed ciphertext CT'. It calculates $M = C_0/(C_0')^{1/t_0}$, print message M, and return "1" as the result of decryption.

   Verify: To test this algorithm, we need the same public parameter *par*, public key tk_{id}^f, public confirmation key vk_{id}^f, control key wk_{id}, decryption result "1/0" and modified ciphertext CT_{id} as input. If the decoding result is "1", and 1 is executed immediately. If not, decompose the CT_{id} and check $C_0'^{t_1} = C_1'$. If this formula is true, the converted CT_{id}' encoded text is calculated by the conversion algorithm. If $CT_{id}' \neq CT_{id}$ it outputs $\perp$, and $\top$ if not. If the equation does not carry, it calculates $\overline{vk}_{id}^A$ as in the *VeriKG* algorithm using tk_{id}^f and wk_{id}, and inspects whether $\overline{vk}_{id}^f = vk_{id}^f$. It returns 1 if the calculation does not carry, and 0 otherwise.

Hypothesis 2: Suppose Rouselkis-Waters is secure with the CP-ABE allocating plan, the proposed ABEPOD plan is secure against ABEPOD [22].

Evidence: The Rouselkis-Waters CP-ABE project has been proven to be secure in outsourced decryption scheme, and as stated in hypothesis 1, the ABEPOD project is safe.

4.3   Integrating Into a Blockchain

Due to the extended inactivity and less performance of blockchain consensus, it is unenviable for the blockchain to handle calculations out of outsourcing. Therefore, it is recommended that most outsourcing work be done using batch technology. This can be done through a hash-based micropayment system where user accounts and cloud functions make a good compromise between outsourcing results and micropayments. There is no need to make final payments on the blockchain until the protocol is complete.

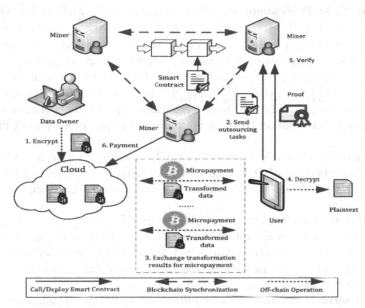

Fig. 2. FEPOD conspire utilizing in payments across blockchain.

Hash-based blockchain micropayment system [23] allows both parties to transact fairly without a trusted third party, each with a longer hash as part of a hash providers service micropayment (e.g., the exchange protocol Hash chain). It works on chains. The FEPOD scheme allows users. Exchange the hash value of the cloud-provided transformation result. Specifically, users generate long chains of hashes like $h_N \rightarrow h_{N-1} \rightarrow \ldots h_1 \rightarrow h_0$ where $N \in Z_p^*, h_0$ is the hash chain root, and $h(\cdot)$ is the hash function such that $h_i = h(h_{i+1})$. The user signs the original source of h_0 and sends a smart contract. When the cloud completes the allocating activity, the client sends the hash chain sequentially from the hash value (signed) to the cloud, starting with the value h_1. So, the processed off-chain block does not perform a transaction which is processed by blockchain that handles final payments between clouds. Using batch techniques in this way can significantly upgrade the ability of some FEPOD schemes. Figure 2 shows how to design a FEPOD deployment on a blockchain. Here, the cloud exchange sends results to a small number of paying users until the final outsourcing activity.

Let us assume, *FEPOD = (FEPOD.Setup, FEPOD.PrivKG, FEPOD.TranKG, FEPOD.VeriKG, FEPOD.Encrypt, FEPOD.Transform, FEPOD.Decrypt, FEPOD.Verify)* is the application encryption outside the FEPOD decryption system. The BFEPOD protocol is described below. It shows how to combine a FEPOD application into a blockchain smart contract (e.g., Ethereum) in 5 steps. The steps 3 and 4 are performed outside of the exchange process chain in the user cloud.

1. **Setup parameters**: A valid KGC calls the *FEPOD.Setup* algorithm to generate public parameters *par* and runs the *FEPOD.PrivKG* algorithm to create the property attribute key sk_{id}^A for the client *id*. Later, KGC exposes common parameters in parallel with the smart contract blockchain.

2. **Multiple Tasks Publication:** The client *id* executes the *FEPOD.TranKG* algorithm to create a tk_{id}^A transaction key and dk_{id} decoding key, and runs the *FEPOD.VeriKG* algorithm to create vk_{id}^A verification key and wk_{id} witness key for numerous paging compute operations. After that user *id* publishes N, then informs the blockchain smart contract, the function of the external account $\left(\left\{CT_A, tk_{id}^A, vk_{id}^A\right\}_1, \ldots, \left\{CT_A, tk_{id}^A, vk_{id}^A\right\}_N\right)$. The CT_A ciphertext generated by the data holder using the *FEPOD.Encrypt* algorithm. Instead, users store a specified amount in a smart contract that they are obligated to pay as the root of the retail chain.

3. **Transform Ciphertexts:** The cloud plays out the re-appropriating calculation undertakings by running the *FEPOD.Transform* calculation on the rethinking calculation task $\left(CT_A, tk_{id}^A, vk_{id}^A\right)$ individually and sends the subsequent changed ciphertext CT_{id} to the client *id*.

4. **Verify Results:** The client *id* interprets CT_{id} by executing the *FEPOD.Decrypt* algorithm. If the decoding is accurate, the client *id* forwards the text hash value of hash string to cloud and action proceeds to step-3. If not, the agreement will expire faster, and the client *id* will release smart contract control key that the operator could use for validation before proceeding to step-5.

5. **Settle Payment:** If convention ends at $n-th$ rethinking calculation task, the instalment will be directed dependent on the $(n-1)-th$ hash esteem on the hash chain (put together by the cloud) and the yield of the *FEPOD.Verify* calculation for the $n-th$ re-evaluating calculation function. Something else, the instalment will be made by the keep going hash esteem on the hash chain.

Hypothesis 3: Suppose the FEPOD system is safe, the hash chain interchange agreement is reasonable, and the blockchain is believed, the BFEPOD protocol is secure.

Evidence: There are three types of enemies in the BFEPOD protocol. A non-secret client (probably the cloud) learns the plaintext of the password, the compromised user gets paid, and the cloud user is compromised. Correct math. Because of the security of the FEPOD plot (portrayed in Principle 1), the blockchain documentation [24, 25], and the reasonableness of the chain hash convention, these assaults are ensured. Along these lines, the BFEPOD convention above is safe.

5 Performance Analysis and Evaluation

5.1 Performance Analysis

In this portion, initially we analyse the presentation of the ABEPOD in principle, and then evaluate the implementation and execution of the suggested ABEPOD technique in the blockchain system.

Table 2 defines the repository costs of ABE plan in, PA-ABE plan in and the suggested ABEPOD plan, where "–" defines "not applicable". k are the multiple attributes related with a personal attribute-key, and l is the multiple attributes in an access structure.

Table 2. Repository costs

	ABE [22]	PA-ABE [12]	ABEPOD
Private Attribute-Key	$2 + 2k$	–	$2 + 2k$
Public Transformation-Key	–	$2 + 2k$	$2 + 2k$
Public Verification-Key	–	–	$2 + 2k$
Transformed Ciphertext	–	2	3
Efficient Verification	–	–	Yes

The initiated ABEPOD project is centred on the CP-ABE project in and compared with another ABEPOD called PA-ABE. Unlike the ABEPOD system, PA-ABE validates the results of outsourced transformation. In particular, the storage and transmission charges of the initiated ABEPOD system are summarized in Table 2 and Table 3 sequentially.

As stated in Table 2, some ABEPOD schemes have additional credit keys and additional cryptographic elements modified with similarly sized switches contrasted to the existing ABEOD plan PA-ABE.

Table 3. Calculus cost

	TranKG User	VeriKG User	Transform Cloud	Decrypt User	Verify Miner
ABE [22]	–	–	–	$\leq(3k+1)\cdot P+k\cdot E$	–
PA-ABE [12]	$(4+3k)\cdot E$	–	$\leq(3k+1)\cdot P+k\cdot E$	E	–
ABEPOD	$(2+2k)\cdot E$	$(5+5k)\cdot E$	$\leq(6k+2)\cdot P+2k\cdot E$	E	E

Table 3 defines the calculus cost of the fundamental ABE plan, the PA-ABE plan [26] and the given *ABEPOD* plan, where "–" means "not applicable". k denotes the multiple attributes related with a personal attribute-key, and l indicates multiple attributes in an access structure. "e" indicates expanding function, and "p" indicates matching function. Table 3 also defines the proposed technique of ABEPOD has the same amount of user decoding (performing outsourced computational work) as the existing ABEOD technique, just adding it to the extraction function.

5.2 Experiment Results

An external cryptographic technique is implemented before performing blockchain-based attribute-based encryption with provable outsourcing decryption (*BABEPOD*), a verifiable external cryptographic protocol for the proposed properties using ABEPOD. It uses a fast Python example framework for application programming. Centred on its charm, it implements all coding algorithms of the ABEPOD plan proposed by Python in about 300 lines of code.

Brilliant agreements are created utilizing around 150 lines of code in density, a language to make shrewd agreements on Ethereum to carry out *BABEPOD* convention. Since Ethereum will not carry the calculation of two-line elliptical curve links and Ethereum source code has changed, miners using PBC version 0.5.14 to verify the conversion results using the Python Web 3 library to communicate with smart contracts via Ethereum. It also implements the BABEPOD protocol in Python, allowing it to interact with Ethereum via Remote Procedure Call (RPC) [27].

Experimentation using a laptop with a 1.2 GHz quad-core Broadcom 64-bit quad-core processor with 1.6 GHz Intel Core i5-8250U processor*4, 24GB RAM, less than 64bit, Ubuntu 18.04 and Raspberry Pi 3B is performed. Raspbian acts as a user with 1GB RAM and Aliyun ecs.c5.x large cloud server. Four processors and 8 GB of RAM serve as the cloud.

First, we run each ABEPOD scheme algorithm on a laptop to test the average computational time to perform various tasks (in the login facility) without using the blockchain, then the outcome is shown in Fig. 3, Fig. 4 and Fig. 5.

Figure 3 shows that the intricacy of creating a private key, public vehicle key, and public credit key is practically direct as for the quantity of highlights. This is reliable with the hypothetical investigation shown in Table 3. The normal intricacy of the bends SS512 (80-bit safety), MNT159 (70-bit safety) and MNT201 (90-bit safety) for 50 exercises is under 60 ms and is truly worthy.

Figure 4 shows the normal computational expense of the ABEPOD chart approval and interpreting calculation. This demonstrates that they were free of the quantity of characteristics broke down in Table 3. For the entrance form with 30 capacities, the intricacy of decrypting and confirmation is under 1 ms on the bends of SS512 (80-piece safety), MNT159 (70-piece safety) and MNT201 (90-piece safety).

Figure 5 shows that computational cost of the encryption and transformation algorithms is almost linear with the multiple features related with the access facility, and the theoretical results in Table 3 can be seen. The proposed scheme is evaluated based on calculation expenses of Encrypt and Transform calculations as far as various qualities in access structures for various elliptical bends in the plan *ABEPOD*. Average computation time curves for SS512 activity (80-bit safety), MNT159 (70-bit safety) and MNT201 (90-bit safety) 50 are achievable in the real world in less than 1 s.

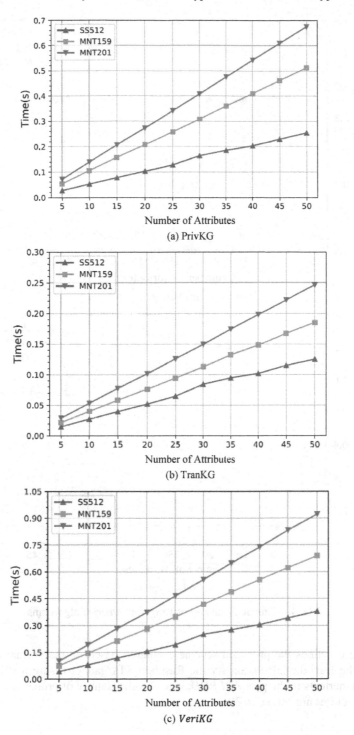

(a) PrivKG

(b) TranKG

(c) *VeriKG*

Fig. 3. Computational cost of (a) *PrivKG*, (b) *TranKG* and (c) *VeriKG* algorithms

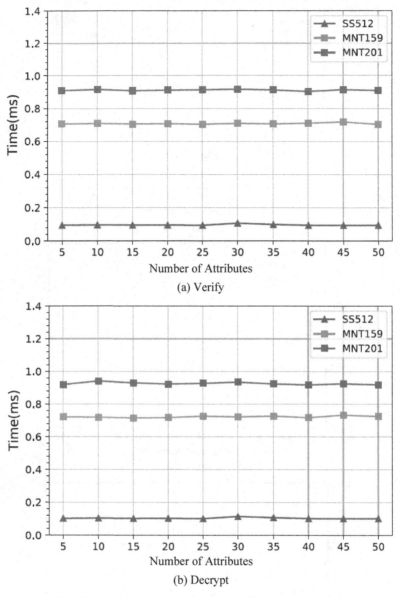

Fig. 4. Computational cost of *Verify* and *Decrypt* algorithms

Figure 6 defines the proposed scheme evaluated based on the normal calculation costs of the Decode calculation for the Raspberry Pi gadget (using a PC) as far as various ciphertexts in the plan *ABEPOD*, where both quantity of properties and area of access structures are settled to 30.

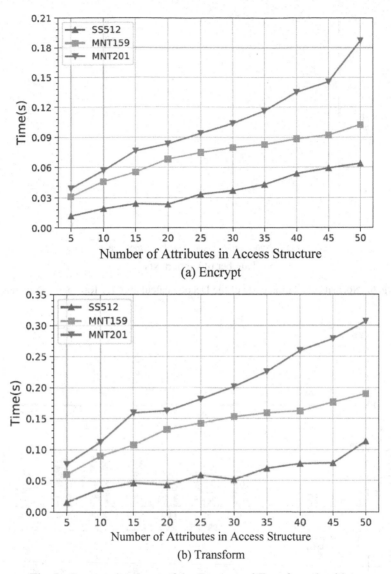

(a) Encrypt

(b) Transform

Fig. 5. Computational cost of the *Encrypt* and *Transform* algorithms

Figure 7 defines the proposed scheme evaluated based on normal calculation costs of conventional *BABEPOD* as far as various ciphertexts, where both quantity of properties and area of access structures are settled to 30.

Figure 8 defines the normal calculation costs of excavator under presumption that the members in convention *BABEPOD* might be noxious when end of convention *BABEPOD* occurs, where the quantity of ciphertexts is settled to 100.

After changing the quantity of client ascribes and the length of the login construction to 30, we test the normal intricacy of decoding Raspberry Pi 3B ciphertext for different

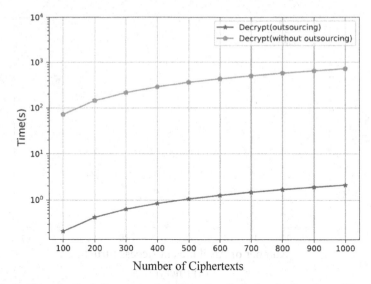

Fig. 6. Normal calculation costs of the Decode calculation for Raspberry Pi gadget

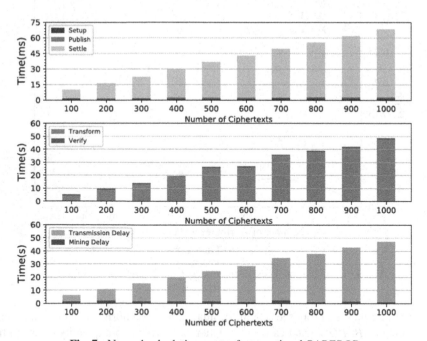

Fig. 7. Normal calculation costs of conventional *BABEPOD*

content encodings. The Raspberry Pi 3B has cloud computing capabilities and smart contract capabilities to outsource processing. The shape is different. The computation

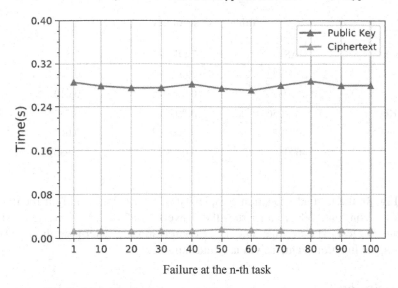

Fig. 8. Normal calculation costs of excavator under presumption

time of decrypted ciphertext is about 50 times slower than that of decrypted ciphertext without outsourcing.

We then evaluate the performance of each step of the BABEPOD protocol and evaluate its effectiveness. In this experiment, we used Ethereum's PoA fast consensus algorithm and set the number of features and the size of the login facility to 30. The Fig. 8 shows the average cost of each operation calculated assuming that all organisms in the BABEPOD protocol are correct. This demonstrates that setup, job submission and processing are very cost-effective, taking not more than 75 ms for 1000 total coding tests. Also, the conversion procedure takes time, but the encryption procedure is too shorter. However, using a more powerful server can reduce the conversion time to the cloud. Figure 8 likewise shows the computational expense of moving transformation results (off-chain) among clients and the cloud. For this situation, the postponement is relative to the quantity of scrambled messages and is controlled by the organization transmission capacity. The blockchain mining process creates a constant BABEPOD delay. This is about 12 s in our experience. Overall, the absolute latency of the BABEPOD protocol is around 1.5 min for 1000 ciphers and 0.09 s for ciphers and is good for real world utility.

Table 4 describes the costs of gas for multiple functions in BABEPOD where Gas price = 1 GWEI, 1 ETH = 228 USD.

It then tests the BABEPOD protocol to determine if the cloud or user is malicious. Figures 5.6 shows the average computational cost of miners in two examples of the BABEPOD protocol, and Table 4 summarizes the cost of Ethereum gas at various stages of the BABEPOD protocol. Here the text of the public key and the words Next Password is updated. Normal validation or update of transmitted ciphertext. Since there is no need for the cloud or the user to be harmless, miners have to change the password and pay, as well as compare the hash value the cloud provides to the user. If the validation extracts the *VeriKG* algorithm to reconstruct the validation public key (represented by the common

Table 4. Costs of gas

Operation	GAS	ETH	USD
Setup	2202342	0.002202	0.5021
Publish	44903	0.000045	0.0102
Public Key	3317756	0.00318	0.7564
Ciphertext	103497	0.000103	0.0071
Settle	40004	0.00004	0.00092

key in Fig. 8), the normal validation complexity will vary significantly if the multiple attributes is constant. When miner runs the conversion algorithm, converted text also completes the test. The average amount of computation to determine the number of cryptographic functions remains virtually unchanged.

6 Conclusion

Because of the benefits across public key encryption, functional encryption is an encryption system intended to ensure information security and protection in numerous new applications, for example, distributed computing administrations. The FE idea with sufficient external decryption (FEOD) capabilities for practical use was introduced to ease the burden on users by delegating most accounts to third parties. However, an unresolved issue in previous FEOD agreements is how payments are handled between outsourced users and third parties performing outsourced account work. In this paper, presented a unique scheme called publicly verifiable External Payment Decryption Scheme (FEPOD) that allows third parties to pay for services via blockchain-based cryptocurrencies. By demonstrating the security of the common architecture of FEPOD, it is described through the process of integrating FEPOD on the blockchain, displaying the FEPOD prototype on the blockchain, and evaluating its actual performance in terms of efficiency is described.

References

1. Liu, J.K., Chu, C.K., Zhou, J.: Identity-based server-aided decryption. In: Parampalli, U., Hawkes, P. (eds.) Information Security and Privacy. 16th Australasian Conference, ACISP 2011, Melbourne, Australia, 11–13 July 2011, Proceedings 16, vol. 6812, pp. 337–352. Springer, Heidelberg (2011). https://doi.org/10.1007/978-3-642-22497-3_22
2. Li, K., Wang, J., Zhang, Y., Ma, H.: Key policy attribute-based proxy re-encryption and RCCA secure scheme. J. Internet Serv. Inf. Secure. **4**(2), 70–82 (2014)
3. Qin, B., Deng, R.H., Liu, S., Ma, S.: Attribute-based encryption with efficient verifiable outsourced decryption. IEEE Trans. Inf. Forensics Secur. **10**(7), 1384–1393 (2015)
4. Liu, J., Li, W., Karame, G.O., Asokan, N.: Toward fairness of cryptocurrency payments. IEEE Secur. Priv. **16**(3), 81–89 (2018)
5. Zhang, Y., Deng, R.H., Liu, X., Zheng, D.: Blockchain based efficient and robust fair payment for outsourcing services in cloud computing. Inf. Sci. **462**, 262–277 (2018)

6. Banasik, W., Dziembowski, S., Malinowski, D.: Efficient zero-knowledge contingent payments in cryptocurrencies without scripts. In: Askoxylakis, I., Ioannidis, S., Katsikas, S., Meadows, C. (eds.) European Symposium on Research in Computer Security, pp. 261–280. Springer, Cham (2016). https://doi.org/10.1007/978-3-319-45741-3_14

7. Parno, B., Howell, J., Gentry, C., Raykova, M.: Pinocchio: nearly practical verifiable computation. Commun. ACM **59**(2), 103–112 (2016)

8. Cui, H., Deng, R.H., Li, Y., Qin, B.: Server-aided revocable attribute-based encryption. In: Askoxylakis, I., Ioannidis, S., Katsikas, S., Meadows, C. (eds.) ESORICS 2016. LNCS, vol. 9879, pp. 570–587. Springer, Cham (2016). https://doi.org/10.1007/978-3-319-45741-3_29

9. Lai, J., Deng, R.H., Guan, C., Weng, J.: Attribute-based encryption with verifiable outsourced decryption. IEEE Trans. Inf. Forensics Secur. **8**(8), 1343–1354 (2013)

10. Jakobsson, M., Wetzel, S.: Secure server-aided signature generation. In: Kim, K. (ed.) PKC 2001. LNCS, vol. 1992, pp. 383–401. Springer, Heidelberg (2001). https://doi.org/10.1007/3-540-44586-2_28

11. Lim, C.H., Lee, P.J.: Server (Prover/Signer)-aided verification of identity proofs and signatures. In: Guillou, L.C., Quisquater, J.J. (eds.) Advances in Cryptology — EUROCRYPT 1995. EUROCRYPT 1995. LNCS, vol. 921, pp. 64–78. Springer, Heidelberg (1995). https://doi.org/10.1007/3-540-49264-X_6

12. Li, J., Li, J., Chen, X., Jia, C., Lou, W.: Identity-based encryption with outsourced revocation in cloud computing. IEEE Trans. Comput. **64**(2), 425–437 (2015)

13. Fuchsbauer, G.: WI is not enough: zero-knowledge contingent (service) payments revisited. In: Proceedings of the 2019 ACM SIGSAC Conference on Computer and Communications Security, pp. 49–62 (2019)

14. Applebaum, B., Ishai, Y., Kushilevitz, E.: From secrecy to soundness: efficient verification via secure computation. Work **716835**, 0627781 (2010)

15. Gennaro, R., Gentry, C., Parno, B.: Non-interactive verifiable computing: outsourcing computation to untrusted workers. In: Rabin, T. (eds.) Advances in Cryptology – CRYPTO 2010. LNCS, vol. 6223. Springer, Heidelberg (2010). https://doi.org/10.1007/978-3-642-14623-7_25

16. Goldwasser, S., Kalai, Y.T., Rothblum, G.N.: Delegating computation: interactive proofs for muggles. J. ACM (JACM) **62**(4), 1–64 (2015)

17. Parno, B., Raykova, M., Vaikuntanathan, V.: How to delegate and verify in public: verifiable computation from attribute-based encryption. In: Cramer, R. (ed.) TCC 2012. LNCS, vol. 7194, pp. 422–439. Springer, Heidelberg (2012). https://doi.org/10.1007/978-3-642-28914-9_24

18. Papamanthou, C., Tamassia, R., Triandopoulos, N.: Optimal verification of operations on dynamic sets. In: Rogaway, P. (ed.) CRYPTO 2011. LNCS, vol. 6841, pp. 91–110. Springer, Heidelberg (2011). https://doi.org/10.1007/978-3-642-22792-9_6

19. Bentov, I., Kumaresan, R.: How to use bitcoin to design fair protocols. In: Garay, J.A., Gennaro, R. (eds.) CRYPTO 2014. LNCS, vol. 8617, pp. 421–439. Springer, Heidelberg (2014). https://doi.org/10.1007/978-3-662-44381-1_24

20. Buterin, V.: A next-generation smart contract and decentralized application platform. White Pap. **3**(37), 2-1 (2014)

21. Wood, G.: Ethereum-a secure decentralised generalised transaction ledger. Ethereum Project Yellow Pap. **151**, 1–32 (2014)

22. Rouselakis, Y., Waters, B.: Practical constructions and new proof methods for large universe attribute-based encryption. In: Proceedings of the 2013 ACM SIGSAC Conference on Computer & Communications Security, pp. 463–474 (2013)

23. Wan, Z.G., Deng, R.H., Lee, D., et al.: MicroBTC: efficient, flexible and fair micropayment for bitcoin using hash chains. J. Comput. Sci. Technol. **34**, 403–415 (2019)

24. Juels, A., Kosba, A., Shi, E.: The ring of Gyges: investigating the future of criminal smart contracts. In: Proceedings of the 2016 ACM SIGSAC Conference on Computer and Communications Security, pp. 283–295 (2016)
25. Kosba, A., Miller, A., Shi, E., Wen, Z., Papamanthou, C.: Hawk: the blockchain model of cryptography and privacy-preserving smart contracts. In: 2016 IEEE Symposium on Security and Privacy (SP), pp. 839–858 (2016)
26. Cui, H., Yi, X., Nepal, S.: Achieving scalable access control over encrypted data for edge computing networks. IEEE Access **6**, 30049–30059 (2018)
27. A Python Interface for Interacting with the Ethereum Blockchain and Ecosystem (2019). Accessed
28. Kushwaha, S.S., Joshi, S., Singh, D., Kaur, M., Lee, H.N.: Systematic review of security vulnerabilities in Ethereum blockchain smart contract. IEEE Access **10**, 6605–6621 (2022)
29. Kamidoi, Y., Yamauchi, R., Wakabayashi, S.I.: A protocol for preventing transaction commitment without recipient's authorization on Blockchain and it's implementation. IEEE Access **9**, 24390–24405 (2021)
30. Hong, L., Zhang, K., Gong, J., et al.: Blockchain-based fair payment for ABE with outsourced decryption. Peer-to-Peer Netw. Appl. **16**, 312–327 (2022)

Multi-environment Audio Dataset Using RPi-Based Sound Logger

Gaurav Govilkar$^{(\boxtimes)}$, Kader B. T. Shaikh , and N. Gopalkrishnan

Vivekanand Education Society's Institute of Technology, Mumbai, India
{2019gaurav.govilkar,kader.shaikh,n.gopalkrishnan}@ves.ac.in

Abstract. Rapid urbanization has led us being surrounded by many machines in our homes, workplaces, and neighborhood. These machines create assorted noises that have disturbing ramifications on the mental and physical health of human beings. A proper analysis of these sounds is essential for medical professionals, especially ENT specialists to study the noise effects on patients. Though AI (Artificial Intelligence) and ML (Machine Learning) technologies have presented complex algorithms to aid research and development in sound analysis and categorization, the paucity of appropriately labeled sound datasets limits the extent of the accuracy of these models. To bridge this gap a database is proposed with a collection of audio recordings of systems operating in ten diverse environments ranging from domestic appliances, automobiles, HVAC systems, industrial machines, construction equipment, etc. encountered by a person on a day-to-day basis. A portable sound logging device is developed using Raspberry Pi 3B+ and INMP441 MEMS microphone breakout board. I2Smic, PyAudio, Librosa, and several python libraries are used in system development. Labeling or annotation of the collected dataset is done manually by experts. All audio files are sampled at 44.1 kHz, single-channel, 16-bit audio wav file format. The database can be used for speech recognition, acoustic classification, audio anomaly detection, and the development of diagnostic systems in otology studies and allied medical research.

Keywords: Raspberry Pi development board · Python audio libraries · Portable sound logger · Environmental sound dataset · Open access sound database · Otology-Audiology Database

1 Introduction

Florence Nightingale recognized noise as a health hazard in 1859 when she wrote "Unnecessary noise is the most cruel abuse of care which can be inflicted on either the sick or the well". Noise pollution is one of the most significant concerns of modern times [1, 2]. Perpetual exposure to noise above certain levels (85 dBs and above) [3] can lead to harmful effects on human ears and mind. Engineering a system for the classification of audio signals assists in analyzing noise pollution and its effects on human health. Though ML and AI technologies have improved

© The Author(s), under exclusive license to Springer Nature Switzerland AG 2024
R. K. Challa et al. (Eds.): ICAIoT 2023, CCIS 1929, pp. 85–97, 2024.
https://doi.org/10.1007/978-3-031-48774-3_6

swiftly, only a few multi-environment audio databases are available that have a comprehensive set of audio signals from domestic appliances, automobiles, HVAC systems, industrial machines, construction equipment, etc. that humans are exposed to on a day-to-day basis [4]. With the heterogeneous nature of the applications that are being developed in this domain, there is an exigency of an application-specific dataset.

To cater to this, a portable sound logging system is designed. The sound logger is developed using Raspberry Pi 3B+ running on the Raspbian operating system. INMP441 a high-precision omnidirectional MEMS microphone breakout board is used to collect audio data. I2Smic and PyAudio python libraries are used to interface the breakout board and RPi.

The sound logging system can be accessed on a mobile phone or tablet using desktop mirroring software such as VNC Viewer over a network thus making the system portable.

The sound logging model collects audio from various on-field sources, like sounds generated by domestic appliances, industrial machines, musical instruments used in public festivals, crowd noise in large gatherings, and quotidian noises of traffic, etc, along with usual environmental sounds. 330 min of recording from 10 different real-life scenarios are recorded and presented in WAV format with signed 16-bit resolution and a sampling rate of 44.1 KHz.

2   Review of Similar Public and Private Databases

Barker [5] proposed an audio dataset under the 5th CHiME (2018) challenge for Automatic Speech Recognition (ASR). The dataset discernibly consisted of speech data wherein the audio was recorded at 20 different parties under similar circumstances. All the parties lasted for two hours at least with conversational audio from four participants. Six distant Microsoft Kinect recording arrays along with Soundman Binaural microphones worn by participants were used to record this dataset.

Ryant [6] as an extension to the work of Barker [5] recorded other single-channel and multichannel conversational audio samples along with using the CHiME-5 dinner party corpus (multichannel audio data) for diarization in the DIHARD II challenge. The single-channel audio tracks included 2-h recordings of various public places including restaurants, courtrooms, interviews, etc. All audio was distributed as 16 kHz, mono-channel FLAC files. Recording arrays and binaural microphones were used as recording media.

AudioSet Ontology [7] made a vast collection of sound events organized in a hierarchical manner. The dataset attempts to cover sounds ranging from human and animal vocals to environmental sounds, to musical and miscellaneous sounds. This dataset was mainly used for constructing Automatic Event Detection systems. Each ontology entry contains a description of the sound event, based on Wikipedia, Wordnet, or written by the author.

ASVSpoof 2017 version 2.0 [8] proposed a database that is used for spoof identification. The database contains recorded audio sets of YouTube videos in assorted environments taking into consideration various playback and recording devices. These devices range from smartphone-based record systems to head-sets and microphones. The spoofed audios were played in the background using speakers and this setup was recorded to develop spoof detection systems.

DCASE 2018 challenge (Task 1) [9] proposed a dataset comprising audios from 10 acoustic scenes in various public places (airports, subways, etc) in dif-ferent cities. 5–6 min of audio, recorded in 2–3 sessions of a few minutes each, were split into segments with a length of 10 s that were provided in individual files. Recording hardware included four different devices, viz, Soundman elec-tret binaural microphone along with 3 smartphones. The audio recorded on the binaural microphone was an in-ear mechanism (mimicking that the voice is reach-ing the human auditory system) recorded at 48 kHz sampling rate and 24-bit resolution.

Dohi [10] presented an anomalous sound detection (ASD) dataset for DCASE 2022 Task 2 challenge. Dataset consists of five distinguished machine sounds, three sets of factory noise data, and three different domain shift scenarios for each machine type. The audio was recorded using a TAMAGO-03 microphone having a 16-bit resolution with a sampling rate of 16 kHz. The recordings (wav file format) were done in soundproof environments with a record length of 10 s each and were supported by short descriptions.

Kumari [11] did a survey on multimedia datasets in 2021 as a compara-tive study on various audio-visual datasets for Anomaly Detection and other applications. The concluding statement of the paper stated that datasets for heterogeneous anomaly are far less compared to specific anomaly datasets and emphasized a strong need to develop more audio-visual datasets for generic scene surveillance.

All the above databases are open source and they cater to certain specific applications. However, this specificity limits the scope of these datasets. Whilst catering to specific applications, the datasets have limited descriptions. On the contrary, the proposed database comprises a wide range of different audio along with a detailed description of the System Under Test (SUT), generated audio, surrounding noise, etc. While writing the description for the audio, attention to detail is not compromised, making the dataset more comprehensive, giving the analysts a better sense of the data, and aiding their application/models to produce more precise and accurate results.

Datasets [7,8] use an algorithm-based annotation and labeling, which is prone to errors whereas the proposed database has been labeled by the experts to achieve error-free results.

Many databases [5–7] offer FLAC, mp3 files whereas files in the proposed database are in WAV format for achieving better bit-depth and sample rate. It is easier to edit the WAV file format using open-source sound editors [12].

3   Experimental System for Sound Recording

3.1   Hardware Description

Raspberry Pi 3B+ [13] development board with INMP441 [14] MEMS Microphone breakout board is used to develop a sound logger. Raspberry Pi board offers GPIO PCM pins that are capable of high-quality audio recording using Inter-IC sound (I2S) interface protocol [15]. INMP441 MEMS microphone has an inbuilt ADC, anti-aliasing filters, and I2S interface. A single piece of INMP441 generates a mono recording. The I2S standard uses three pins, viz, Serial Data pin to record the data, Serial Clock pin to keep the sampling of data in sync (SCK), and Word Select pin to determine whether an input/output is in the left channel or right channel. Figure 1 shows the interconnection between Raspberry Pi and INMP441.

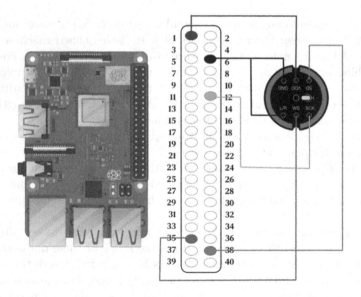

Fig. 1. Raspberry Pi and INMP441 connection diagram

3.2   Software Description

Raspberry Pi 3B+ works on Raspbian Operating System [13]. Python is used to develop the software for device interface, communication, control, data collection, storage, labeling, and other tasks related to the sound logger. Along with common libraries such as NumPy, matplotlib, scipy, etc., the following python libraries are used to read, save, and analyze audio input from INMP441 MEMS microphone breakout-board on to Raspberry Pi:

- I2Smic: The Raspberry Pi (RPi) needs to be prepared for I2S communication by creating/enabling an audio port in the RPi OS system. This audio port will then be used to communicate with MEMS microphones and record mono/stereo audio [16].

- Pyaudio: Python library that provides bindings for PortAudio the cross-platform audio input/output stream library [17].

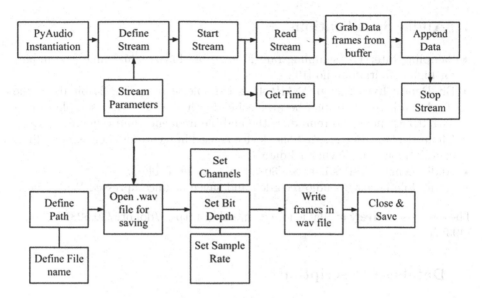

Fig. 2. Data Collection Algorithm

3.3 Data Collection Algorithm

Figure 2 shows the block diagram representation of the data collection algorithm and the algorithm is as follows:

1. Identify Device Index: Device index of the external breakout board is identified using the I2Smic library.
2. Define recording parameters: Parameters such as sampling rate, chunks, channels, recording length, device index (identified in step 1), etc. are defined.
3. Create a PyAudio instant: PyAudio instant is created, which gathers system requirements for PortAudio [17].
4. Define Pyaudio Stream: To record or play audio there is a need to define the stream. After the definition, the stream is started.
5. Grab (collect) audio data: The data is grabbed as frames from the buffer. These frames are appended to the previous frames to form a 'stream' of data. The data is recorded for a given time with specified audio specifications (parameters) defined in the initial step. Later, the stream is stopped.
6. Save the data: Timestamp of the recording is collected and we define the filename, and location (path) for the data to be saved. The WAV file is opened and the recorded frames are written in the file with the appropriate number of channels, bit depth, and sampling rate. The data is saved in WAV file format at the specified location.

7. Close instance & garbage collection: After saving the data in step 5, PyAudio instant is closed and the garbage collection is initiated.
8. Data Transfer to external storage device: Recorded data is transferred to the laptop/desktop from the Raspberry Pi memory using the VNC software [18].

4   Audio Specifications

- Sampling rate: The sampling rate is set to 44.1 kHz. This value is standard for high-definition audio [19].
- Bit Depth: Even though INMP441 is capable of recording 24-bit data, the bit depth is set to 16-bits because a bit depth of 16-bit for a sample rate of 44.1 kHz is enough to reproduce the audible frequency and dynamic range.
- Channels: Only one microphone device is used in recording, hence, recordings contain mono/single channel data.
- Audio Length: Varies between 30–60 min. Refer Table 1.
- Audio File Format: Audio recorded and saved as uncompressed WAV Files.

The database is freely available for download at https://doi.org/10.5281/zenodo.7493820.

5   Database Description

- **Air compressor.** The Air compressor unit (Fig. 3a) provides clean air at high pressure. This machine creates a high operating noise that is irritating. The air compressor sound was recorded by keeping the logger at a distance of 50 cm from the Air compressor unit. The change in the pressure of the output air can be compared with the intensity of the sound produced by the compressor.
- **Pump/Motor.** The pump under study (Fig. 3b) is a 2900 RPM, 400V, 3-phase Monoblock pump used for redundancy application. This pump is a substitute for three submersible pumps. The pump is activated on failure in any primary pump. The monoblock pump has a higher flow rate and is used to drain the underground tank on the failure of the submersible pumps. The audio is recorded from a distance of 30 cm from the source of the sound.
- **Kitchen Exhaust Fan.** SUT is a typical exhaust fan installed in a traditional kitchen. The fan was kept ON for 30 min. The recording system was kept very close to the fan (Less than 10 cm away), making it possible to record a clear and distinct sound of the kitchen exhaust. The fan is over 20 years old and thus is louder than the latest models. Given the passé nature of the machine, it has a blade that is slightly bent which causes that blade to brush one of the metal bars that hold the exhaust fan in place, though it doesn't occur quite often.

Table 1. Description of collected sound datasets

Environment	Type	Approx. distance of recording device from the sound source (m)	Duration (mins)	Notes	Background Noise
Air Compressor	Industry	0.5	60	Mechanical sound of electric motor, belt, wheezing due to pressure release	–
Pump	Industry	0.3	30	Mechanical sound of pump	–
Kitchen Exhaust Fan	Domestic	0.1	30	Sound of motor and blades, blade sound	–
Washing Machine	Domestic	0.3	30	Sequential sounds of water pouring, drain, dryer, etc.	–
Concrete Mixer	Construction	20	30	Sounds of Mixer, Pulley, Motor, etc.	Other construction sounds like clinging of metals, dumping of goods, conversation amongst site workers
Dhol	Instrument	50	30	Beating of Drums, Tasha	Traffic noise
Laptop cooling fan	Domestic/ Workplace	0.2	30	Maximised rpm of the CPU and GPU Fans	–
Disturbances in Local Train	Public place	–	30	Sounds of train acceleration and deceleration, Occasional sound of breaks	Occasional sound of announcements on the train stations, indistinct chatter
Canteen Noises	Public place	–	30	Distinct and Indistinct chatter conversational sounds	Sound of clanging of metals, cutlery.
Traffic Noises	Public place	–	30	Honking, acceleration and ignition sounds of vehicles	Indistinct chatter

(a) Air Compressor　　　(b) Monoblock Pump　　　(c) Concrete Mixer

Fig. 3. Systems under test

- **Washing Machine.** The machine under study is a top-load washing machine. The recording was done for 30 min with the machine in full action. The actions of the machine recorded by the system include filling the washing tub, rinsing, washing, spinning (the dryer), and draining the water. All these processes are repeated twice in the recording. The filling of the tub is fairly less noisy as most of the tub is already full of clothes and the noise of water hitting the surface is minimum. In the rinsing process, the noise fairly increases due to the whirling of the water. The washing action is similar to rinsing but an increase in the speed of whirling leads to increased noise. The spinning action produces substantial noise initially when the dryer tub begins to gain momentum. This process is accompanied by the noise made by the washing machine's body shaking vigorously for about 3 to 4 s before gaining maximum speed. At maximum spinning action, the noise is comparatively lower and constant. As the spin is decreased the noise again picks up and wanes down with a final 'jerk' sound. The drain of the water makes a gurgling sound. As mentioned earlier, this cycle is repeated again before the washing machine completely stops. Since the machine is a new purchase and of the latest type, the noise produced is fairly lower. For older, semi-automatic machines more noise could be observed.
- **Concrete Mixer.** The machine under study is a cement mixer used to prepare concrete (Fig. 3c). Recording from this machine not only includes the blaring noise of a typical concrete mixer but also includes various other on-site noises like a construction lift, its motor, and the pulley. Normally the lift (also called 'Builder's Hoist') mechanism noise is masked by the loud din of the concrete mixture, yet that noise (of the lift mechanism) can be prominently heard in certain places. The recording width is 30 mins and has a mixture of noises that are generated simultaneously unlike the sequential noise change in the case of the washing machine. The voices can, however, be distinguished if the on-field process is understood. Other minor sounds in the recording also include metals hitting against each other along with other subsidiary constructional disturbances.

- **Dhol.** The traditional Indian drums called Dhol (often clubbed with 'Tasha', a smaller drum) are a ritual during festivities in the country. The noise produced by a 'dhol' is quite large than the traditional drums used on stage. The recording of the banging of these large drums was done for 30 min from a place around 50 m away from the source. The sound of the drums is not continuous but can be heard at intervals. Drum sounds are present in 60–70 percent of the recording. The sound of Tasha Drums is present at a few instants only while at other instants only the Dhol sound is audible, the combination of the two voices is observed as well. They can be distinguished based on the bass produced. Tasha drums have lower bass than the Dhol drums [20].
- **Laptop Cooling Fan.** With gaming laptops that achieve higher performances for games with heavy graphics coming in vogue, additional and high-end cooling systems are required to be installed. Under normal conditions, these fans operate at regular speeds but when operated in the 'Gaming Mode' the cooling fans are pumped to greater speeds to achieve smoother and lag-free gameplay. The noise created by multiple cooling fans operating at the same time with their maximum capacity is irritating to human ears. The cooling fan noise was recorded for 30 min. The speed of the fan was set to maximum manually.
- **Disturbances in Local Train.** Trains are a popular medium of transport in Indian metropolitan cities like Mumbai. Given the colossal number of travelers using the local trains on a daily basis, this environment is usually very noisy. The recording was done on a train routing from the less crowded suburbs of Mumbai to more crowded regions of the city resulting in less noisy initial parts of the recording. The disturbances increase towards the end of the recording as the train gets crowded. Halting of the train at subsequent stations can be figured out by the announcements inside the train as well as on the platform as the train stops. Another factor that identifies the slowing down of trains is the noise of the brakes. Speeding of the train is identified by a characteristic sound.
- **Canteen Noises.** This recording is done on the premises of a college canteen during the lunch break. There are around 100–150 people in the canteen while the audio was being recorded. The specimen contains noises from the crowd, people discussing along with distant shrieking. Most of this recording is conversations and murmurs. A seldom & indistinct sound of the clanging of metals can be heard from the kitchens of the canteen. Sounds of cutlery are also recognized, though not very distinct.
- **Traffic Noise.** The audio is recorded in the peak traffic hours of the evening near a traffic signal. The traffic signal is roughly 200 m away from the site of the recording. The audio comprises honking, acceleration, and ignition sounds produced by vehicles of distinct types. The honking can be heard whenever the signal has just turned green. Honking is followed by acceleration and ignition sounds. A distant chatter can be treated as background noise, though its magnitude is less.

6   Application Test Cases

- **Otology and Audiology.** Effects of surrounding noises and sounds on the human auditory system can be analyzed using the dataset by suitable feature extraction [21]. The dataset can aid audiology researchers and ENT specialists to anatomize the auditory effects of miscellaneous sounds on a human. A spectrogram constructed using the proposed dataset is shown in Fig. 4. Using this figure audiologists can decipher the intensity (amplitude) for various frequencies along with the duration of exposure thus suggesting the extent of exposure to sound in a particular environment. The combined spectrogram shows the contrast between the extent of exposures for various sound environments discussed in this paper.

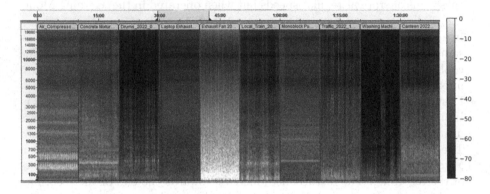

Fig. 4. Spectrogram with data from diverse environments

- **Environmental Noise Control Analysis.** The dataset can be used by environmentalists to obtain a detailed report on 'noisy' parameters in heterogeneous environments by extracting sound specifications of the audio [22]. Correspondingly, environmentalists can suggest suitable measures to keep noise pollution under control in similar environments. The extent of similarity can be defined on the basis of the comprehensive description of the scenarios in which the audio was recorded.
- **Speech Recognition.** With the indispensability of speech recognition technology in almost every sector, the major challenge has been the detection of speech in noisy environments [5,6]. This dataset can assist the experts in analyzing the audio trends in various environments to achieve a better performance of their speech recognition models.
- **Anomaly Detection & Machine Analysis.** With Industry 4.0 & 5.0 taking over every sector, industries are developing novel techniques like audio anomaly detection for on-site machine analysis [4,23–25]. The dataset includes samples of industry-recorded data that can aid the process of developing algorithms for acoustic anomaly detection.

- **Acoustic Classification.** In applications where manual or optical scanning is not possible, sound classification provides optimum results. It is possible to determine, classify and label the types of sound from noisy audio [26]. For such applications, the dataset provides a set of sounds properly defined and described.
- **Event Detection & Localization.** A particular event can be detected by the occurrence of a typical sound in a noisy environment. Detection of such events can be used for safety, event classification, and other purposes [27,28]. The dataset provides a well-established set of various sounds that will aid in event detection and localization.

7   Conclusion

The proposed dataset is a compilation of ten audio sets from diverse environments. These audio sets are recorded using a portable sound logger developed using Raspberry Pi 3B+ and INMP441 MEMS microphone breakout board. Application software is developed using various python libraries. Details of system hardware and software are discussed in the paper. In breadth discussion of all SUTs, their surrounding environment, and generated acoustics are summarized. The primary aim of releasing this database is to foster research in the field of medicine, especially otology along with other areas such as speech recognition, event detection, anomaly detection, etc.

References

1. Mukherjee, K., Deb, N., Roy, A.D., Dash, P.: Impact of noise pollution on human health in Barasat Urban Area, West Bengal. In: Sustainable Urbanism in Developing Countries on Proceedings, pp. 375–393. CRC Press (2022)
2. Thompson, R., et al.: Noise pollution and human cognition: an updated systematic review and meta-analysis of recent evidence. Environ. Int. **158**, 106905 (2022). https://doi.org/10.1016/j.envint.2021.106905
3. Ali, S.A.: Industrial noise levels and annoyance in Egypt. Appl. Acoust. **72**(4), 221–225 (2011). https://doi.org/10.1016/j.apacoust.2010.11.001
4. Shaikh, K.B.T., Jawarkar, N.P., Ahmed, V.: Machine diagnosis using acoustic analysis: a review. In: 2021 IEEE Conference on Norbert Wiener in the 21st Century (21CW), pp. 1–6. (2021). https://doi.org/10.1109/21CW48944.2021.9532537
5. Barker, J., Watanabe, S., Vincent, E., Trmal, J.: The fifth 'CHiME' speech separation and recognition challenge: dataset, task and baselines. arXiv preprint arXiv:1803.10609 (2018). https://doi.org/10.48550/arXiv.1803.10609
6. Ryant, N., et al.: The second dihard diarization challenge: dataset, task, and baselines. arXiv preprint arXiv:1906.07839 (2019). https://doi.org/10.48550/arXiv.1906.07839
7. Gemmeke, J.F., et al.: Audio set: an ontology and human-labeled dataset for audio events. In:2017 IEEE International Conference on Acoustics, Speech and Signal Processing (ICASSP), pp. 776–780. IEEE (2017). https://doi.org/10.1109/ICASSP.2017.7952261

8. Delgado, H., et al.: ASVspoof 2017 version 2.0: meta-data analysis and baseline enhancements. In: Odyssey 2018-The Speaker and Language Recognition Workshop (2018)

9. Mesaros, A., Heittola, T., Virtanen, T.: A multi-device dataset for urban acoustic scene classification. arXiv preprint arXiv:1807.09840 (2018). https://doi.org/10.48550/arXiv.1807.09840

10. Dohi, K., et al.: MIMII DG: sound dataset for malfunctioning industrial machine investigation and inspection for domain generalization task. arXiv preprint arXiv:2205.13879 (2022). https://doi.org/10.48550/arXiv.2205.13879

11. Kumari, P., Bedi, A.K., Saini, M.: Multimedia datasets for anomaly detection: a review. arXiv e-prints, arXiv-2112 (2021). https://doi.org/10.48550/arXiv.2112.05410

12. Neyaz, A., Varol, C.: Audio steganography via cloud services: integrity analysis of hidden file. Int. J. Cyber-Secur. Digit. Forensics **7**(1), 80–87 (2018)

13. Raspberry Pi Foundation. https://www.raspberrypi.com/products/raspberry-pi-3-model-b-plus/. Accessed 20 Jan 2023

14. INMP441 Datasheet. https://invensense.tdk.com/wp-content/uploads/2015/02/INMP441.pdf. Accessed 20 Jan 2023

15. NXP Homepage, I2S Protocol User Manual. https://www.nxp.com/docs/en/user-manual/UM11732.pdf. Accessed 20 Jan 2023

16. Maker's Portal I2SMic Homepage. https://github.com/makerportal/rpii2s. Accessed 20 Jan 2023

17. SCAIL MIT PyAudio Homepage. https://people.csail.mit.edu/hubert/pyaudio/docs/. Accessed 20 Jan 2023

18. MIT Edu Homepage VNC Technology. http://web.mit.edu/cdsdev/src/howitworks.html. Accessed 20 Jan 2023

19. Smith, S.W., et al.: The Scientist and Engineer's Guide to Digital Signal Processing. California Technical Publishing, San Diego (1997)

20. Ballengee, C.: From Dhol-Tasha to Tassa: tradition and transformation in Indian Trinidadian Tassa drumming. Roczniki Humanistyczne **70**(12), 121–136 (2016). https://doi.org/10.18290/rh227012.8

21. Lokwani, P., Prabhu, P., Nisha, K.V.: J. Otol. (2022). https://doi.org/10.1016/j.joto.2022.08.001

22. Li, J., Dai, W., Metze, F., Qu, S., Das, S.: A comparison of deep learning methods for environmental sound detection. In: 2017 IEEE International Conference on Acoustics, Speech and Signal Processing (ICASSP), pp. 126–130. IEEE (2017). https://doi.org/10.1109/ICASSP.2017.7952131

23. Koizumi, Y., et al.: Description and discussion on DCASE2020 challenge task2: unsupervised anomalous sound detection for machine condition monitoring. arXiv preprint arXiv:2006.05822 (2020). https://doi.org/10.48550/arXiv.2006.05822

24. Cao, H., Yu, J., Wang, Y., Zhang, L., Kim, J.: A fault diagnosis system for a pipeline robot based on sound signal recognition. Sensors **22**(9), 3275 (2022). https://doi.org/10.3390/s22093275

25. Bondyra, A., Kołodziejczak, M., Kulikowski, R., Giernacki, W.: An acoustic fault detection and isolation system for multirotor UAV. Energies **15**(11), 3955 (2022). https://doi.org/10.3390/en15113955

26. Barchiesi, D., Giannoulis, D., Stowell, D., Plumbley, M.D.: Acoustic scene classification: classifying environments from the sounds they produce. IEEE Signal Process. Mag. **32**(3), 16–34 (2015). https://doi.org/10.1109/MSP.2014.2326181

27. Mesaros, A., Heittola, T., Eronen, A., Virtanen, T.: Acoustic event detection in real life recordings. In: 2010 18th European Signal Processing Conference on Proceedings, pp. 1267–1271. IEEE (2010)
28. Zhuang, X., Zhou, X., Hasegawa-Johnson, M.A., Huang, T.S.: Real-world acoustic event detection. Pattern Recognit. Lett. **31**(12), 1543–1551 (2010). https://doi.org/10.1016/j.patrec.2010.02.005

A Comparative Analysis of Android Malware Detection Using Deep Learning

Diptimayee Sahu$^{(\boxtimes)}$ ⓘ, Satya Narayan Tripathy ⓘ, and Sisira Kumar Kapat ⓘ

Department of Computer Science, Berhampur University, Berhampur , India
ds.rs.cs@buodisha.edu.in

Abstract. The fear of Android malware infection is building up with its user acceptance in the global market space. The existing anti-malware techniques are archaic to signature-based detection which limits the detection scope against newly crafted malware. Several machine learning, as well as deep learning approaches, has been proposed so far to combat against the rising threat of the android mobile world. Most of these have relied on static features due to its lower cost. In this paper, a Deep Neural Network model is proposed and evaluated it with static, dynamic and hybrid features respectively to prepare the comparative statement. In this experiment we used the OmniDroid dataset [1] which consists of 22,000 Android Package Kits (APKs) including both benign and malware APKs. For analysis 7955 static features and 4805 dynamic features are extracted using proper selection criteria. The efficacy of the proposed models on different data types is compared in terms of accuracy and loss. This experiment achieved a higher accuracy of 99.66% with dynamic features and a lower accuracy of 85.74% with combined features. Similarly, we achieved a minimal loss of 0.06 with dynamic analysis and a higher loss of 0.61 with static analysis. This confirms that the efficacy of dynamic analysis is more prominent in An- droid malware detection.

Keywords: Android malware detection · dynamic analysis · artificial neural network · deep learning · deep neural network

1 Introduction

The adoption of digitalization makes Smartphone an important aspect of everyday life. Android operating system shares the maximum user acceptance among Smartphone users with 75% of the global market share [2]. The growing platform of Android implies Google play to add more number of applications. Google play store has more than 2.9 million Applications in 2020 [2]. The use of Smartphone and intrusion of malicious apps in Smartphone is growing in parallel. Malicious Apps refer to a piece of software with malevolence intent such as; financial losses, data theft and privacy disclosure etc. According to a report by Kaspersky [3] the recent COVID-19 pandemic situation encountered many banking Trojans, spyware and adware droppers etc. A study says the AV-TEST Institute reports more than 350,000 novel malware every day [4]. The rising number of malware applications attaining analysts and researchers concern on a wide scale across the world. Various approaches have evolved so far to come up with an efficient solution using different methodologies [5, 6, and 7].

R. K. Challa et al. (Eds.): ICAIoT 2023, CCIS 1929, pp. 98–110, 2024.
https://doi.org/10.1007/978-3-031-48774-3_7

Android application comes in a packed format basically as a zip file with a .apk extension. Every APK file includes two important files; AndroidManifest.xml, classes.dex. Baskaran et al. [8] in their survey categorized malware detection techniques into static, dynamic and hybrid. Static detection is done by gathering the information without executing the application while dynamic detection is done after running the application. Static detection is done by extracting features such as; permissions, intents, actions, services and other metadata recorded from the AndroidManifest.xml file and API calls from a source code file 'classes.dex' which contains the entire bytecode of the application in the native Dalvic Executable (DEX) format of the Dalvik Virtual Machine. AndroidManifest.xml file has sufficient information to discriminate the common malware applications [9] while using information collected at runtime is efficient to improve the performance of detection systems [10].

Static analysis is hassle-free as it does not need one dedicated environment to run the applications. It inspects the malware concerning to metadata present in the .apk files, signatures, code structure and patterns observed. The code structure can be represented as control flow graphs. Considering only available static features is not always feasible to detect malware since malware writers can easily obfuscate the benign applications just by adding a few lines of redundant code.

Dynamic analysis needs a dedicated sandbox environment for executing the malicious applications to understand it's behaviour. A sandbox environment is a virtual simulated environment that ensures the safety of the user system from harm by the execution of malicious applications where the analysis of the applications can be performed more thoroughly with no fear of threat to user system. The applications run in a virtual environment for a certain period and run-time information is collected after executing the application. The collected information after the execution of an application is then compared with the information before the execution of the application to encounter the exact behavioural analysis.

The existing Android malware detection systems rely on the application's signature. It can be ineffective for detecting evasion threats as the signatures are not re- corded yet. In addition, the evasion techniques adopted by the attackers to evade detection make the attacks hard to detect. To counteract these concerns, several experiments are performed on various feature sets using deep learning and performed a comparative analysis of the static, dynamic and combined approach.

Deep learning has a history of performing equally better on a large number of input features [11]; hence deep learning techniques are considered on a substantial number of features of our dataset in this experiment.

This paper focuses to develop a deep learning framework using Deep Neural Network (DNN) techniques on various identical features. Package-level information such as permissions, Opcodes, API calls, System Commands are considered and dynamic feature includes System call sequences. A comparative analysis is also performed to analyze the efficacy of the proposed system with other deep learning approaches of the literature survey. Based on the comparison it is made out that the proposed system has a better performance in malware detection.

Hereafter the paper is arranged as Section 2 lists the literature survey for the proposed work, Section 3 explains the proposed methodology for performing the experiments, Section 4 narrates the proposed framework, Section 5 gives details of the experimental setup where the experiments are conducted, Section 5 explains the result and discussion based on the experiments performed, Section 7 describes the comparative analysis of the work followed by Section 8 which includes the conclusion that summarizes our work.

2 Literature Survey

Chaba et al. [5] proposed a three steps approach detection techniques which include system call log generation, chi-square filtering method to create the dataset and implementation of the machine learning algorithm on the dataset. They used NB, RF and Stochastic Gradient Descent (SGD) algorithms on the Waikato Environment and validated the correctness and quality of their dataset by obtaining 93.75% accuracy on NB Classifier and 93.84% on RF Classifier and 95.5% on SGD algorithm.

Rodrigo et al. [6] proposed a hybrid detection model using OmniDroid dataset which contains 22,000 samples. They performed feature selection using Pearson Correlation and selected 840 features. They used a neural network to train and validate the model and achieved 85.8% accuracy. Later they relabeled the dataset by changing the threshold and achieved 92.9% accuracy.

Oliveira et al. [7] proposed a hybrid android detection model which consists of 3 different models Chimera-S, Chimera-R, Chimera-D using DNN techniques. Chimera-S is a static detection model that uses 200 features which include 100 Intents and 100 Permissions. Chimera-R uses DEX byte-codes as images and Chimera-D is a dynamic analysis model which uses system call sequences. Finally, they concatenated the models and passed it to the final Chimera's DNN classifier and achieved an accuracy of 90.9%.

Baskaran and Ralescu [8] in their survey listed static analysis features that include permissions, API calls, intents, actions, services, etc. which can be extracted from AndroidManifest.xml file. Dynamic features include network traffic, battery usage, IP address, etc. which can be extracted while the application is running. The hybrid analysis includes both static and dynamic features. Based on these three different types of features malware detection techniques can also be categorized into static detection, dynamic detection and hybrid detection respectively.

Sagar et al. [12] surveyed various detection models where most of the models use permissions and API calls for static analysis and dynamic behaviors for dynamic analysis. Every android '.apk' file contains 'androidmenifest.xml' in which all permissions that an application will need are listed. An android malware detection model can be designed by extracting that permission and carefully applying classification to it Budiarto et al. [13] proposed a hybrid approach considering both static and dynamic features. They used Principal Component Analysis (PCA) technique on 215 static and 46 dynamic features. Their experiments apply some of the machine learning algorithms such as Naive Bayes(NB), Random Forest(RF), K-Nearest Neighbor (K-NN), Multi-Layer Perceptron (MLP), Gradient Boost (GB) etc. on top 10 selected features and come up with GB performed better among other algorithms.

Bayazit et al. [14] proposed a detection model on static features of the CICInves-sAndMal2019 dataset that records 1126 benign apps and 396 malware apps. They considered 8115 permission, and intent features and evaluated their model using Recurrent Neural Network (RNN) algorithms such as Long Short-Term Memory (LSTM), Bidirectional LSTM (BiLSTM) and Gated Recurrent Unit (GRU) where BiLSTM achieved a better accuracy rate of 98.85%.

Shatnawi et al. [15] proposed a framework and performed a comparative analysis on the performance of static, dynamic and hybrid approaches using several classifiers such as Extreme Gradient Boosting (XGBoost), GB, DT, RF. Based on the result obtained where all the analysis achieved accuracies above 94%, they concluded that static analysis alone would be efficient and cost less.

Aboaoja et al. [16] considered API call sequences of android applications to propose a malware dynamic detection model against android, and used a box-whisker algorithm to produce the malicious behaviour dataset. Their framework is evaluated on RF, SVM, CART, NB, KNN, ANN, LR, and XGBoost machine learning algorithms. Their final dataset consists of 3848 benign samples and 7208 evasive malware where XGBoost classifier outperformed with values 0.967, 0.040, 0.971, 0.978 and 0.975 with respect to accuracy, False Positive Rate (FPR), detection rate, precision and F1 score.

Shen et al. [17] have proposed a technique called complex flow analysis for which they focused on app behavior related to sensitive information and then analyzed the API sequences in detail. After analyzing the API sequences they found that by considering the sent sensitive data is not enough to categorize the applications as malware and benign. In their model they used Multi flow analysis technique. They have extended the present blue seal to find complex flows and used a two-class SVM classifier which obtains an accuracy of 0.913 with 1 gram size.

Ki et al. [18] proposed a model by analyzing API call sequences. They collected 2,727 kinds of APIs. They used the longest common subsequence (LCS) to find the common sequence patterns. They created a database of signatures and for each new program the LCS is matched with the signatures. Their model achieved precision, recall and F1 score as 1, 0.998, 0.999 respectively.

3 Proposed Methodology

The proposed methodology includes several steps that are explained in the following sections. The data preprocessing is important in order to shape the dataset as per the required format that in turn increase the efficiency of the detection framework.

3.1 Dataset

OmniDroid dataset [1] is being used in this paper. It is a comprehensive benchmark dataset with an extensive number of manifold features consisting static and dynamic types. The relevant features are considered from the dataset for analysis purpose. The dataset has 22000 applications out of which 11000 are benign and 11000 are malicious.

From the dataset 7955 distinct static features are considered which include permissions (5500), Opcodes (224), API calls (2128), System Commands (103) and 4805 dynamic features. The dynamic feature includes the System call sequences because the System call sequence represents the actual behavioral pattern of an application. Two malware samples or two benign samples have many similarities in their System call sequences pattern; hence we have considered System call sequences as dynamic feature in the proposed study. The above features are accepted based on the literature survey of this paper.

3.2 Data Preprocessing

The dataset is represented as a $i \times j$ matrix; where 'i' imply to total number of rows (APKs), 'j' imply to total number of columns (individual features) with a label column which represents the class of the individual records; 0 (for benign) and 1(for malware).

$$The\ value\ of\ Label\ Column = \begin{cases} 0\ for\ benign \\ 1\ for\ malware \end{cases} \quad (1)$$

Data cleaning is the initial task performed to normalize the dataset by eliminating missing, duplicate, and null values from the dataset. Here row reduction is performed to remove the duplicate entries in the dataset. In this study a two stage data transformation technique is performed to increase the acceptability of the data by an algorithm. Python MinMaxScaler and StandardScaler functions are used to standardize the features. MinMaxScaler scales the features between a certain range [0, 1] and StandardScaler scales the data to unit variance.

3.3 Feature Selection

To select the significant features Information Gain (IG) feature selection algorithm is used. It ranks the features with an IG score. It is an entropy based feature evaluation method that calculates the IG score for individual features in the dataset by subtracting the weighted entropies of individual features from the entropy of the output column that can be represented as Eq. (2).

$$H(y) = -(class(b) * \log_2(class(b)) + class(m) * \log_2(class(m))) \quad (2)$$

where, $H(y)$ is entropy of output class and y is the class label (1 and 0). Class (b) is the predicted probability of the feature being for all the class labeled as 0 ($y = 0$) and Class (m) is the predicted probability of the feature malware for all the class labeled as 1 ($y = 1$) and $\log_2(class(b))$ is the log probability of the feature being benign and $\log_2(class(m))$ is the log probability of the feature being malware.

The IG value represents the importance of the feature in the dataset. Fig. 1 and Fig. 2 represent the IG ranking of different features, where x- axis stands for the features in descending order of calculated IG values and y-axis stands for IG values. The selection of features for final experiment depends on the IG ranking, and the features for the experiment are selected based on the IG score. The percentage of selected features can be observed from Table 1.

Table 1. Number of Selected Features.

Name of feature	Total	Selected
Static features	7955	1591 (20%)
Dynamic features	4805	961 (20%)
Combined features	13760	2552

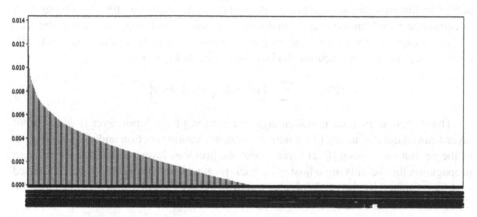

Fig. 1. Ordering of the feature based on IG score calculated per feature basis of dynamic features.

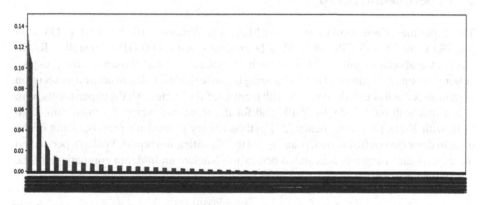

Fig. 2. Ordering of the feature based on IG score calculated per feature basis of static features.

4 Proposed Framework

For detection, our proposed framework uses a combination of static features and dynamic features. The resulting feature vector includes permission features, Opcode features, API call features, System command features and System call sequences. The proposed framework consists of three major processes for the detection such as; feature selection, model generation and model evaluation. In this section, we introduce a DNN approach for the prediction and evaluation of malware existence in an application. This section discusses several essential concepts, the system workflow, and the evaluation parameters.

DNN has a generic architecture; the input layer consists of dummy neurons that receive individual features as input, and one output layer which predicts the output followed by number of hidden layers with artificial neurons which projects the data into different dimensions. The model is determined by its arguments; the maximum number of hidden layers in the network, number of neurons in each hidden layer, suitable activation function for hidden layers, batch size (number of samples per batch), suitable activation function for the output layer, etc. and the performance for forward propagation is computed based on the weight matrices assigned to each input including the bias vector on each hidden layers. The output of one layer is input to the next layer. The output computed by each neuron can be represented as Eq. (3).

$$Zj^i = \sigma(\sum k\left(w_{jk}^i\right) + z_k^{l-1}) + bias_k^l \qquad (3)$$

The computation at each hidden layer for neuron j from input layer (l-1) to output layer l can be defined as Eq. (3): where 'σ' is an activation function and w^l is the weight of the neuron j to neuron 'k' at layer l from the previous layer (l-1). After the forward propagation the weights are adjusted by back-propagation with gradient descent based on loss estimated by the model during training and the model efficacy is evaluated on the test set of data.

5   Experimental Setup

The experiments are conducted on the Microsoft Windows 10 Pro.(64-bit) OS with Intel(R) Core(TM) i5 CPU @ 1.80 GHz processor with 8.00 GB of installed RAM. Google Colaboratory with a hosted Jupiter notebook service was chosen for data pre- processing, feature selection and model building because this provides an arbitrary execution of python codes through the browser with many useful libraries. All the experiments were performed with the "Tesla T4" GPU and for the front-end wrapper TensorFlow 2.8.0 [19], with Keras [20]. Scikitlearn [21] python library is used for preprocessing of the data, to draw the confusion matrix and to get the classification report. We have performed the experiments in two phases stated bellow to visualize an in-depth comparison of the model performance.

$$precision = \frac{True\ Positives}{True\ Positives + False\ Positives} \qquad (4)$$

$$recall \frac{True\ Positives}{True\ Positives + False\ Negatives} \qquad (5)$$

$$F1\ score = 2 \times \frac{precision \times recall}{precision + recall} \qquad (6)$$

$$accuracy = \frac{TP + TN}{TN + TP + FP + FN} \qquad (7)$$

where True Positives (TP) represents the amount of malware that are detected as malware. False Positive (FP) represents the amount of benign apps that are wrongly categorized as malware. True Negative (TN) represents the amount of benign apps that are correctly classified as benign. False Negative (FN) represents the amount of malware apps those are wrongly classified as benign. Precision is the probability for a malware app to be classified as malware. Recall helps to understand how the model is able to identify the data. F1-score is the harmonic mean of precision and recall. Accu- racy measures the overall performance of the detection model.

6   Result and Discussion

Several experiments were carried out on the proposed framework for performance evaluation. The experiments cover using static features, dynamic features and combin- ing both. A comparative analysis of the proposed system is performed for malware detection and evaluation according to the performance metrics described as follows.

6.1   Analysis Using Static Features

This section hereafter presents and discusses the results obtained using static features of OmniDroid dataset [1]. The total extracted features are 7955 which includes permissions, API calls, Opcodes, System Commands. The top 20% of the features are selected after feature selection. So total of 1591 static features are used for the performance evaluation of the model. A two-stage data transformation is performed with the help of MinMaxScaler and StandardScaler. The whole data is parted into two parts such as training and testing set. The training set consists of 80% whereas the testing set con- sists of 20% of whole data. The proposed DNN model is constructed using Keras sequential model with TensorFlow support and represented by Fig. 3.

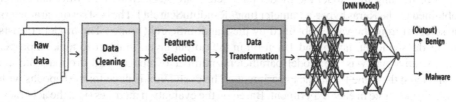

Fig. 3. Proposed System Architecture

In our proposed framework the rectified linear unit 'ReLU' is used in every hidden layer to deal with the vanishing gradient problem and on the output layer 'sigmoid' is used as an activation function to predict the classification output. For tuning the model parameters Random search hyper-parameter tuner is used. Adam is used as the optimization function. Binary_crossentropy is used to measure the error. The experiments are performed on all the feature sets using the algorithm given bellow.

ALGORITHM: Experiment for Model Generation

Input	:	**Omnidroid Dataset**
Output	:	**Accuracy and Loss**
Step 1	:	Perform feature selection based on calculated $IG(f_i)$
Step 2	:	$\{X, Y\}$ (Static features + Dynamic features)
Step 3	:	$S_{sample} \rightarrow (X_{train}, Y_{train})$
Step 4	:	Split the dataset into 2 parts:
		1. Training set (80%)
		2. Test set (20%)
Step 5	:	$T_i = number\ of\ columns\ in\ X_{train}$
Step 6	:	*Build_model(DNN)*
Step 7	:	1. Input layer (Input dimension = T_i)
		2. Hidden layers (selected by RandomSearch()) with (Percentage of validation set + batch size + number of epochs + optimizer + activation function + loss optimizer + learning_rate)
		3. Output layer (One output with sigmoid activation function)
Step 8	:	Evaluation matrices(accuracy, val_accuracy, loss, val_loss)
Step 9	:	Return (val_accuracy, val_loss)

For training the model the model parameters are selected based on the parameter obtained by the Keras hyper-parameter tuner RandomSearch(). The best set of parameters is selected from the values obtained by the RandomSearch() tuner. The model consists of 13 dense hidden layers and 1 neuron in the output layer. In order to make a proper comparative analysis the model parameters, number of epochs and dropout rate are considered the same for all the experiments. The evaluation is done for 400 epochs with a 0.2 dropout rate in each experiment. Based on the evaluation matrices (i.e., the accuracy, loss) the mean accuracy obtained by the model is 89.04 % and the loss calculated by the model is 0.61 as presented in Figs. 4 and 5

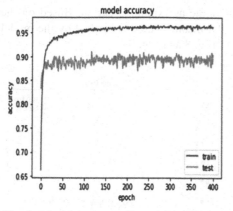

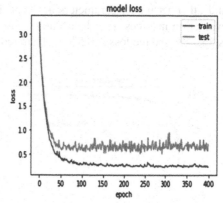

Fig. 4. Model Accuracy Score using Static Features.

Fig. 5. Model Loss Score using Static Features.

6.2 Analysis Using Dynamic Features

This section hereafter presents and discusses the results obtained using dynamic features of the OmniDroid dataset [1]. The total extracted features are 4805 which includes System call sequences. The top 20% of the features which is 961 dynamic features are selected after feature selection for performance evaluation of the model and all other model parameters are kept the same as explained earlier. Based on the evaluation matrices (i.e., the accuracy, loss) the mean accuracy obtained by the model is 99.66% and the loss is 0.06 as illustrated in Fig. 6 and Fig. 7.

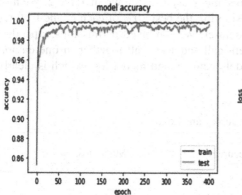

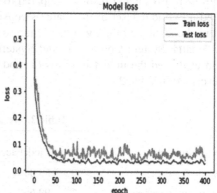

Fig. 6. Model Accuracy Score using Dynamic Features.

Fig. 7. Model Loss Score using Dynamic Features.

6.3 Analysis Using a Combination of Static and Dynamic Features

Out of all extracted features after feature selection top 20% of the static features and the top 20% of the dynamic features are combined for performance evaluation of the model and all other model parameters are kept the same as explained earlier. Based on the evaluation matrices (i.e., the accuracy, loss) the mean accuracy obtained by the model is 85.74% and the loss is 0.57 as illustrated in Fig. 8 and Fig. 9.

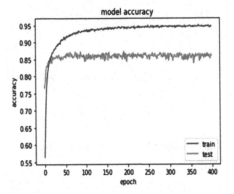
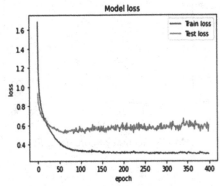

Fig. 8. Model Accuracy Score using Combination of Features.

Fig. 9. Model Loss Score using Combination of Features.

7 Comparison Analysis

Based on all the result analysis it is observed that static features are easy to use in a detection model whereas dynamic features require extra resources for environmental setup and after feature extraction, a lot of manual work is needed to prepare the dynamic dataset. But as the dynamic data is the real-time behavioural representation of the applications it gives better results compared to static analysis as shown in Table 2. We also used a hybrid dataset of the features considering the best 20% of each of the features from the static and the dynamic feature dataset. We combined permissions, Opcodes, API calls, System Commands and System call sequences all together in one dataset and evaluated the model in terms of model accuracy and model loss which is clearly described in Table 2.

Table 2.   Accuracy and Loss.

Name of feature	Model accuracy	Model loss
Static features	89.04%	0.61
Dynamic features	99.66%	0.06
Combined features	85.74%	0.57

Recent researches like Rodrigo et al. [6] and Oliveira et al. [7] used the Omnidroid dataset in their research work. The comparative analysis of this research work with [6] and [7] is represented in Table 3.

Table 3. Comparison with Similar Research Work.

Name of the research work	Year	Accuracy
Rodrigo et al. [6]	2021	91.1%
Oliveira et al. [7]	2020	90.9%
Proposed Framework	–	99.66%

8 Conclusion

The increase of Android malware attacks in the last few years needs a coherent solution to the current situation. In this proposed work, we have performed a comparative analysis and based on the result obtained, we used the dynamic features to set up a DNN model with the potential to spot malwares from android applications. Several experiments were conducted and the test results reveal that the proposed system achieves a ceiling of 99.66% accuracy with a minimal loss of 0.06. The inclusion of dynamic features for model training signifies more concise and precise malware detection as compared to the other analysis performed in this work. Finally, the efficiency of the proposed system is compared with some other works of the literature survey considering accuracy as a parameter for comparison. The comparison shows that our pro- posed system gives better accuracy as compared to some recent works. For future work, we look forward to implement other deep learning techniques on varied features of android applications to provide an efficient system for android malware detection.

References

1. OmniDroid Dataset. http://aida.etsisi.upm.es/download/omnidroid-dataset-csv-features-v1/. Accessed 21 May 2020
2. Android statistics (2022). https://www.businessofapps.com/data/android-statistics/. Accessed 2 Aug 2022
3. Mobile malware evolution (2020). https://securelist.com/mobile-malware-evolution-2020/101029/. Accessed 20 Nov 2022
4. Malware (2022). https://www.av-test.org/en/statistics/malware/. Accessed 12 Feb 2022
5. Chaba, S., Kumar, R., Pant, R., Dave, M.: Malware Detection Approach for Android systems Using System Call Logs. arXiv:1709.08805 (2017). doi: https://doi.org/10.48550
6. Rodrigo, C., Pierre, S., Beaubrun, R., Khoury, F.E.: BrainShield: a hybrid machine learning-based malware detection model for android devices. Electronics **10**(23), 2948 (2021). https://doi.org/10.3390/electronics10232948
7. Oliveira, A.S., Sassi, R.J.: Chimera: An Android Malware Detection Method Based on Multimodal Deep Learning and Hybrid Analysis. TechRxiv, Preprint (2020). https://doi.org/10.36227/techrxiv.13359767.v1
8. Baskaran, B., Ralescu, A.: A study of android malware detection techniques and machine learning. In: 27th Modern Artificial Intelligence and Cognitive Science Conference (MAICS 2016), pp 15–23 (2016)
9. Arp, D., Spreitzenbarth, M., Hubner, M., Gascon, H., Rieck, K.: Drebin: effective and explainable detection of android malware in your pocket. In: NDSS Symposium (2014). ISBN 1–891562–35–5. https://doi.org/10.14722/ndss.2014.23247

10. Rastogi, V., Chen, Y., Jiang, X.: Catch me if you can: evaluating android anti-malware against transformation attacks. IEEE Trans. Inf. Forensics Secur **9**(1), 99–108 (2014)
11. Deep learning performance breakthrough - Servers & Storage. https://www.ibm.com/blogs/systems/deep-learning-performance-breakthrough/. Accessed 2 July 2020
12. Sabhadia, S., Barad, J., Gheewala, J.: Android malware detection using deep learning. In: 3rd International Conference on Trends in Electronics and Informatics (ICOEI). IEEE (2019)
13. Hadiprakoso, R.B., Kabetta, H., Buana, K.S.: Hybrid-Based malware analysis for effective and efficiency android malware detection. In: International Conference on Informatics, Multimedia, Cyber and Information System (ICIMCIS). IEEE Xplore (2020). https://doi.org/10.1109/ICIMCIS51567.2020.9354315
14. Bayazit, E.C., Sahingoz, O.K., Dogan, B.: A deep learning based android malware detection system with static analysis. In: International Congress on Human-Computer Interaction, Optimization and Robotic Applications (HORA). IEEE Xplore (2022). doi:https://doi.org/10.1109/HORA55278.2022.9800057
15. Shatnawi, A.S., Jaradat, A., Yaseen, T.B., Taqieddin, E., Al-Ayyoub, M., Mustafa, D.: An android malware detection leveraging machine learning. Wireless Commun. Mobile Comput. Hindawi (2022). https://doi.org/10.1155/2022/1830201
16. Aboaoja, F.A., Zainal, A., Ali, A.M., Ghaleb, F.A., Alsolami, F.J., Rassam, M.A.: Dynamic extraction of initial behavior for evasive malware detection. Mathematics. **11**(2), 416 (2023). https://doi.org/10.3390/math11020416
17. Shen, F., Vecchio, J.D., Mohaisen, A., Ko, S.Y., Ziarek, L.: Android Malware Detection using Complex-Flows. In: 2017 IEEE 37th International Conference on Distributed Computing Systems (ICDCS), Atlanta, GA, USA, pp. 2430–2437 (2017). https://doi.org/10.1109/ICDCS.2017.190
18. Ki, Y., Kim, E., Kim, H.K.: A novel approach to detect malware based on API call sequence Analysis. Inter J. Distributed Sensor Netw. Hindawi Publishing Corporation (2015). https://doi.org/10.1155/2015/659101
19. TensorFlow. https://www.tensorflow.org/. Accessed 13 June 2020
20. Keras. https://keras.io/getting_started/. Accessed 13 June 2020
21. scikit-learn, Machine Learning in Python. https://scikit-learn.org/stable/. Accessed 13 June 2020

Optimization of Virtual Machines in Cloud Environment

Kamal Kant Verma[1]([✉]) [iD], Ravi Kumar[1] [iD], Shivani Chauhan[1] [iD], Sagar Gulati[2] [iD],
Brij Mohan Singh[1] [iD], and Mridula[1] [iD]

[1] COER University, Roorkee, India
kkv.verma@gmail.com, drmridfce@coer.ac.in
[2] iNurture Education Solutions Pvt. Limited, Bangalore, India

Abstract. The proposed VMP-LR (Virtual Machine Placement-Load Rebalancing) has three main components, namely, Resource Request Handling Component, Placement Component and Load Monitoring Component. VMP-LR is designed using a two-phase methodology, where the first phase handles the tasks involved with Resource Request Handling and Placement Components, while Phase II handles the tasks of Load Monitoring Component. The algorithms proposed in both the phases are designed to handle multiple resource requests. During non-rush hours (low traffic), as the number of requests is minimal, the VMP-LR uses a simple enhanced round-robin method to place virtual machines (VMs) to physical machines (PMs). During rush hours (heavy traffic), the three queues created are handled using three separate hybrid scheduling and load- balancing algorithms to perform placement operation efficiently and accommodate high resource demands. To solve this issue, the load monitoring component is used. For this purpose, a hybrid algorithm that combines ant colony optimization (ACO) with an artificial bee colony (ABC) algorithm is used. The proposed algorithms are implemented using CloudSim Simulator and evaluated using seven performance metrics. They are throughput, response time, resource utilization rate, power usage, load unbalancing rate, SLA Violation rate and migration rate. The average SLA violation rate of VMP-LR is 0.6% less as compared with FF (1.26%), BF (1.1%), RR (1.36%), GA (0.98%), ACO (1.16%) and PSO (1.01%). This shows that VMP-LR is a much-improved version among existing load balancing algorithm.

Keywords: Monitoring · Virtual machine development · CloudSim Simulator

1 Introduction

A virtual machine is a digital model of a physical computer. Virtual machine software requires updates and system monitoring since it may execute program and operating systems, store data, connect to networks, and perform other computer operations. Virtual machine software may be used to manage several VMs that are hosted on a single physical computer, frequently a server. This increases overall efficiency by allowing for the flexible distribution of computational resources (computer storage and network)

R. K. Challa et al. (Eds.): ICAIoT 2023, CCIS 1929, pp. 111–135, 2024.
https://doi.org/10.1007/978-3-031-48774-3_8

among VMs as needed. The fundamental building blocks for the sophisticated virtualized resources we use today, such as cloud computing, are provided by this architecture. Cloud computing is described as "a model for convenient, on-demand network access to a shared pool of configurable computing resources (e.g., networks, servers, storage, applications, and services) that can rapidly be provisioned and released with little management effort or service provider interaction," by NIST [1]. From this definition, cloud computing can be considered as network accesses (or requests) to potential resources from a shared pool, which are provided in an on-demand basis in a manner that requires less response time, minimal effort from the management, minimal interaction from the service provider. In cloud computing, there are three basic service models: software as a service (SaaS), platform as a service (PaaS), and infrastructure as a service (IaaS) [2]. These three services can all be set up in one of four possible cloud configurations: private, public, community, or hybrid. This section presents four deployment models [3] used by cloud systems such as Private Cloud Deployment Model, Public Cloud Deployment Model, Hybrid Cloud Deployment Model, Community Cloud Deployment Model. It has emerged as a crucial technology for data centers and the cloud, contributing to scalability, dynamic topologies, and the ability to move virtual machines between PMs for load balancing. There are three different virtualization techniques: client (or desktop) virtualization, server virtualization, and storage virtualization. (see Fig. 1).

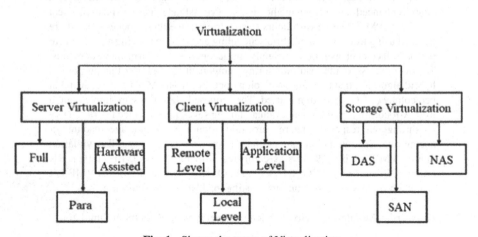

Fig. 1. Shows the types of Virtualization

There are three types of data storage used in virtualization: DAS (Direct Attached Storage), NAS (Network Attached Storage), and SAN (Storage Area Network). DAS is a traditional data storage method in which the storage drive is directly attached to the server machine. NAS is a shared storage mechanism connected over a network. NAS is used for file sharing, device sharing, and backup storage between machines. A SAN is a storage device that is shared with various servers over a high-speed network. A hypervisor is a software package that controls work access to the physical hardware of a host computer. Furthermore, as clouds offer an on-demand pay-as-you-go business model, users can demand creation and termination of any number of VMs according to their requirements.

1.1 Load Rebalancing

Most of the popular cloud providers offer different categories of VMs with the speci fication for each type of resource. These VM instances differ in their individual re source capacity, where some instances are larger than others and some instances have a relatively higher capacity of one type of resource compared to other resources. Be cause of the above properties, complementary resource demands across different re source dimensions are common in cloud data centers. Load imbalance arises when the existing services are stopped either by the cloud customers or in the event of host power cycling. This causes fragmentation that leads to degradation in server resource utilization, where a new request cannot be accommodated often. As VM workloads change frequently, eventually in a dynamic cloud environment, even the well- optimized placement choices are not sufficient to maintain a balanced load. This sce nario can be managed through the use of load rebalancing algorithms. An example of a fragmented and rebalanced load is shown in Fig. 2(a) and Fig. 2(b).

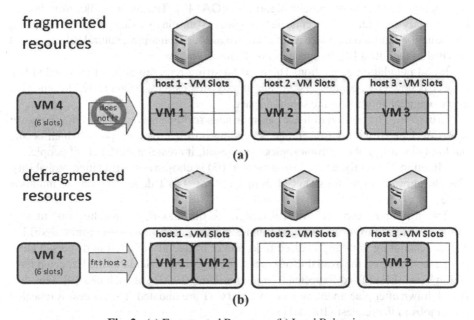

Fig. 2. (a) Fragmented Recourses (b) Load Balancing

Figure 2(a) and Fig. 2(b) represent the effects of load rebalancing.

Moving services between hosts to guarantee equitable resource distribution is known as load balancing. Moving and condensing the VMs into the less physical hardware resources is a way to lower the number of active PMs [4].

2 Literature Review

Recently, several studies have successfully used ant colony optimization (ACO) meta-heuristics as solutions for VM placement and VM consolidation. This section presents the basic concepts of ACO and how ACO is used in scheduling and load balancing along with a survey of studies that have used the ACO algorithm for placement, load balancing, and rebalancing. The movement of ants in an ACO-based load-balancing algorithm can belong to two types: (i) forward movement in which the ants maintain shifting ahead till they find out PMs which are each overloaded and underloaded (ii) backward movement in which the traversal will reverse and go backward if an ant encounters an overloaded PM after visiting an underloaded PM to determine whether it is still underloaded. It will evenly allocate resources if it is still underloaded. If an underloaded PM cannot be found, then a PM with maximum foraging pheromone will be selected for redistribution of resources.

On the other hand, [5] and [6] have provided a detailed working description and compared performance evaluation of several existing prominent scheduling and load balancing methods. The methods compared include round robin algorithm, greedy algorithm, Backward Speculative Placement, Dynamic Priority Based Scheduling Algorithm, genetic algorithm, and power save algorithm. An evolutionary biology- inspired learning strategy is offered by genetic algorithms (GA) [7]. The most wellknown class of evolutionary algorithms, GA replicate biological evolution by using processes including reproduction, mutation, crossover (also known as recombination), natural selection, and survival of the fittest [8]. Its specifics are discussed below.

Initial population generation. The fixed bit string representation of each individual answer is how GA operates. As a result, binary strings have been created to represent every potential solution in the solution space.

Crossover. The portion of these chromosomes residing on one side of the crossover site is exchanged with the other side depending on the crossover point, which occurs randomly in this pool of chromosomes. As a result, it creates a fresh pair of people.

Mutation. Currently, a very low number (0.05) is chosen as the mutation probability. The chromosomal bits are switched from 1 to 0 or 0 to 1 depending on the mutation value.

The algorithms used in VMP-LR design are the max-min algorithm, first-fit algorithm, best-fit algorithm, and round-robin algorithm. The typical Max-min algorithm begins scheduling with unprocessed tasks and is frequently utilized in dispersed environments. Based on the available resources, the algorithm calculates the Expected Execution Time (EExT) and Expected Completion Time (ECoT) of each task [9]. This task is withdrawn after placement, and the ExT and CoT are updated. The process is repeated for all jobs in the request [10] and [11].

Step 1: Start with a list of unscheduled resource, U.

Step 2: Determine the set of minimum completion times for U.

Step 3: Choose the next task with a maximum minimum completion time and assign it to the machine that provides the minimum completion time.

Step 4: Remove the newly mapped task from U Step 5: Repeat steps 2–4 until U is empty.

The First Fit algorithm [12] is one of the popular simple solutions that can help minimize the number of servers used. This algorithm follows the criterion: "Place are-quest in the first (called as lowest-indexed) bin which will fit, that is, if there are any partially filled PMj with level (PMj) + s(Resourse Request (RRi)) < = 1, then place RRi in the lowest indexed bin [16]. Otherwise, start a new bin to meet the VM requirement and place RRi as its first item [17]. Here 's' indicates the size of the resource.

The best-fit algorithm is also a popular bin-packing algorithm [15]. The best-fit algorithm [13] also called match-making algorithm, is similar to the first-fit algorithm in the manner that it filters out all the PMs that do not meet the resource request character-istics. To get the finest VM-to-PM mapping possible, the best-fit algorithm conducts a thorough search by assessing how well the PMs are matched for the location rather than selecting the first PM that meets the criterion [18]. The goal of the traditional round robin scheduling technique [19] is to distribute the workload across all PMs evenly [20]. This approach cycles through allotting one VM to each node [14]. Prior to going on to the request for the following PM, the algorithm first assigns a request to each PM. The pre-empted operation is then moved to the back of the ready queue and executed during the subsequent time slice or quantum [21]. This placement ensures that the scheduler moves on to the next task without waiting for a PM's resources to run out.

3 Proposed Methodology

The research analyzes the VM requests concerning three dimensions: CPU Using this multi-dimensional resource characteristic [22] and [23], the proposed scheduling algorithm can achieve high efficiency and greater utilization of resources resolving the first issue identified [24]. The Placement Component efficiently maps VMs to PMs using traffic and load-aware scheduling methods [25]. The efficiency with which these algorithms handle large, medium, and low resource requests have also been improved. The VM manager will constantly monitor all of these related parts. Two research steps go into the design of the proposed VMP-LR [26] represented in Fig. 3

Resource handling and placement components, the research methodology's initial phase suggests algorithms for building request queues and techniques for effectively scheduling them. The three unique queues that are formed are High Resource Queue (HRQ), Medium Resource Queue (MRQ), and Low Resource Queue (LRQ).

The second stage of the study technique focuses on load balance to boost the place-ment work and further optimize resource utilization. Resource fragmentation brought on by an increased load imbalance lowers server resource usage [27]. A load imbalance can also occur if a host power cycle occurs or cloud users or customers discontinue existing services. Load rebalancing methods can be used to control this situation [28]. The experiments are created in the following six stages.

Stage 1: Review the suggested scheduling and load balancing algorithm. Stage 2: Evaluate the effect of queuing algorithm.

Stage 3: Selection of algorithms according to traffic intensity.

Stage 4: Analysis of scheduling and load balancing algorithm to queues can be mapped to any of the suggested algorithms.

Stage 5: Analysis of the rebalancing algorithm.

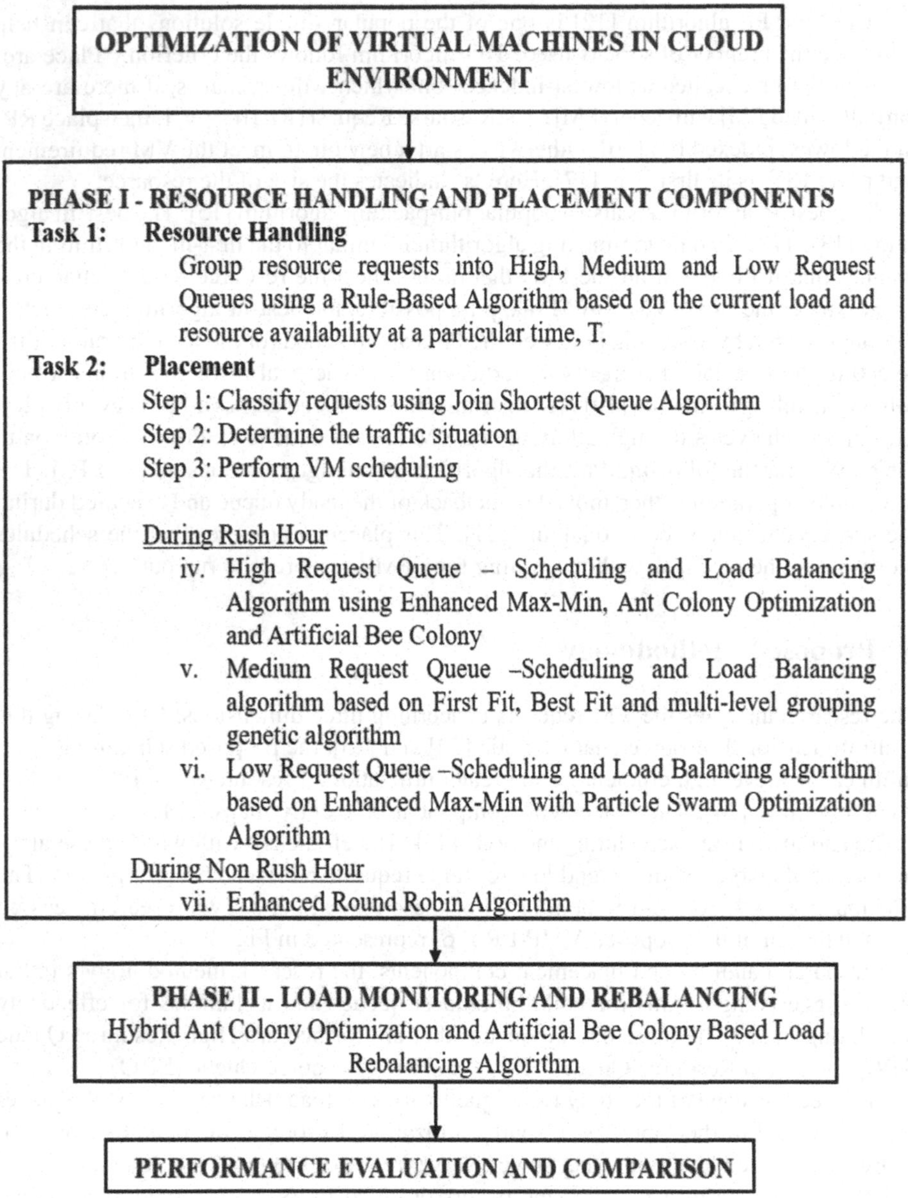

```
┌─────────────────────────────────────────────────────┐
│   OPTIMIZATION OF VIRTUAL MACHINES IN CLOUD          │
│                  ENVIRONMENT                         │
└─────────────────────────────────────────────────────┘
```

PHASE I - RESOURCE HANDLING AND PLACEMENT COMPONENTS
Task 1:   Resource Handling
Group resource requests into High, Medium and Low Request Queues using a Rule-Based Algorithm based on the current load and resource availability at a particular time, T.

Task 2:   Placement
Step 1: Classify requests using Join Shortest Queue Algorithm
Step 2: Determine the traffic situation
Step 3: Perform VM scheduling

<u>During Rush Hour</u>
iv. High Request Queue – Scheduling and Load Balancing Algorithm using Enhanced Max-Min, Ant Colony Optimization and Artificial Bee Colony
v. Medium Request Queue –Scheduling and Load Balancing algorithm based on First Fit, Best Fit and multi-level grouping genetic algorithm
vi. Low Request Queue –Scheduling and Load Balancing algorithm based on Enhanced Max-Min with Particle Swarm Optimization Algorithm

<u>During Non Rush Hour</u>
vii. Enhanced Round Robin Algorithm

PHASE II - LOAD MONITORING AND REBALANCING
Hybrid Ant Colony Optimization and Artificial Bee Colony Based Load Rebalancing Algorithm

PERFORMANCE EVALUATION AND COMPARISON

Fig. 3. Proposed Methodology

Stage 6: Evaluation of VMP-LR after incorporating the enhanced Algorithms Space is here.

Let P or PM be the collection of all physical machines in a cloud system's data center and P = {p1, p2,..., pm}, where m is the number of actual machines overall and pi represents a specific physical machine. 'i' ($1 \leq I \leq m$). Similarly, let V or VM be the set of virtual machines on each Pi and Vi = {vi1, vi2,..., vin} where n is the total number of VMs in PM 'i'.

Using the PM load variation law, it is possible to divide time (t) into a total of K consecutive time intervals., that is t = [(t1-t0), {t2-t1),..., (tk-tk-1)]. The time interval 'k' is denoted as (tk-tk-1). Assuming that the VM load is constant throughout the day, then $V_{in}(j, k)$ denotes a load of Vin in P_j at time interval k. Thus, the average load of $Vin(j, k)$ on Pj in a time interval T is defined as Eq. (1).

$$\text{Average Load(AL)} = \frac{1}{T} \sum_{k=1}^{k} V_{in}(j, \ k) * (t_k - t_{k-1}) \tag{1}$$

Using Eq. (2) a load of PM for the last T intervals can be estimated by adding all the loads of VMs on that PM.

$$\text{Load of } P_j(T) = \sum_{j-1}^{m} AL(V_{in}(T)) \tag{2}$$

After placement, the VMP-LR has resource information for V, from which, PM load can be re-estimated using Eq. (3).

$$\text{Load of} P_j(T) = P_j(T) + AL((V_{in}T)) \tag{3}$$

The cloud system works on a heterogeneous resource requirement environment, where the queuing models based on single resource might not be sufficient. The requests inside each queue are given equal importance. The usage of the multi-queuing model increases both customer satisfaction and provides profitability. It can also reduce starvation when combined with dynamic scheduling [29]. A cloud system generally uses a queue manager to generate and manage the queues. Three types of queues, small queue contain 40% of the requests, medium queue contains the next 40% of the requests and long queue contains the remaining 20% of the request are used during scheduling and load balancing [30].

3.1 Proposed Queuing Model

The main goal of the proposed queuing model is to group the resource requests into High, Medium and Low Request Queues using an algorithm based on the current load and resource availability at a particular time, T. The proposed queuing model is designed as a multi- queue model with multiple data centers, each having multiple PMs that can serve a set of VMs. The algorithm 1 of the resource handing component is shown below:

Algorithm 1

Input: ReqID, Req (Resource Requirement), Primary Queue
Output: HRQ, MRQ and LRQ
    Step 1: Estimate the current load
    Step 2: Estimate Resource Available
    Step 3: Handle Requests (Admission Control)
    Step 4: Estimate Thresholds at time t
    Step 5: Classify each request in the primary queue into any one of the secondary
           queues (HRQ, MRQ, and LRQ)
    Step 6: Sort each queue in descending order of R
    Step 7: Update load, resource availability, thresholds after each placement
    Step 8: Repeat steps 1 – 7 For each new request, classify request to anyone
           queue and link it with any one of the corresponding shortest queues.

The next step of the queuing model examines the requests in the primary queue and moves them to HRQ, MRQ or LRQ. This step gives equal priority to all the three resources and uses a rule-based algorithm 2.

Algorithm 2

For each datacenter, i,
    If $RD_t(Re_i) > $ T2(Re_j) then Add ReqID, to $RD_t(Re_i)$ THRQ
    If $RD_t(Re_i)$ <=T2 and >=T1 Add ReqID, $RD_t(Re_i)$ to TMRQ
    Then If $RD_t(Re_i)$< T1 then Add ReqID, to $RD_t(Re_j)$ TLRQ
    If count (ReqID) in THRQ >=2 then Move ReqID, $RD_t(Re_j)$ to HRQ; Loop;
    If count (ReqID) in TMRQ>=2 then Move ReqID, $RD_t(Re_j)$ to MRQ; Loop;
    If count (ReqID) in TLRQ >= 2 then Move ReqID, $RD_t(Re_j)$ to LRQ; Loop;
    else
Move ReqID, $RD_t(Re_j)$ to LRQ
End for

3.2 Proposed Placement Algorithm

The first step of the proposed algorithm is to perform pre-scheduling using the improved max-min algorithm. After the initialization of pheromone trials, the threshold level of PMs (value in the [0,1]), the fitness of the population is estimated using Eq. (4).

$$F_{ij} = \frac{\sum_{i=1}^{NHRQ} RC_{ij}}{VC_j} \tag{4}$$

where *Fij* is the capacity of VMj with bee number of I or the fitness of the bee population of I in *VMj*.

The revised max-min algorithm employed in the proposed Scheduling and Load Balancing Algorithm using Improved Max-Min, ACO and ABC algorithm (SLAM2A) results in an average efficiency improvement of 21.8 percent compared to the standard Scheduling and Load Balancing Algorithm based on ACO (SLAACO) method when considering the number of PMs utilized. The average Scheduling and Load Balancing Algorithm (SLA) violations rate of SLAACO algorithm is 1.16%, which has been reduced to 0.9% by SLAM2A algorithm. As a consequence of the lowered load imbalance rate and migration rate, the power consumption rate of the proposed algorithm is also reduced to 2.25% compared to the conventional algorithm.

The proposed Scheduling and Load Balancing Algorithm based on First Fit, Best Fit algorithm and multi-level grouping (SLAFBG) is designed using an amalgamation of algorithms: First-Fit, Best-Fit, and Genetic Algorithms. The algorithm uses a Load Percentage Threshold (LPT) threshold, which acts as the deciding criterion during placement. The LPT is initially set to 50%, which indicates half- full PMs that can safely accommodate all basic RRs. As a result of integrating the benefits of First-Fit and Best-Fit for scheduling, while a multi-level grouping GA is used to execute load balancing, the suggested algorithm's efficiency has increased when compared to its traditional techniques with all performance measures. The two basic components of SLAMP algorithm are pre-scheduling and scheduling with load balancing. This problem can be stated as "Find a mapping instance M such that scheduling a RR to a PM, the total makespan based on EExT is minimized". The objective function to reduce the execution time is given using Eq. (5) and (6).

$$\text{MinF} = \sum_{i=1}^{n} \sum_{j=1}^{m} = E_x E_{tij} 1 - \sum_{i=1}^{n} RA(PM_i(\text{Re}) - \mu(\text{Re})) \tag{5}$$

Subject to

$$\sum_{J-0}^{M} RR(\text{Re}) \leq \text{RLREt} \tag{6}$$

where n is the total number of VM requests, m is the total number of PMs available for mapping and EExT is the expected execution time, (Re) is C, R or B and μ represents the mean of the resource (C, R, B) estimated.

The standard round-robin approach cycles through allocating one virtual machine to each node, recording the location of the previous scheduler visit and starting from that point the next time a new request is received. The proposed enhanced round robin algorithm solves these issues. The algorithm simultaneously performs two tasks. The first is to place a new request to an appropriate PM and the second is to monitor PMs to find situations when it can be shut down. In the algorithm, steps 1 to 8 take care of task 1, while steps 9 to 11 take care of the second task. The proposed method enhances the working of the conventional round robin algorithm based on two conditions. On the other hand, when it is greater than Tn, the distribu tion of remaining VMs to other PMs is performed to save power. The period 'T' is estimated automatically using the SR register. The SR register stores the time difference of EExT of VMs and CoT of VMs of a PM. Analyzed and compared [31] various existing VM rebalancing algorithms. On the other hand, [32] presented a new VM placement and rebalancing algorithm for data intensive applications of cloud systems. VMs were mapped to PMs with minimum access time, and migration took place when the access time was greater than a predefined time threshold value.

4 Results and Discussions

The revised max-min algorithm employed in the proposed SLAM2A algorithm results in an average efficiency improvement of 21.8 percent when compared to the standard Scheduling and Load Balancing Algorithm based on ACO (SLAACO) method when taking into account the number of PMs utilized. Similarly, while considering the response time metric, on average, the proposed SLAM2A algorithm maximizes the speed by 36.4ms when compared to SLAACO algorithm. The average SLA violations rate of SLAACO algorithm is 1.16%, which has been reduced to 0.9% by SLAM2A algorithm. As a consequence of the lowered load imbalance rate and migration rate, the power consumption rate of the proposed algorithm is also reduced to 2.25% compared to the conventional algorithm. The comparison of Scheduling and Load Balancing Algorithm based on First Fit (SLAFF), Scheduling and Load Balancing Algorithm based on Best Fit (SLABF), Scheduling and Load Balancing Algorithm based on GA (SLAGA), Scheduling and Load Balancing Algorithm based on First Fit, and Best Fit algorithm and multi-level grouping Genetic (SLAFBG) algorithm over No. of active physical devices, SLA Violation Rate, Power Usage, Migration rate parameters are given in the Fig. 4.

The comparison of SLAPSO and SLAMP algorithm is given in the Fig. 5. The effect of introducing the queuing algorithm along with Join Shortest Queue (JSQ) algorithm is studied and analyzed in Stage 2 experiments and the result of this analysis is presented in this section. In the figures showing this effect, WO indicates the proposed algorithm without preprocessing algorithm and W represents the proposed algorithm with the preprocessing algorithm. By contrasting the outcomes of the suggested methods with and without the addition of preprocessing algorithms, the significance of the preprocessing stage in VM placement has been examined. The efficiency gain (%) of SLAFBG algorithm over the conventional algorithms has been mentioned in the Table 1.

The effect of preprocessing on the performance of SLAM2A has been shown in the Fig. 6. The performance has been measured over No of active PMs, SLA violation rate, power usages and migration rate parameters.

Stage 3 experimental results shows the selection of algorithms according to traffic intensity. The results also show that with high traffic intensity, the performance of the SLAM2A, SLAFBG, and SLAMO algorithms stabilizes. This demonstrates the benefit of these algorithms in situations with high traffic volumes and suggests that they are better suited for rush hour. Figure 7 to Fig. 11 shows the suggested VM placement algorithms for the chosen performance measures, including the number of PMs used, response time, SLA violation rate, resource use, power usage and load imbalance rate.

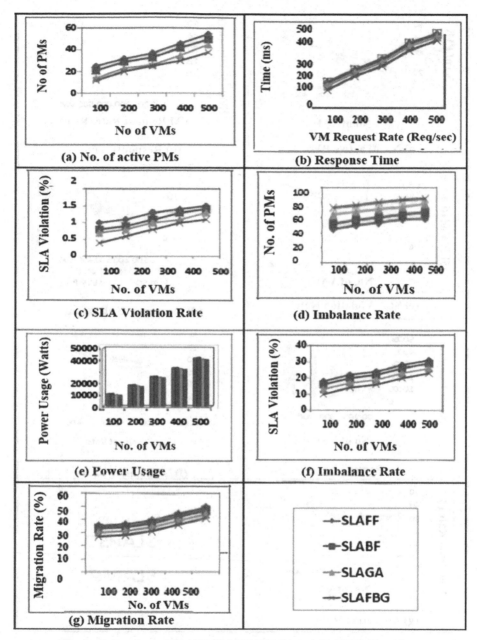

Fig. 4. Comparison of SLAFF, SLABF, SLAGA and SLAFBG Algorithms.

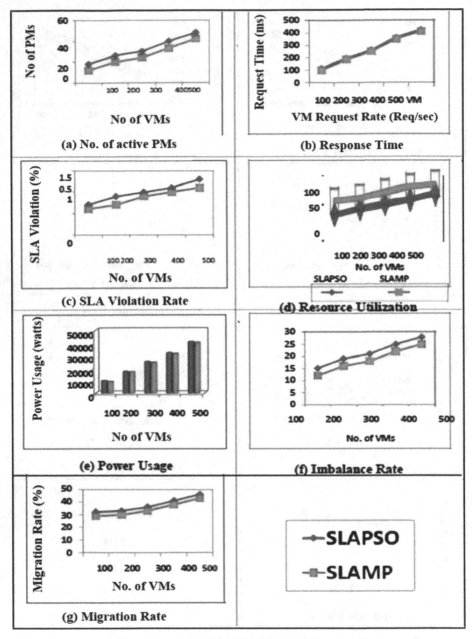

Fig. 5. Comparison of SLAPSO and SLAMP Algorithms.

Stage 4 experimental results identifying and mapping proposed algorithms to HRQ, MRQ or LRQ. The main aim of stage 4 experiments is to identify the algorithm that can efficiently handle the HRQ, MRQ and LRQ requests. As these queues are used only during rush hour and from the previous section, it was decided that the SLAM2A

Table 1. Efficiency gain (%) of SLAFBG Algorithm over the Conventional algorithms.

Performance Metric	First Fit	Best Fit	Genetic Algorithm
No. of Active PMs	35.89	28.57	12.58
Response Time	14.15	15.41	8.52
SLA Violation Rate	38.09	29.10	20.40
Resource Utilization	33.25	24.42	10.58
Power Usage	5.76	6.67	2.76
Load Imbalance Rate	32.52	26.55	15.31
Migration Rate	19.70	15.54	8.43

and SLAFBG algorithms are more suitable in handling rush hour traffic, the queue mapping analysis is performed only for these three algorithms. The request per second was adjusted in stages of 100 during the exercise, ranging from 100 to 500 indicate the selection of the suggested algorithms based on resource request intensity and the number of active PMs employed, response time, migration rate, resource utilization, power usage, SLA violation rate and load imbalance rate. The performance analysis of SLAM2A, SLAFBG, and SLAMP algorithm with high, low and medium requests based on the number of active PMs have been shown in Fig. 12(a) to Fig. 12(c).

Similarly, the performance analysis of SLAM2A, SLAFBG, and SLAMP algorithm with high, low and medium requests according to response time have been shown in Fig. 13(a) to Fig. 13(c).

The performance analysis of SLAM2A, SLAFBG, and SLAMP algorithm with high, low and medium requests according to SLA Violations have been shown in the Fig. 14(a) to Fig. 14(c).

The performance analysis of SLAM2A, SLAFBG, and SLAMP algorithm with high, low and medium requests according to resource utilization have been shown in the Fig. 15(a) to Fig. 15(c)

The performance analysis of SLAM2A, SLAFBG, and SLAMP algorithm with high, low and medium requests according to load imbalance rate have been shown in Fig. 16(a) to Fig. 16(c).

From the experimental results mentioned in Fig. 12(a) to Fig. 16(c), it can be deciphered that with all the performance metrics, the SLAM2A works well with high resource requests, SLAFBG exhibits better performance with medium resource requests and SLAMP algorithms' performance is higher with low resource requests. Hence, it is decided to use the SLAM2A to handle requests in HRQ, SLAFBG to handle the request in MRQ and SLAMP to handle requests in LRQ.

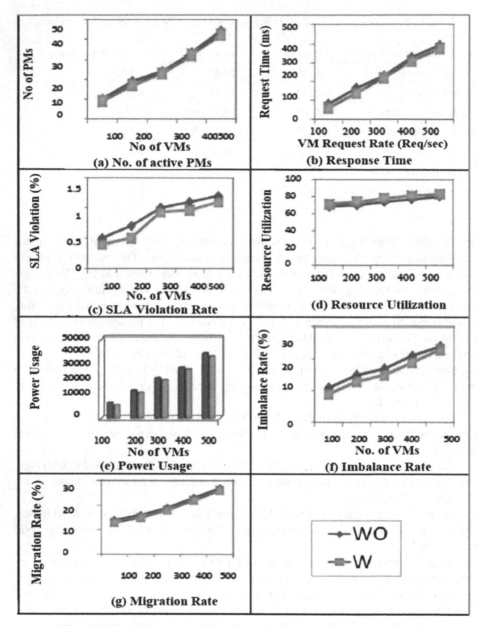

Fig. 6. Effect of Preprocessing on the Performance of SLAM2A Algorithm

The ACO algorithm uses modified objective functions, modified heuristic infor
mation, and pheromone vaporization rules and is combined with the ABC optimiza
tion algorithm. VM requests per second are changed from 100 to 500 in 100 increments
to analyze the low and high demand algorithms. SLAM2A using the proposed Load
Rebalancing using ACO and ABC Algorithm (LR2A) rebalancing algorithm is 18.18%

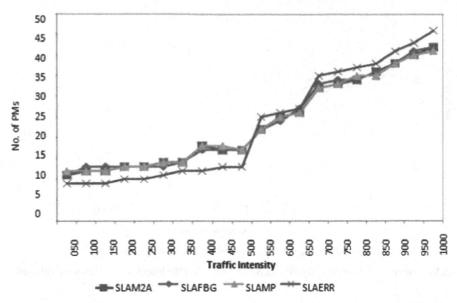

Fig. 7, Selection of proposed algorithms according to traffic intensity and No. of PMs

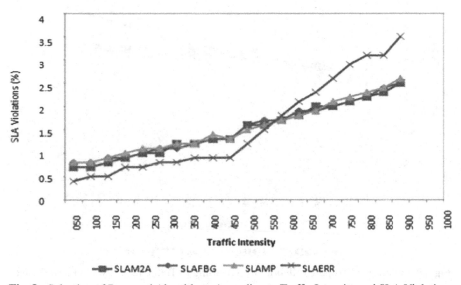

Fig. 8, Selection of Proposed Algorithms According to Traffic Intensity and SLA Violations

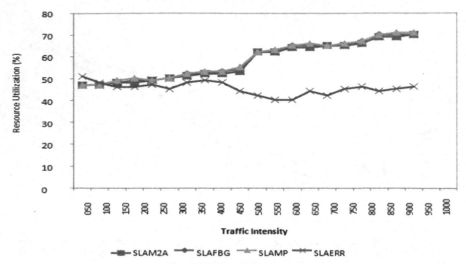

Fig. 9, Selection of Proposed Algorithms According to Traffic Intensity and Resource Utilization.

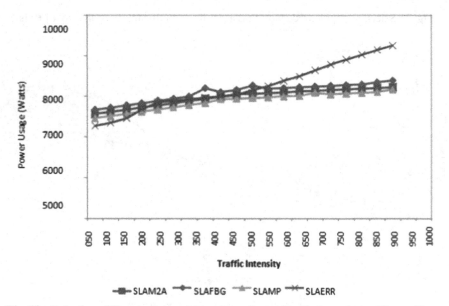

Fig. 10. Selection of Proposed Algorithms According to Traffic Intensity and Power Usage

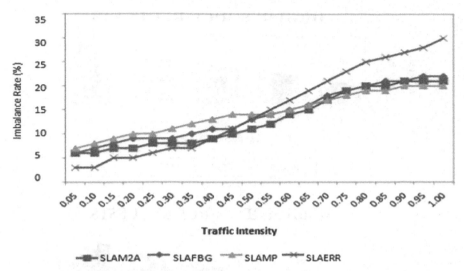

Fig. 11. Selection of Proposed Algorithms according to Traffic Intensity and Load Imbalance Rate

better than SLAM2A using the traditional Load Rebalancing using ACO Algorithm (LRACO) rebalancing method when considering average load balancing. Compared to SLAFBG with LRACO, SLAFBG with LR2A shows an average efficiency improvement of 18.84%. Similarly, SLAMP and Scheduling and Load Balancing Algorithm based on Enhanced Round Robin (SLAERR) with LR2A improve performance by 6.98 percent and 10.75 percent, respectively, over their respective counter algorithms with the traditional ACO algorithm.

On the other hand, SLAM2A, SLAFBG, SLAMP, and SLAERR using LR2A improved by 8.67%, 6.29%, 5.06%, and 4.85% compared to using the standard ACO algorithm, according to a survey of average migration assessments is showing. The load imbalance and VM Migration rate of rebalancing algorithms are given in Fig. 17 and Fig. 18 respectively.

From all above-mentioned results of stage 4 experiments, it can be deciphered that with all the performance metrics, the SLAM2A works well with high resource requests, SLAFBG exhibits better performance with medium resource requests and SLAMP algorithms' performance is higher with low resource requests. Hence, it is decided to use the SLAM2A to handle requests in HRQ, SLAFBG to handle the request in MRQ and SLAMP to handle requests in LRQ.

HIGH RESOUCE REQUESTS

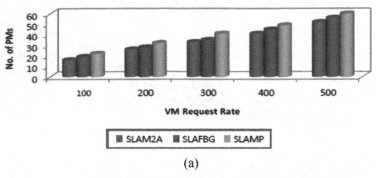

(a)

MEDIUM RESOURCE REQUESTS

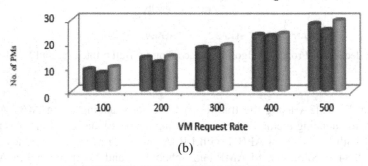

(b)

LOW RESOURCE REQUESTS

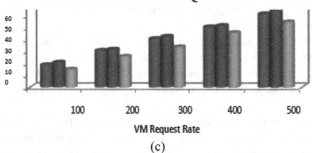

(c)

Fig. 12. (a) to (c) Performance Analysis of SLAM2A, SLAFBG, and SLAMP with High, Medium, Low Requests according to No. of Active PMs Used.

HIGH RESOURCE REQUESTS

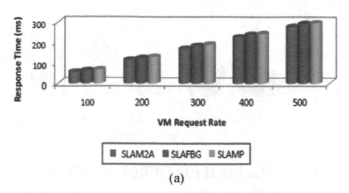

(a)

MEDIUM RESOURCE REQUESTS

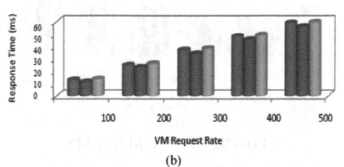

(b)

LOW RESOURCE REQUESTS

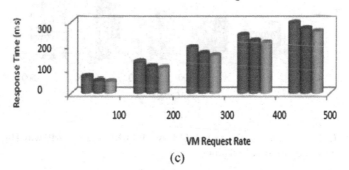

(c)

Fig. 13. (a) to (c) Performance Analysis of SLAM2A, SLAFBG and SLAMP with High, Medium, Low Requests according to Response Time.

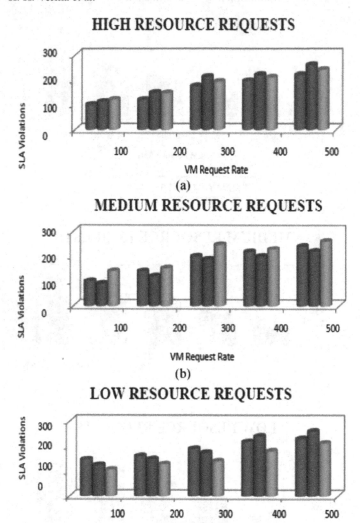

Fig. 14. (a) to (c) Performance Analysis of SLAM2A, SLAFBG and SLAMP with High, Medium, Low Requests according to SLA Violations.

HIGH RESOURCE REQUEST

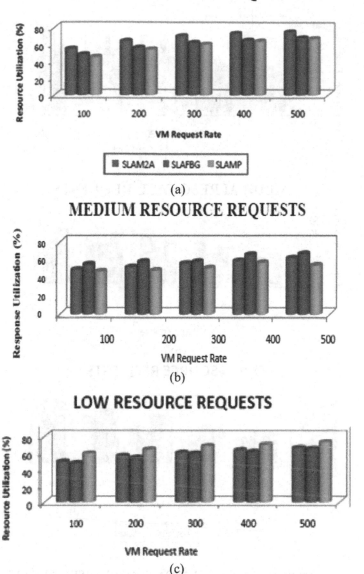

(a)

MEDIUM RESOURCE REQUESTS

(b)

LOW RESOURCE REQUESTS

(c)

Fig. 15. (a) to (c) Model Structure of Analysis of SLAM2A, SLAFBG and SLAMP with High, Medium, Low Requests according to resource utilization.

HIGH RESOURCE REQUESTS

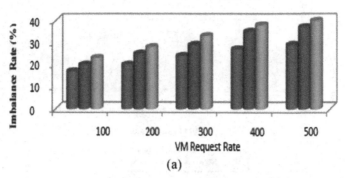

(a)

MEDIUM RESOURCE REQUESTS

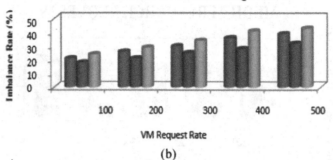

(b)

LOW RESOURCE REQUESTS

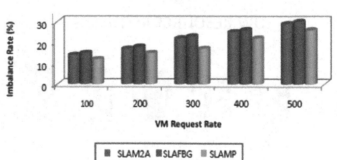

■ SLAM2A ■ SLAFBG ■ SLAMP

(c)

Fig. 16. (a) to (c) Performance Analysis of SLAM2A, SLAFBG and SLAMP with High, Medium, Low Requests according to load imbalance rate.

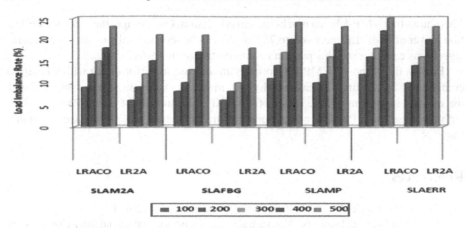

Fig. 17. Load Imbalance Rate of Rebalancing Algorithms

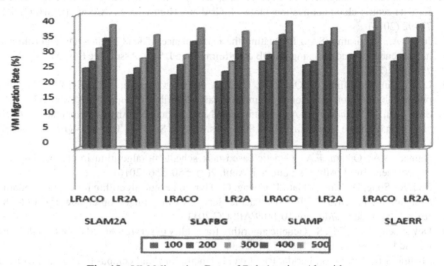

Fig. 18. VM Migration Rate of Rebalancing Algorithms

5   Conclusion

Thus, from the results of Stage 1 to Stage 4 experiments it can be deciphered that all the proposed scheduling and load balancing algorithms (SLAM2A, SLAFBG, SLAMP, and SLAERR) are improved versions to their conventional counterpart algorithms. Including queuing and JSQ algorithms with the proposed algorithms has improved the VM placement process. The linking of the proposed LR2A re- balancing algorithm with the proposed algorithms has also increased the efficiency of VM placement operation. The average SLA violation rate of VMP-LR is 0.6% which is a small rate when compared with FF (1.26%), BF (1.1%), RR (1.36%), GA (0.98%), ACO (1.16%) and PSO (1.01%).

This shows that VMP-LR is a much- improved version. On average, the resource utilization power of VMP-LR increases to 76.8%. As can be seen from the results, the existing algorithms cannot reach this peak in resource utilization power.

Finally, the proposed VMP-LR portrays an increase in performance with all metrics compared to the existing algorithms. The system shows an efficiency gain compared with the existing algorithms in the range of 17–43% with number of active PMs, 10–16% with response time, 15–62% with resource utilization and 9–40% with load imbalance rate.

References

1. Mell, P., Grance, T.: The NIST definition of cloud computing (2011)
2. Rani, B.K, Padmaja Rani, B., Vinaya Babu, A.: Cloud computing and inter-clouds–types, topologies and research issues. Proc. Comput. Sci. **50**, 24–29, (2015)
3. Mandal, A.K., Changder, S., Sarkar, A., Debnath, N.C.: Architecting software as a service for data centric cloud applications. Inter. J. Grid and High-Perform. Comp. (IJGHPC) **6**(1), 77–92 (2014)
4. Joshi, A., Munisamy, S.D.: Evaluating the performance of load balanc- ing algorithm for heterogeneous cloudlets using HDDB algorithm. Inter. J. Syst. Assurance Eng. Manag. **13**(1), 778–786 (2022)
5. Li, J., Qiu, M., Ming, Z., Quan, G., Qin, X., Gu, Z.: Online optimization for scheduling preemptable tasks on IaaS cloud systems. J. Parallel Distribut. Comput. **72**(5), 666–677 (2012)
6. Saravanan, N., Mahendiran, A., Subramanian, N.V., Sairam, N.: An im- plementation of RSA algorithm in google cloud using cloud SQL. Res. J. Appl. Sci. Eng. Technol. **4**(19), 3574–3579 (2012)
7. Hamad, S.A., Omara, F.A.: Genetic-based task scheduling algorithm in cloud computing environment. Int. J. Adv. Comput. Sci. Appl. **7**(4), 550–556 (2016)
8. Zhu, K., Song, H., Liu, L., Gao, J., Cheng, G.: Hybrid genetic algorithm for cloud computing applications. In: 2011 IEEE Asia- Pacific Services Computing Conference, pp. 182–187. IEEE (2011). https://doi.org/10.1109/APSCC.2011.66
9. Devarasetty, P., Reddy, S.: Genetic algorithm for quality of service- based resource allocation in cloud computing. Evol. Intel. **14**, 381–387 (2021)
10. Mishra, S.K., Sahoo, B., Parida, P.P.: Load balancing in cloud compu- ting: a big picture. J. King Saud Univ. Comput. and Inform. Sci. **32**(2), 149–158 (2020)
11. Elzeki, O.M., Reshad, M.Z., Elsoud, M.A.: Improved max-min algo- rithm in cloud computing. Inter. J. Comput. Appli. **50**(12), 22–27 (2012)
12. Usmani, Z., Singh, S.: A survey of virtual machine placement techniques in a cloud data center. Proc. Comput. Sci. **78**, 491–498 (2016)
13. Dósa, G., Sgall, J.: First fit bin packing: a tight analysis. In 30th International symposium on theoretical aspects of computer science (STACS 2013). Schloss Dagstuhl-Leibniz-Zentrum fuer Informatik (2013)
14. Patel, S., Bhatt, M.: Implementation of load balancing in cloud computing through round robin & priority using cloudSim. Inter. J. Rapid Res. Eng. Technol. Appli. Sci. **3**(11) (2017)
15. Albers, S., Khan, A., Ladewig, L.: Best fit bin packing with random or- der revisited. Algorithmica **83**, 2833–2858 (2021)
16. Kapoor, S., Dabas, C.: Cluster based load balancing in cloud computing. In: 2015 Eighth International Conference on Contemporary Computing (IC3), pp. 76–81. IEEE (2015)

17. Anand, A., Lakshmi, J., Nandy, S.K.: Virtual machine placement optimization supporting performance SLAs. In: 2013 IEEE 5th International Conference on Cloud Computing Technology and Science, vol. 1, pp. 298- 305. IEEE (2013)
18. Madhumala, R.B., Tiwari, H., Devaraj, V.C.: An improved virtual machine placement in cloud data center using particle swarm optimization algorithm. Inter. J. Adv. Res. Eng. Technol. **11**(8), 760–768 (2020)
19. Maniyar, B., Kanani, B.: Review on round-robin algorithm for task scheduling in cloud computing. J. Emerging Technologies Innovative Res. **2**(3), 788–793 (2015)
20. Gibet Tani, H., El Amrani, C.: Smarter round robin scheduling algorithm for cloud computing and big data. J. Data Mining Digital Human. (2018). https://doi.org/10.46298/jdmdh.3104
21. Panda, S.K., Bhoi, S. K.: An effective round robin algorithm using min- max dispersion measure. arXiv preprint arXiv:1404.5869, (2014)
22. Yao, Z., Papapanagiotou, I., Callaway, R.D.: Multi-dimensional scheduling in cloud storage systems. In: 2015 IEEE International Conference on Communications (ICC), pp. 395–400. IEEE (2015). doi: https://doi.org/10.1109/ICC.2015.7248353
23. Gupta, H., Verma, K.K., Sharma, P.: Using data assimilation technique and epidemic model to predict tb epidemic. Inter. J. Comput. Appli. **128**(9), 1–5 (2015)
24. Tomar, A., Pant, B., Tripathi, V., Verma, K.K., Mishra, S.: Improving QoS of cloudlet scheduling via effective particle swarm model. In: Machine Learning, Advances in Computing, Renewable Energy and Communication: Proceedings of MARC 2020, pp. 137–150 (2022)
25. Masdari, M., Nabavi, S., Ahmadi, V.: An overview of virtual machine placement schemes. cloud computing. J. Netw. Comput. Appli. **66** (2016). https://doi.org/10.1016/j.jnca.2016.01.011
26. Ndayikengurukiye, A., Ez-Zahout, A., Omary, F.: An overview of the different methods for optimizing the virtual resources placement in the cloud computing. J. Phys. Conf. Ser. **1743**(1), 012–030 (2021. https://doi.org/10.1088/1742-6596/1743/1/012030
27. Wenting W., Kun W., Kexin W., Huaxi G., Hong S.: Multi-resource balance optimization for virtual machine placement in cloud data centers. Comput. Elect. Eng. **88** (2020)
28. Aggarwal, G., Motwani, R., Zhu, A.: The load rebalancing problem. In: Proceedings of the Fifteenth Annual ACM Symposium on Parallel Algorithms and Architectures, pp. 258–265 (2003)
29. Karthikeyan, S., Seetha, H.S., Manimegalai, R.: Nature inspired optimization techniques for cloud scheduling problem. In: Proceedings of International Conference on Energy Efficient Technologies for Sustainability, St. Xavier's Catholic College of Engineering, TamilNadu, India (2018)
30. Mishra, S.K., Bibhudatta, S., Priti, P.P.: Load balancing in cloud computing: a big picture. J. King Saud Univ. Comput. Inform. Sci. **32**(2), 149–158 (2020)
31. Beloglazov, A., Jemal, A., Rajkumar, B.: Energy-aware resource alloca- tion heuristics for efficient management of data centers for cloud computing. Futur. Gener. Comput. Syst. **28**(5), 755–768 (2012)
32. Masdari, M., Sayyid, S.N., Vafa, A.: An overview of virtual machine placement schemes in cloud computing. J. Netw. Comput. Appl. **66**, 106–127 (2016)

A Construction of Secure and Efficient Authenticated Key Exchange Protocol for Deploying Internet of Drones in Smart City

Dharminder Chaudhary$^{(\boxtimes)}$ [iD], Tanmay Soni[iD], Soumyendra Singh[iD],
and Surisetty Mahesh Chandra Gupta[iD]

Department of Computer Science and Engineering, Amrita School of Computing
Amrita Vishwa Vidyapeetham, Chennai, India
manndharminder999@gmail.com

Abstract. The concept of a smart city is increasing because of the demand for intelligent drones. So, the Internet of Drones came into picture, providing several benefits/services for daily life. Services that IoD offers are monitoring, FANET (Flying Ad-Hoc Networks), management of any infrastructure, and IoT (Internet of Things). These services can be deployed in the smart city environment. Still, communication among drones is a significant concern. For communicating, drones use the insecure channel, and there is a risk of security threats while sending critical information. They are also prone to physical capture attacks because of their usage in an environment devoid of human beings. Regarding communication and computation, drones are resource constrained, so it is not feasible to implement public key cryptography because it requires more power to perform those actions. A recent protocol for the internet of drones is also analyzed in this article. Therefore, this article presents a lightweight authentication protocol that is reliable to the users and meets their demands. The performance analysis ensures the efficiency of the proposed protocol.

Keywords: Authentication · Key Agreement · Internet of Drones · Physical Unclonable Function

1 Introduction

The technologies are emerging with novel innovations being implemented simultaneously. Fields like IoT (Internet of Things), FANET (Flying Ad-hoc Networks), and 5G communication are also following the same trend, which led to the development of smart cities [1–5]. But there are certain challenges being faced by it like maintaining and storing a large amount of data. This data is collected by the sensors and IoT, which is later stored in their memory. One way to overcome this challenge is by combining UAV, FANET, and IoT into IoD

R. K. Challa et al. (Eds.): ICAIoT 2023, CCIS 1929, pp. 136–150, 2024.
https://doi.org/10.1007/978-3-031-48774-3_9

[6]. Each one plays their individual role even after getting combined. FANET, when used in UAVs, gives fast speed, low latency, and back-end services to the user. In drones also, when IoT is deployed, it gathers critical information from various difficult scenarios and carries that information forward with the help of FANET. So this led to the rapid demand for FANET-based IoD [7,8]. Their features include monitoring, surveillance, logistic transportation, and providing relief during critical rescue operations. The services which are available to use by IoD are possible because of the portability, flexibility, and rapid deployment. For example, if any natural calamity happens, the sensors present in the IoD can gather all the necessary information. Affected people can be located by their body temperature with the help of thermal sensors [9]. After this, medical assistance will be provided to them once the data collected by IoD is shared with them. Even after having many advantages still, there are certain issues that need to be addressed and resolved for better functioning of the IoD. FANET-based IoD uses public channels for sharing information which leads to the violation of privacy [10]. The data, which was gathered by the drones, can be used by the attacker for unethical activities if it is not secured properly. For example, an adversary can get the details of sensitive localities by taking pictures and recording videos and can even use them to transport illegal substances. Apart from this, adversaries can intrude into the privacy of someone else by clicking pictures with the help of a camera present in drones. If we talk of the extreme scenario, the adversary can gather information stored in the drone by capturing it and then pretend to be the captured drone using the information stored in it. Along with this, there is a boost in demand for services which is there in smart cities. And another challenge it faces is the lightweight property. IoD has a set of conditions or constraints like computation power and consumption of energy. This was the main objective for the evolution of a lightweight authentication scheme. It is difficult to perform highly complex computations, and any computation that a system is performing must be solved within a certain threshold time interval, or else the performance of the schema will be affected, and the desired result will not be achieved. So to ensure the reliable and efficient working of IoD in smart cities, secure and lightweight authentication and key agreement scheme is introduced [11]. For many years, researchers are proposing a schema that is secure and follows the condition of efficient Authentication Key Agreement (AKA) protocols for IoD so that they can be implemented in smart cities [12–14]. Some of them asserted that their scheme is reliable, efficient, untraceable, and capable of withstanding various attacks. Unfortunately, their assertion proved to be wrong when their schema faced security attacks. Elliptic Curve Cryptography and other existing AKA schemes can not be implemented in IoD-based smart cities because of their lightweight property.

2   Related Works

For many decades, several authentication and key agreement schemes have been developed in order to improve the security and privacy of IoT [15–17]. "Password-based single factor AKA scheme" was introduced by Lamport [18]. But Lamport

[18] was not successful because it was not resistant to password guessing attacks in offline mode because only password privacy and security was considered. So to improve this, "smart card based two-factor authentication and key agreement scheme and password" was introduced by Das [19]. But Das's [19] scheme also has a drawback which was explained by Nyang's and Lee's [20]. The drawback was prone to guessing passwords and capturing sensor node attacks. So the "Secure and efficient two-factor authentication and key agreement scheme" was introduced by Nyang and Lee [20] to overcome the drawback faced by Das's [19] scheme. To introduce the new feature of providing user anonymity and privacy, He et al. [21] proposed an "enhanced two-factor based authentication and key agreement scheme ." But He et al.'s [21] proposed a scheme was unable to establish the session key and also provide secure mutual authentication. This vulnerability was analyzed by Kumar and Lee [22]. So, it was concluded that two factors, AKA schemes for IoT [18–22] are prone to multiple security attacks. In previous years, several scholars have been introducing secure, and lightweight three-factor authentication and key agreement schemes based on biometric [23–27]. They were introducing these schemes so that IoT-enabled drones can overcome the vulnerability, security challenges and privacy challenges faced by the previous two factor-based authentication and key agreement scheme. The three-factor AKA-based scheme was introduced by Wazid et al. [27]. He described several security requirements for the different types of issues faced by IoD environments. But Wazid et al.'s scheme have one issue that it cannot guarantee perfect backward secrecy and independent aliasing. This was pointed out by Alladi et al. [24]. So he introduced the two stages of lightweight authentication and key agreement scheme for "software-defined network-based unmanned aerial vehicle." But Alladi et al. [24] have one issue, which was raised by Beebak et al. [25], that forgery, offline password guessing, and replay cannot be prevented. In addition to this, the confidentiality of the data and forward secrecy can not be guaranteed. To prevent this security issue "temporary login based anonymous lightweight three-factor authentication and key agreement scheme in the internet of drones," also known as TCALAS, was proposed by Srinivas et al. [26]. But Srinivas's scheme also faced one issue that it is prone to impersonation attacks and can also be traced easily. This was analyzed by Ali et al. [23]. Ali et al. [23] introduced an "enhanced authentication and key agreement scheme for the internet of drone-based smart city environments" to overcome the issue faced by the scheme developed by Srinivas et al. [26]. Still, for some reason, Ali et al. [23] scheme was vulnerable to forgery, server spoofing attacks, and session key disclosure. Similar to these schemes, others also tried to introduce new security features between the IoD and users, but those features have some vulnerabilities. The schemes [23–27] are still prone to security attacks in future. Recently, several public key cryptography-based authentication and key agreement schemes [28–31] have been introduced or proposed in the IoD based smart cities environment in order to enhance the level of security and privilege controls as compared to the previously defined schemes (three factor-based AKA schemes). "ECC based certificate less authentication and key agreement

scheme for the IoD-based smart city environment" was proposed by Won et al. [31]. But this scheme has some vulnerabilities, like it lacks anonymity of the user and formal security is not discussed. Another scheme, "Authentication and key agreement scheme to provide services in the internet of drone environments based on homomorphic encryption," was introduced by Cheon et al. [28]. But this scheme is prone to insider attacks and session key discloser. Another scheme [29] was proposed stating, "Secure and efficient authentication and key agreement framework for mobile sinks used in IoD environment based on bilinear pairing." But in several attacks like impersonation attacks and perfect forward secrecy is not guaranteed. In addition to these, this scheme cannot provide real-time services because of the high communication computation costs required by the bilinear pairing. This was pointed out by Nikooghadam et al. [30] who proposed a "Secure and lightweight authentication and key agreement scheme for the IoD-based smart city environments based on ECC," which is highly secure against several security compromised attacks. Later his scheme failed to be resistant against security attacks like replay, impersonation attacks, and insider and there was no mutual authentication between two or more devices. This vulnerability in the Nikooghadam et al. scheme [30] was discovered by Ali et al. [32]. These public key cryptography based authentication and key agreement schemes [28–31] can be prone to physical drone capture attacks. If it happens, then the adversary can extract all the sensitive information from it and impersonate the captured drone. These schemes are not made to solve complex problems because it requires high communication and computation cost.

3   Motivation and Contribution

The IoD is an emerging area for researchers, security and privacy are significant concerns for communication among drones. Many authentications and key exchange protocols have been proposed [23,26,27,30,31], but it was found that either they were not secure against possible attacks like password guessing, anonymity, Man in the middle [MITM], malicious insider, user/server impersonation, etc., or they are not efficient in terms of computation and communication. We have also studied a recent protocol [33] and found a vulnerability. According to the protocol to which we have referred,$\{id_i, rpw_i, r_i\}$ was submitted to the control server through a secure channel, but there is no meaning to send the random number r_i to the control server as it can be used by the attacker to retrieve the secret credential with certain attacks. Therefore, one need a secure and efficient authenticated key exchange mechanism for the IoD environment. This article proposes the required mechanism, which attains most of the security attributes. A performance analysis of the proposed protocol is also done, with relevant protocols, and found that the proposed protocol takes less computation cost. Therefore this protocol can be implemented for communication among IoD.

4   Threat Model

The most widely used threat model for finding the security of secure, lightweight authentication protocol for the IoD is Dolev and Yao [34] threat model. The actions which can be performed by the malicious attacker or adversary (MA) are discussed here. The Dolev and Yao model states that attackers can compromise, delete, eavesdrop, inject some codes and modify some of the data shared through the public medium. A malicious attacker can also perform a powerful analysis attack on smartphones by acting as a legal user [35]. This leads to the compromise of sensitive information present in the smartphone. Apart from this, the attacker can do capturing the physical device. After this, the login details can be extracted, and the attacker can behave like a legal user. The attacks performed by the attacker after getting the login details are forgery and impersonation attacks. Another scheme that is more secure and efficient than Dolev and Yao's(DY) threat model is Canetti and Krawczyk(CK), model [36]. It is also known as the CK threat model. The standard CK model is de facto for all AKA schemes. The CK threat model states that the attacker (MA) has all the capabilities mentioned in the DY model. But in addition to it, the adversary can also compromise sensitive information by performing attacks based on session hijacking. Also, a MA can perform the attack on CK [36] model like an ephemeral secret leakage attack. So it is essential for any scheme that it does not reveal the data even after the session hijacking is performed. The scheme must be resistible enough to protect the data of other devices connected to it.

5   Physical Unclonable Functions

When we talk about security, there are hardware devices like sensors and the IoD. To protect them from the malicious adversary or attacker, physical unclonable functions(PUF) [37] is used. Generally, PUF produces single unique output for the input which is given to it. For example, the fingerprint sensor present in the devices uses PUF. The unique feature of the PUF is that the secret key is not saved, and public key cryptography is not used for authentication purposes. Apart from this, making the exact copy of the previously known PUF is difficult because they are designed by nanoscale variation at the time of production of integrated circuit chips. PUF are durable, unique and tamperproof. They protect the devices connected from side-channel attacks and cloning. The mathematical way of representing PUF is $O = PUF(c)$, where O is output; PUF is a physical unclonable function, and c is a challenge given to it. Properties of PUF are (1) architecture of PUF decides the output, (2) produced output is unclonable and unique, and they are easy to implement on any device. In case there is any modification of PUF during manufacturing, the output will also be affected. The user and device authenticity is verified by a PUF before the connection is established [38]. If we implement these features into our scheme, it will result in an efficient and durable model.

6   Network Model

The internet of the drone-enabled smart city will have the network model. The suggested scheme has three major structural elements: control server(cs), drone(d), and mobile user(mu). A brief description of each in detail is as follows:

1. Control Server(cs): As the name suggests, it refers to the ground station server. They act as a medium between the drone and the mobile user. It allows the mobile user to communicate with the drones. Apart from this, mobile users can monitor the drones provided that the mobile users must be authorized. To authorize the entity's control server assigns the credentials to mobile users and drones and registers them. One can say that the control server is the link between mobile users and drones. The mobile user and the drone can also mutually authenticate themself in the public channel through which they are connected. This authentication is possible with the help of the control server only. The database of the control server is not accessible by the malicious adversary(MA). So the data stored in the database is secured.
2. Mobile user(mu): The users who carry mobile devices like smartphones, tablets, etc., are referred to as mobile users. In the registration phase, the credential is assigned to the mobile user by the control server. Once the credentials are given, mutual authentication is performed for mobile users and drones. After this, the session key is established between them for secure communication and data sharing.
3. Drone(d): Similar to the mobile user, in the registration phase, the credential is assigned to the drone by the control server. Once the credentials are given, they are eligible for deployment in flying areas. The drone which is deployed is controlled by the control server in order to send data collected by it using the sensors. The data, which is collected, is then sent to mobile users.

7   Proposed Secure and Efficient Authenticated Key Exchange Protocol for Deploying Internet of Drones in Smart City

This section discusses the proposed authenticated key exchange for deploying the IoD in the smart city. The protocol is divided into four phases (i) initialization, (ii) drone registration, (iii) user registration, and (iv) authentication and key agreement phases, respectively. Table 1 shows all the most important notations and terminologies.

7.1   Initialization

In this phase, the public parameters of the system, such as physical unclonable function PUF(.) and fuzzy extractor functions like generator gen(.) and reproduction Rep(.), are issued. In addition, the control server also chooses the Z_p and msk, which belongs to Z_p. The control server does certain pre-assignment

Table 1. Notations

Symbol	meaning
mu_i	Mobile user
d_j	Drone
cs	control server
bio_i	biometric of mobile user
id_i, pw_i	identity and password of mobile user
did_j	identity of drone
r_1, r_2, r_3	random nonces
t_i	timestamp
sk	session key between mobile user and drone
msk	maskter key of control server
$h()$	hash function
$PUF()$	physical unclonable function
$\oplus$	XOR operation
$\|$	Concate operation

operations, like the secret credentials assigned to a drone for registering and authenticating before they can operate in their flying areas. The Drone identity (did) is chosen for each drone by the control server. Then the drone identity (did) is sent to the respective drones. The drones save their identity in the secure database. Once initialized, the system goes to the next phase of registration.

7.2   Drone Registration

After registering, the drones receive their secret credentials. Drones use the credentials for authentication purposes. Drone registration is elaborated in three steps as follows:

1. DRP 1: The drone d_j selects its did_j and arbitrary number n_j. Then drone d_j submit its identity and random number as a single entity $\{n_j, did_j\}$ to the control server through the secure transmission channel.
2. DRP 2: After receiving the message, the control server checks if $did_j^* = did_j$. If true, then a random challenge set c_j is selected by the control server. The response $Res_j = PUF(c_j)$ is calculated with that challenge c_j. Two element set r_j and δ_j are calculated when the response is passed to the PUF $Gen(.)$ function with a condition that $(r_j, \delta_j) = (Gen(Res_j))$. Then, the control server calculates $z_j = h(did_j\|msk)$, $n_j = z_j \oplus h(did_j\|n_j)$ and $e_j = \delta_j \oplus h(z_j\|n_j\|did_j)$, which sent $\{n_j, e_j\}$ to the drone. At last, the control server stores $\{z_j, (c_j, r_j)\}$ as a single entity in its memory.
3. DRP 3: Once the message is received, the secret credential $\{n_j, e_j\}$ is stored by the drone in its memory.

7.3   Mobile User Registration

After registering, the mobile users mu_i receive their secret credentials. Then mobile users mu_i use the credentials for authentication. Mobile user registration is described briefly in 3 steps mentioned below:

1. URP:1 mu_i represents mobile user, selects a mobile user identity id_i, password pw_i and random number r_i. After this, the mobile user mu_i computes a random password $rpw_i = h(pw_i\|r_i)$and sends identity and random password $\{id_i, rpw_i\}$as a single entity to the control server through the secure transmission channel.
2. URP:2 After receiving the message, the control server calculates a random identity $rid_i = h(id_i\|rpw_i\|msk_i)$, $x_i = h(msk_i\|rid_i\|rpw_i)$ and saves $\{rid_i, x_i\}$ as a single entity in its memory. After this, cs the fetches did_j and submit $\{did_j, rid_i, x_i\}$ to mobile user through a secure medium.
3. URP:3 Once the message is received, the mobile user calculates $\beta_i^* = \beta \oplus h(id_i\|pw_i)$, $x_i^* = x_i \oplus h(id_i\|pw_i\|r_i)$, $rid_i^* = rid_i \oplus h(pw_i\|id_i\|rpw_i)$, $did_j^* = did_j \oplus h(id_i\|rpw_i\|pw_i)$, and $c_i^* = h(rid_i\|pw_i\|x_i\|r_i)$. Then, gateway gw_i changes $\{rid_i^*, x_i^*\}$ with $\{rid_i, x_i\}$ after that it saves $\{\beta_i^*, c_i^*, did_i^*\}$ in the mobile user device which is connected to it.

7.4   Authentication and Key Agreement Process

In the Authentication process, Control Server authenticates the mobile user and drone to establish a session key. A detailed description of AKA process of our proposed model is stated below:

1. AKP-1: A user enters its secret id_i and pw_i in the mobile devices. $r_i=$, $rpw_i = \text{h}(pw_i \| r_i)$, $x_t = x_i^* \oplus \text{h}(id_i\|pw_i\|r_i)$, $rid_i = rid_i^* \oplus \text{h}(pw_i \| id_i\| rpw_i)$, $did_j = did_j^* \oplus \text{h}(id_i \| rpw_i \| pw_i)$, and $c_i^* = \text{h}(rid_i \| pw_i \| x_i \| r_i)$, and verifies $c_i^* \overset{?}{=} c_i$. If its false then the session is aborted by mu_i if not, chooses a arbitrary numpty r_1 and a timestamp t_1, and calculates $m_1 = (r_1\|did_j) \oplus \text{h}(rid_i\|x_i\|t_1)$ and $auth_us = \text{h}(rid_i\|r_1\|x_i\|t_1)$, and over a public channel the message $\{m_1, auth_us, rid_i, t_1\}$ is sent to CS.
2. AKP-2: After receiving the messages, a timestamp t_2 is generated, and freshness of $|t_2 - t_1| \leq \triangle t$ is verified and where maximum time delay for transmission is denoted by $\triangle t$ and message reception time is denoted as t_2. If the date and time of the message are valid, Control Server calculates $(r_1 \| did_j)$ $= m_1 \oplus \text{h}(rid_i \| x_i \| t_1)$ and $auth_{us}^* = auth_{us}$. If it's true, the control server fetches $(c_j, r_j) \leftarrow did_j$, and chooses an arbitrary numpty r_2. Then, control server calculates $z_j = \text{h}(did_j \| z_j \| t_2)$ and $auth_{sd} = \text{h}(did_j \| r_2 \| z_j \| \| r_j \| t_2)$ and sends $\{c_j, m_2, auth_{sd}, t_2\}$ to the drone.
3. AKP-3: Once the message is received, current timestamp t_3 is selected by d_j and validity of $|t_3 - t_2| \leq \triangle t$ is checked by d_j. If its true, d_j retrives $\{n_j, e_j\}$ in the memory and calculates $z_j = n_j \oplus \text{h}(did_j \| b_j)$ and $\delta_j = e_j \oplus \text{h}(z_j \| b_j \| did_j)$. After this drone computes $r_j = \text{rep}(\text{PUF}(c_j), \delta_j)$,$(r_1 \| r_2) = m_2 \oplus \text{h}(did_j \|$

$r_j \parallel z_j \parallel t_2$) and $auth_{sd}^*$=h($did_j \parallel r_2 \parallel r_3 \parallel r_j \parallel t_3$) and $auth_{du}$=h($r_1 \parallel r_3 \parallel r_j \parallel did_j \parallel sk$). Finally, Over a public channel message $m_3, auth_{ds}, auth_{du}, t_3$ to control server by d_j.

4. AKP-4: After receiving the messages, The timestamp t_4 is generated by CS and freshness of $|t_4 - t_3| \leq \triangle t$ is verified. If it is valid, CS calculates r_3=$m_3 \oplus$h($r_j \parallel r_2 \parallel t_3$) and $auth_{ds}^*$=h($did_j \parallel r_2 \parallel r_3 \parallel r_j \parallel t_3$) and verifies if $auth_{ds}^* \stackrel{?}{=} auth_{ds}$. Whenever affliction is valid, control server calculates m_4=($r_2 \parallel r_3 \parallel r_j \oplus$)h($rid_i \parallel did_j \parallel r_1 \parallel x_i \parallel t_4$) and $auth_{su}$=h($rid_i \parallel r_1 \parallel r_2 \parallel x_i$). Finally, CS sends $m_4, auth_{du}, auth_{su}, t_4$ to mu_i.

5. AKP-5: Upon receiving the messages, mu_i a timestamp t_5 is selected and $|t_5 - t_4| \leq \triangle t$ is verified. If they are same then mu_i calculates ($r_2 \parallel r_3 \parallel r_j$)=$m_4 \oplus$h($rid_i \parallel r_1 \parallel r_2 \parallel x_i$) and validate whether $auth_{su}^* \stackrel{?}{=} auth_{su}$. If the validation fails mu_i closes the current session alternatively mu_i calculates sk = h($r_1 \parallel r_3 \parallel r_j$) and $auth_{ds}^* \stackrel{?}{=} auth_{ds}$. The session key sk will be established successfully if its valid and mobile user and drone are authenticated mutually.

8   Security Analysis

A security analysis is done to prove the security of our proposed scheme. To analyze the security of our proposed model or scheme, the informal security analysis is discussed. In this, it is to prove that the model is immune to specific attacks like revealing anonymity, disclosing session key agreements, etc.

8.1   Impersonation Attacks

This attack happens when a malicious attacker or adversary tries to impersonate a legal or authorized user mu_i by intercepting the communication sent through the public medium. But while performing this attack, the malicious adversary must know the $\{m_1, auth_{us}, rid_i, t_1\}$ and $\{m_4, auth_{du}, auth_{su}, t_4\}$. So an attacker can't get the messages because the attacker fails to get the arbitrary numpty r_1 and private credential x_i. At last, it can be said that the model is immune to impersonation attacks.

8.2   Replay Attack

Suppose the adversary intercepts $\{m_1, auth_{us}, rid_i, t_1\}$, $\{c_j, m_2, auth_{sd}, t_2\}$, $\{m_3, auth_{ds}, auth_{du}, t_3\}$, and $\{m_4, auth_{du}, auth_{su}, t_4\}$ are sent during the authentication and key agreement phase. Whenever a malicious attacker tries to resent the previously sent messages then our model will check for the current timestamp present on it or not. Here, the messages which are sent have private credentials x_i, and random nonces $\{r_1, r_2, r_3\}$. So our model is immune to replay attack.

8.3   Physical Capture Attack on Drones

Suppose a malicious attacker captures some drones to extract all the sensitive information $\{n_j, e_j\}$ where $n_j = z_j \oplus h(did_j||b_j)$ and $e_j = \delta_j \oplus h(z_j||b_j||did_j)$ present in its memory. But, a malicious adversary cannot calculate the common session key which was established because the adversary does not know the drone's arbitrary number b_j and arbitrary numpty r_2. And each drone has a unique and independent random number and random nonce. This happens because challenge and response (c_j, r_j) are randomly formed. So it is very difficult to find the common session key and it can be concluded that our scheme is immune to physical capture attacks.

8.4   Disclosed Session Key Attack

If the malicious attacker gets the private credentials $\{x_i^*, rid_i^*, c_i^*, did_j^*, \beta_i^*,\}$ after performing a stolen password attack to imitate a legal user mu_i. But the attacker must have random nonces $\{r_1, r_2\}$, response $\{r_j\}$ to get the session key $sk = h(r_1, r_3, r_j)$. But an attacker can't get the random nonces because all the nonces are masked with private credentials $\{x_i, z_j\}$. In addition to it, the attacker also does not know the physically unclonable function's private parameters δ_j. So our model is immune to the disclosed session key attacks.

8.5   Offline Password Guessing Attack

If the malicious attacker ma gets the private credentials after performing the attack mentioned in the threat model section. So malicious attackers can try all possibilities to get the real password pw_i of the legal user mu_i. But, the actual password is hashed with the random number and stored as a random password $rpw_i = h(pw_i||r_i)$. So it is difficult to get pw_i if the arbitrary number r_i is not known to the adversary. So the model is immune to offline password-guessing attacks.

8.6   Man-in-the-middle Attack

As mentioned in the threat model section, a malicious attacker can eavesdrop on $\{m_1, auth_{us}, rid_i, t_1\}$, $\{c_j, m_2, auth_{sd}, t_2\}$, $\{m_3, auth_{ds}, auth_{du}, t_3\}$, and $\{m_4, auth_{du}, auth_{su}, t_4\}$ sent through public medium and tries to perform a man-in-the-middle attack. But the attacker can not get the authentication and confirmation messages because the arbitrary number $\{r_1, r_2, r_3\}$ and secret credentials $\{x_i, z_j\}$ have masked them. In addition to it, a malicious adversary cannot get the session key $sk = h(r_1||r_2||r_3)$ without having the arbitrary numpty $\{r_1, r_2, r_3\}$ and physical unclonable function response parameter r_j. So our model is immune to a man-in-the-middle attack.

8.7   Ephemeral Secret Leakage Attack

As mentioned in the Canetti Krawczyk model, if the malicious adversary ma gets only the private credentials and session states from the other features as mentioned in Dolev and Yao model. Even if the long-term keys $\{r_i, b_j\}$ are known to the attacker, the session key sk is still unknown because the real identity $\{id_i, did_j\}$ and secret value of physical unclonable functions α_i are not known to the attacker. While the other case can be if an attacker gets the short-term keys $\{r_1, r_2, r_3\}$ still the session key sk cannot be revealed because the private credentials $\{x_i, z_j\}$ and secret parameters of the physically unclonable function $\{r_j\}$ are not known to the attacker. So our model is immune to the ephemeral secret leakage attack.

8.8   Anonymity

If malicious adversary ma eavesdrop on the messages, send them during the authentication and key agreement phase. Still, it is difficult for the malicious adversary to get the real identity $\{id_i, did_j\}$ if the master key msk and masked password rpw_i are not known. This happens because $\{id_i, did_j\}$ of mobile user and drone are stored as $rid_i = h(id_i||rpw_i||msk)$ and $did_j^* = did_j \oplus h(id_i||rpw_i||pw_i)$. So our model offers anonymity.

8.9   Mutual Authentication

Our model mutually authenticates all the connected devices. The control server gets the login request $\{m_1, auth_{us}, rid_i, t_1\}$ from the mobile user and it checks $auth_{us}^* \stackrel{?}{=} h(rid_i||r_1||x_i||t_1)$. If it's true, then the control server authenticates the mobile user. Similarly, the drone also verifies the authentication request message $\{c_j, m_2, auth_{ds}, t_2\}$ is sent by the control server and checks authentication $auth_{sd}^* \stackrel{?}{=} h(did_j||t_2||r_j||z_j||r_2)$. If it's true, then the drone authenticates the control server. After this, a confirmation message $\{m_3, auth_{ds}, auth_{du}, t_3\}$ reaches to control server from the drone. Then, the control server checks if it's true or not $auth_{ds}^* \stackrel{?}{=} h(did_j||r_2||r_3||r_j||t_3)$. If it's true, control authenticates the drone. Now the authentication confirmation messages $\{m_4, auth_{du}, auth_{su}, t_4\}$ are sent to mobile users from drones and control servers. If $auth_{su}^* \stackrel{?}{=} h(rid_j||r_1||r_2||x_i)$ and $auth_{du}^* \stackrel{?}{=} h(r_1||r_2||r_j||did_j||sk)$ are true, then the mobile user authenticates the drone and control server. As a result, the model provides mutual authentication to all the devices connected.

9   Performance Analysis

The proposed scheme's authentication and key agreement computation overhead and communication overhead are compared with the previously proposed schemes, as mentioned in related works.

9.1   Computation Time

The overhead computation cost of the proposed scheme has been analysed by comparing it with the previously published schemes [23,26,27,30,31,33]. This comparison is made at the authentication and key agreement phase. For comparison, the testbed experiment results will be used. The measurement of cost is done based on the computation time defined for the cryptographic primitives. The average time required by the cryptographic primitives, particularly for the control server, is represented by the Table 2.

Table 2. Computation Cost

Notations	Scheme	mu_i	cs	d_j	Total Cost
[A]	[27]	$t_{fe} + 16t_h \approx 7.792$ ms	$8t_h \approx 0.44$ ms	$7t_h \approx 2.163$ ms	10.395 ms
[B]	[26]	$t_{fe} + 14t_h \approx 7.174$ ms	$9t_h \approx 0.495$ ms	$7t_h \approx 2.163$ ms	9.832 ms
[C]	[23]	$t_{fe} + 10t_h \approx 5.938$ ms	$3t_{sed} + 7t_h \approx 0.388$ ms	$7t_h \approx 2.163$ ms	8.489 ms
[D]	[31]	$5t_{ecpm} + 5t_h \approx 15.785$ ms	–	$4t_{ecpm} + 2t_h \approx 12.01$ ms	27.795 ms
[E]	[30]	$2t_{ecpm} + 6t_h \approx 7.55$ ms	$8t_h \approx 0.44$ ms	$2t_{ecpm} + 5t_h \approx 7.241$ ms	15.231 ms
[F]	[33]	$t_{fe} + 12t_h \approx 6.556$ ms	$9t_h \approx 0.495$ ms	$t_{fe} + 8t_h \approx 5.32$ ms	12.371 ms
[G]	$ProposedProtocol$	$t_{fe} + 16t_h \approx 7.692$ ms	$8t_h \approx 0.43$ ms	$7t_h \approx 2.162$ ms	10.284 ms

In such cases, t_{bp} , t_{ecpm}, t_{fe}, t_h and t_s is considered. Now the average time required by the cryptographic primitives, particularly for the drones and the mobile user, is represented in Table 2. In such cases, t_{bp} , t_{ecpm}, t_{fe}, t_h and t_s is considered. The result of our computation cost when compared with the previously defined schemes, is represented in Table 2 and Fig. 1. After seeing the result, it can be concluded that the proposed scheme [G] gives less computation time when compared with other schemes [27] denoted by [A], [26] [B], [23] denoted by [C], [31] denoted by [D], [30] [E], and [33] denoted by [F] in the Fig. 1.

9.2   Communication Cost

For finding the communication cost, several factors need to be considered. For example, timestamp size, ciphertext and plaintext of the private key algorithm, any arbitrary number, identity, $h(.)$, and various elliptic curve points like 32, 64, 128, 160, and 320 bits. The message created and sent by the mobile user in the secure lightweight authentication protocol internet of drones is $\{m1, auth_{us}, rid_i, t_1\}$. The total overhead is 512 bits. Similarly, control server also creates some message $\{c_j, m_2 auth_{sd}, t_2\}$ and $\{m_4, auth_{du}, auth_{su}, t_4\}$ and sent them. The total overhead of the control server is 512 bits. Likewise drone also create a message $\{m_3 auth_{ds}, auth_{du}, t_3\}$. The total overhead of the drone is 512 bits. The result of our communication cost compared with the previously defined schemes is represented by the Table 3.

The total overhead of communication cost of the proposed scheme is 2048 bits. After seeing the result, it can be concluded that the proposed scheme gives less communication time when compared with other schemes [23,26,27,30,31].

In addition to it, the security and efficiency are also better than other schemes. In Fig. 1, an analysis of the proposed protocol with relevant ones is shown.

Table 3. Communication cost

Schemes	Message 1	Message 2	Message 3	Message 4	Total Cost
[27]	672 bits	512 bits	512 bits	- bits	1696 bits
[26]	672 bits	512 bits	352 bits	- bits	1536 bits
[23]	480 bits	672 bits	512 bits	- bits	1664 bits
[31]	1952 bits	- bits	- bits	- bits	1952 bits
[30]	832 bits	992 bits	512 bits	- bits	2336 bits
[33]	512 bits	512 bits	512 bits	512 bits	2048 bits
proposed protocol	512 bits	512 bits	512 bits	512 bits	2048 bits

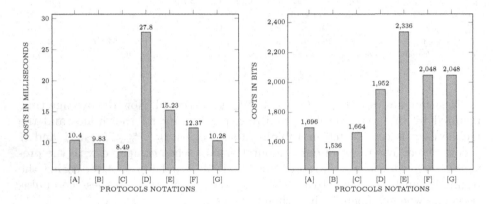

Fig. 1. Illustration of Computations and Communications Costs of Protocols

10 Conclusion

Many authentications, and key exchange protocols were studied, but it was found that either they were not secure against possible attacks like password guessing, anonymity, Man in the middle, malicious insider, user/server impersonation, etc., or they are not efficient in terms of computation and communication. A recent study on protocol [33] was studied and a vulnerability was found. According to the protocol to which we have referred,$\{id_i, rpw_i, r_i\}$ was submitted to the control server through a secure channel but there is no meaning to send the random number r_i to the control server as the attacker can use it to retrieve the secret credential with certain attacks. Therefore, we have designed a secure and efficient authenticated key exchange mechanism for the internet of drones environment. We have also done a performance analysis of the proposed protocol, with relevant protocols, and found that the proposed protocol takes less computation cost. Therefore this protocol can be implemented for communication among the Internet of Drones.

References

1. Das, A.K., Bera, B., Wazid, M., Jamal, S.S., Park, Y.: igcacs-iod: an improved certificate-enabled generic access control scheme for internet of drones deployment. IEEE Access **9**, 87024–87048 (2021)
2. Khan, M.A., et al.: An efficient and secure certificate-based access control and key agreement scheme for flying ad-hoc networks. IEEE Trans. Veh. Technol. **70**(5), 4839–4851 (2021)
3. Li, X., Jianwei Niu, Md., Bhuiyan, Z.A., Fan, W., Karuppiah, M., Kumari, S.: A robust ECC-based provable secure authentication protocol with privacy preserving for industrial internet of things. IEEE Trans. Ind. Inf. **14**(8), 3599–3609 (2017)
4. Mandal, S., Bera, B., Sutrala, A.K., Das, A.K., Choo, K.K.R., Park, Y.: Certificateless-signcryption-based three-factor user access control scheme for IoT environment. IEEE Internet Things J. **7**(4), 3184–3197 (2020)
5. Yu, S., Lee, J., Park, K., Das, A.K., Park, Y.: IoV-SMAP: secure and efficient message authentication protocol for IoV in smart city environment. IEEE Access **8**, 167875–167886 (2020)
6. Long, T., Ozger, M., Cetinkaya, O., Akan, O.B.: Energy neutral internet of drones. IEEE Commun. Maga. **56**(1), 22–28 (2018)
7. Boccadoro, P., Striccoli, D., Grieco, L.A.: An extensive survey on the Internet of Drones. Ad Hoc Netw. **122**, 102600 (2021)
8. Gharibi, M., Boutaba, R., Waslander, S.L.: Internet of drones. IEEE Access **4**, 1148–1162 (2016)
9. Mishra, B., Garg, D., Narang, P., Mishra, V.: Drone-surveillance for search and rescue in natural disaster. Comput. Commun. **156**, 1–10 (2020)
10. Yahuza, M., et al.: Internet of drones security and privacy issues: taxonomy and open challenges. IEEE Access **9**, 57243–57270 (2021)
11. Wazid, M., Das, A.K., Lee, J.K.: Authentication protocols for the internet of drones: taxonomy, analysis and future directions. J. Ambient Intell. Human. Comput. 1–10 (2018)
12. Chaudhry, S.A., Yahya, K., Karuppiah, M., Kharel, R., Bashir, A.K., Zikria, Y.B.: GCACS-IoD: a certificate based generic access control scheme for Internet of drones. Comput. Netw. **191**, 107999 (2021)
13. Cho, G., Cho, J., Hyun, S., Kim, H.: Sentinel: a secure and efficient authentication framework for unmanned aerial vehicles. Appl. Sci. **10**(9), 3149 (2020)
14. Zhang, Y., He, D., Li, L., Chen, B.: A lightweight authentication and key agreement scheme for internet of drones. Comput. Commun. **154**, 455–464 (2020)
15. Gope, P., Hwang, T.: A realistic lightweight anonymous authentication protocol for securing real-time application data access in wireless sensor networks. IEEE Trans. Ind. Electron. **63**(11), 7124–7132 (2016)
16. Park, K., et al.: LAKS-NVT: provably secure and lightweight authentication and key agreement scheme without verification table in medical internet of things. IEEE Access **8**, 119387–119404 (2020)
17. Shen, J., Zhou, T., Wei, F., Sun, X., Xiang, Y.: Privacy-preserving and lightweight key agreement protocol for v2g in the social internet of things. IEEE Internet Things J. **5**(4), 2526–2536 (2017)
18. Lamport, L.: Password authentication with insecure communication. Commun. ACM **24**(11), 770–772 (1981)
19. Manik Lal Das: Two-factor user authentication in wireless sensor networks. IEEE Trans. Wirel. Commun. **8**(3), 1086–1090 (2009)

20. Nyang, D.H., Lee, M.K.: Improvement of das's two-factor authentication protocol in wireless sensor networks. Cryptology EPrint Archive (2009)
21. He, D., Gao, Y., Chan, S., Chen, C., Jiajun, B.: An enhanced two-factor user authentication scheme in wireless sensor networks. Ad Hoc Sens. Wirel. Netw. **10**(4), 361–371 (2010)
22. Kumar, P., Lee, H.J.: Cryptanalysis on two user authentication protocols using smart card for wireless sensor networks. In: 2011 Wireless Advanced, pp. 241–245. IEEE (2011)
23. Ali, Z., Chaudhry, S.A., Ramzan, M.S., Al-Turjman, F.: Securing smart city surveillance: a lightweight authentication mechanism for unmanned vehicles. IEEE Access **8**, 43711–43724 (2020)
24. Alladi, T., Chamola, V., Kumar, N., et al.: Parth: a two-stage lightweight mutual authentication protocol for UAV surveillance networks. Comput. Commun. **160**, 81–90 (2020)
25. Bakkiam David Deebak and Fadi Al-Turjman: A smart lightweight privacy preservation scheme for IoT-based UAV communication systems. Comput. Commun. **162**, 102–117 (2020)
26. Srinivas, J., Das, A.K., Kumar, N., Rodrigues, J.J.: TCALAS: temporal credential-based anonymous lightweight authentication scheme for Internet of drones environment. IEEE Trans. Veh. Technol. **68**(7), 6903–6916 (2019)
27. Wazid, M., Das, A.K., Kumar, N., Vasilakos, A.V., Rodrigues, J.J.: Design and analysis of secure lightweight remote user authentication and key agreement scheme in internet of drones deployment. IEEE Internet Things J. **6**(2), 3572–3584 (2018)
28. Cheon, J.H., et al.: Toward a secure drone system: flying with real-time homomorphic authenticated encryption. IEEE Access **6**, 24325–24339 (2018)
29. Yoney Kirsal Ever: A secure authentication scheme framework for mobile-sinks used in the internet of drones applications. Comput. Commun. **155**, 143–149 (2020)
30. Nikooghadam, M., Amintoosi, H., Islam, S.H., Moghadam, M.F.: A provably secure and lightweight authentication scheme for Internet of Drones for smart city surveillance. J. Syst. Arch. **115**, 101955 (2021)
31. Won, J., Seo, S.-H., Bertino, E.: Certificateless cryptographic protocols for efficient drone-based smart city applications. IEEE Access **5**, 3721–3749 (2017)
32. Ali, Z., Alzahrani, B.A., Barnawi, A., Al-Barakati, A., Vijayakumar, P., Chaudhry, S.A.: TC-PSLAP: temporal credential-based provably secure and lightweight authentication protocol for IoT-enabled drone environments. Secur. Commun. Netw. **2021**, 1–10 (2021)
33. Yu, S., Das, A.K., Park, Y., Lorenz, P.: SLAP-IoD: secure and lightweight authentication protocol using physical unclonable functions for internet of drones in smart city environments. IEEE Trans. Veh. Technol. **71**(10), 10374–10388 (2022)
34. Dolev, D., Yao, A.: On the security of public key protocols. IEEE Trans. Inf. Theory **29**(2), 198–208 (1983)
35. Kocher, P., Jaffe, J., Jun, B.: Differential power analysis. In: Wiener, M. (ed.) CRYPTO 1999. LNCS, vol. 1666, pp. 388–397. Springer, Heidelberg (1999). https://doi.org/10.1007/3-540-48405-1_25
36. Canetti, R., Krawczyk, H.: Universally composable notions of key exchange and secure channels. In: Knudsen, L.R. (ed.) EUROCRYPT 2002. LNCS, vol. 2332, pp. 337–351. Springer, Heidelberg (2002). https://doi.org/10.1007/3-540-46035-7_22
37. Aman, M.N., Chua, K.C., Sikdar, B.: Mutual authentication in iot systems using physical unclonable functions. IEEE Internet Things J. **4**(5), 1327–1340 (2017)
38. Gao, Y., Al-Sarawi, S.F., Abbott, D.: Physical unclonable functions. Nat. Electron. **3**(2), 81–91 (2020)

Comparative Analysis of Quantum Key Distribution Protocols: Security, Efficiency, and Practicality

Neha Agarwal[✉] [iD] and Vikas Verma [iD]

Vivekananda Global University, Jaipur 302012, India
meet2neha.261@gmail.com

Abstract. Conventional cryptography commonly relies on the complexity of mathematical algorithms and the impractical amount of time required to crack the method, which ensures the strength of security in key distribution. However, if the process for distributing secret keys is inaccurate, it will be unsuccessful. As a recent solution to the key distribution problem, Quantum Key Distribution (QKD) has recently attracted a lot of research interest. QKD is a method for securely distributing encryption keys using the properties of quantum mechanics. There are several different protocols for QKD, each with its own advantages and limitations. In this comparative analysis, the most commonly used QKD protocols, including BB84, B92, E91, and SARG04 are examined and compared. Also, their proposed work, generation rates, and experimental feasibility are analyzed. The future directions and challenges for each of these protocols are discussed in this paper. The goal of this study is to provide a comprehensive understanding of the various QKD protocols and their strengths and weaknesses.

Keywords: Cryptography · Encryption · Quantum Cryptography · Quantum Key Distribution · Secret Key

1 Introduction

Data communication security is a complicated process that involves individuals, networks, and applications, all of which are interconnected by a variety of latest technologies. Information systems are therefore extremely susceptible to attacks and unauthorized intrusions, whether the data is accidental or malicious. To secure information transmission over such networks, cryptography is a technique. Cryptography and security are essential components of our daily network communications. In cryptography [1] data are encrypted and decrypted using mathematical tools. It allows users to store sensitive information or send it through unsecured networks (like the Internet) so that only the intended recipient can understand it. Classical data encryption cannot provide complete security for legal parties due to the weaknesses of existing networking techniques. Most of the traditional cryptographic algorithms [2] depend upon mathematical models and computational assumptions in the network communication environments. Due to this reason, they are actually not safe and easily accessible by many attackers.

Quantum Cryptography (QC) has attracted the attention of information security experts in recent years. In order to secure and transmit data in a way that cannot be intercepted, QC [3] employs the inherent features of quantum physics. In contrast to conventional cryptographic systems, QC uses physics rather than mathematics as the primary component of its security concept. Modern cryptography relies heavily on key distribution methods because they enable the use of more effective cryptography algorithms. QC makes use of key distribution methods known as Quantum Key Distribution (QKD). QKD is a secure form of communication for sharing encryption keys that are only known to share parties. It exchanges cryptographic keys in a verifiable manner that ensures security using principles from quantum physics.

The conventional key distribution uses mathematics to protect the data, but QKD uses a quantum system that relies on fundamental natural laws to do so. The capacity to detect the presence of eavesdroppers is a novel capability that QKD [4] has which conventional cryptography techniques don't possess. It is possible to identify every eavesdropper activity as an error. The security offered by the QKD system has been demonstrated to be resistant to adversary attacks, even with infinite computing power. Attackers are prevented from simply copying the data in the same way that they can today by the no-cloning theorem, which states that it is impossible to make identical copies of an unknown quantum state. Furthermore, the system undergoes alterations that enable the intended individuals to detect any interference or unauthorized access by an attacker. This procedure is resistant to increased processor power.

In order for QKD [5] to function, numerous light particles, or photons, must be sent between parties over fibre optic cables. The photons sent constitute a stream of ones and zeros, and each photon has a random quantum state. Qubits, which consist of a continuous stream of ones and zeros, are the binary system's counterpart of bits. A photon passes via a beam splitter at the receiving device, which compels it to take whichever path it chooses at random into a photon collector. The receiver then sends information on the order of the photons in response to the original sender, and the sender compares that information with the emitter, which would have delivered each photon, separately. The remaining bits of a particular series of bits after photons in the incorrect beam collector are eliminated. The key to encrypting data can then be created using this bit sequence. During an error-correction phase and other post-processing procedures, any mistakes and data leakage are eliminated. Another post-processing phase known as delayed privacy amplification eliminates any knowledge an eavesdropper might have acquired about the ultimate secret key. Figure 1 shows the outline of the generation of the secret key from the plaintext. In QKD protocols, a secret key is generated through a process that begins with the sender converting classical bits into quantum bits. These qubits are then transmitted through a quantum channel to the receiver. During transmission, the sender and receiver perform operations on the qubits to ensure the security of the key. Once the transmission is complete, the sender and receiver use a classical system to confirm the validity of the key and correct any errors that may have occurred during transmission [5]. Overall, the process involves a combination of quantum and classical systems to create a secure and reliable secret key.

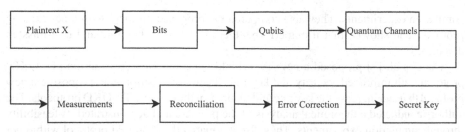

Fig. 1. Generation of the secret key in quantum key distribution protocols

2   Comparative Analysis

In a modern communication environment, QKD is an emerging solution for safeguarding sensitive data during transmission. In order to develop a secure connection for files dependent on different simulator conditions, numerous scientists have concentrated on the simulation of QKD. There are various categories for categorizing cryptographic algorithms, and these categories will be based on the number of keys used for encryption and decryption, as well as their application and use (Fig. 2).

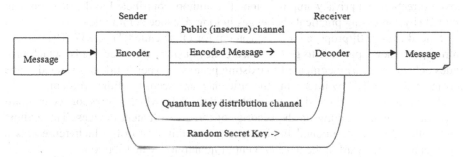

Fig. 2. Working of QKD

Hamouda et al. [2] conducted a comparative study of various cryptographic algorithms. Their contribution is in providing an overview and comparison of different cryptographic algorithms and evaluating them based on factors such as security, performance, efficiency, and ease of implementation. The study helps in selecting the appropriate algorithm for specific applications based on their requirements.

Abushgra et al. [5] emphasize the importance of security in QKD protocols and discusses different types of attacks that can compromise their security. Additionally, the paper covers different techniques for improving the performance of QKD protocols. Overall, the review provides a comprehensive overview of the state-of-the-art research in QKD protocols and is useful for researchers in the field of quantum cryptography.

Wang et al. [6] proposed a new Coherent One-Way Quantum Key Distribution (COW-QKD) protocol based on non-orthogonal states. Their contribution is in proposing a new protocol that can increase the secret key rate and distance for QKD. The authors conducted a theoretical analysis of the protocol and demonstrated its feasibility through simulation experiments. They also compared the proposed protocol with other existing protocols and showed that it outperforms them in terms of key rate and transmission distance.

Guan et al. [7] proposed a new Twin-Field Quantum Key Distribution (TF-QKD) protocol with improved security and key rate. Their contribution is in proposing a new protocol that can increase the key rate and security of the existing TF-QKD protocols. The authors conducted a theoretical analysis of the protocol and demonstrated its feasibility through simulation experiments. They also compared the proposed protocol with other existing protocols and showed that it outperforms them in terms of key rate and security. Specifically, the authors introduced a new procedure of error estimation and correction to reduce the errors caused by the fluctuation of the phase modulator. Additionally, the authors used a modified reconciliation algorithm to enhance the security of the protocol.

Bhattacharya et al. [8] provided a detailed theoretical analysis of their proposed approach and show that it has several advantages over existing Measurement Device Independent Quantum Key Distribution (MDI-QKD) protocols, such as higher key rates and improved security against certain types of attacks. They also provide a numerical simulation of their proposed protocol to demonstrate its practicality and feasibility.

Jouguet et al. [9] demonstrated the experimental implementation of continuous variable QKD over long distances, which was achieved using a high-performance system for error correction and privacy amplification. The authors contributed to the development of QKD by addressing practical challenges in long-distance communication.

Sajeed et al. [10] proposed a new protocol for QKD called TF-QKD. The authors introduced a new approach to generate secret keys using correlated fields, which provides enhanced security compared to existing protocols. Their work contributed to the advancement of QKD by improving the efficiency and security of the protocol.

Tang et al. [11] introduced the concept of MDI-QKD, which allows for secure communication without relying on the security of the measurement devices. The authors demonstrated the experimental implementation of this protocol, which represents a significant step forward in the development of quantum key distribution.

Sit et al. [12] proposed an MDI-QKD protocol that uses high-dimensional states, which provides increased security compared to previous protocols. The authors demonstrated the feasibility of this protocol experimentally, contributing to the development of secure quantum communication protocols.

Huang and Wang [13] proposed a continuous-variable MDI-QKD protocol that improves the security of existing protocols. The authors demonstrated the feasibility of their protocol experimentally, which represents a significant contribution to the development of secure quantum communication protocols.

Sharma and Kumar [14] reviewed quantum computing and its various applications, such as quantum public key cryptography, quantum key distribution, and quantum authentication. This paper also demonstrated that QC is not only secure but also has claimed to demonstrate the intention of traditional cryptography. The sender and receiver can recognize eavesdropping and take necessary action due to the qualities it has obtained from quantum physics. The second goal is that nobody can crack the quantum key.

Gheorghies et al. [15] examined three distinct types of protocols: classical, QKD, and blockchain based protocols, with examples from each category. Also discussed were the specifics and difficulties of each protocol, as well as potential solutions and the effects of these protocols. This paper also proposed an outline of PKI solutions in the context of quantum computing and blockchain.

Kour et al. [16] attempted to introduce QC, QKD protocols, and QC applications in this study. It provided information on several QKD mechanisms. In order to attain a higher level of security, these protocols can be used in conjunction with encryption technology.

Al-Shabi [17] conducted a comparative analysis of the most significant algorithms in terms of speed (implementation) and security (special keys). This paper covers a number of significant algorithms used for the encryption and decryption of data across all fields. The comparison of symmetric and asymmetric algorithms demonstrates that the former is quicker than the latter. Advanced Encryption Standard (AES) is the most dependable algorithm in terms of speed encryption, decoding, the length of the key, structure, and usability, according to past studies and comparison results.

Panhwar et al. [18] presented the features of different symmetric and asymmetric algorithms, including triple Data Encryption Standard (DES), AES, and DES, which are addressed in this study [18] in relation to their application in mobile computing based work solutions.

Bharathi et al. [19] compare and contrast the various block cipher algorithms like DES, RC6, BLOWFISH, and UR5 while also conducting a literature review on each method. Blowfish and UR5 have 8 rounds, and RC6 has 12 rounds. In the study, performance metrics for the encryption process are analyzed in light of security concerns.

QKD is different from symmetric encryption, asymmetric encryption, and hash encryption in several ways. Table 1 shows the comparison of QKD over different encryptions in terms of security, key distribution, quantum resistance and use cases. There are Several different algorithms [15] that have been proposed for use in QKD systems. Bennett and Brassard proposed the initial QKD protocol in 1984, and the first successful QKD deployment took place in 1989. Table 2 shows the comparison of different QKD protocols [20] based on their authors, year, proposed work, advantages and disadvantages. QKD is considered to be more advanced and secure than conventional encryption techniques for several reasons:

- Security: QKD is based on the principles of quantum mechanics and is considered to be one of the most secure forms of encryption. It is resistant to eavesdropping, which is a major concern in conventional encryption techniques because any attempt to eavesdrop on the key distribution process will be detected.
- Quantum-resistance: QKD is considered to be quantum-resistant, meaning that it is secure against attacks by quantum computers, whereas conventional encryption methods such as symmetric and asymmetric encryption, will be broken by the power of Quantum computers.
- Key Distribution: In QKD, the key distribution process is secure and does not rely on a secure initial channel, which is required in conventional encryption techniques. This makes it more suitable for use in environments where a secure initial channel is not available.

- No need for trust: In QKD the parties do not need to trust the communication channel or the devices used in the key distribution process, as the security is based on the laws of physics, which can be trusted.
- Versatility: QKD can be used in combination with other encryption methods to enhance the security of the overall system. For example, a one-time pad encryption key generated by QKD can be used in a symmetric-key encryption algorithm such as AES.

Table 1. Comparison of using QKD over Symmetric Encryption, Asymmetric Encryption and Hash Encryption

Parameter	Symmetric Encryption	Asymmetric Encryption	Hash Encryption	QuantumKey Distribution (QKD)
Security	Can be broken if the secret key is compromised	Can be broken if a private key is compromised	Not meant for encryption but for integrity check	Considered to be one of the most secure forms of encryption
Key Discontribution	The secret key must be shared securely	Public key can be shared openly, the private key must be kept secret	Not applicable	Allows for secure key distribution without the need for a secure initial channel
Quantum-resistance	Not quantum-resistant	Not quantum-resistant	Not applicable	Quantum-resistant
Use-cases	Data encryption, VPN, and Disk encryption	The digital signature, Secure communication	Data integrity, the Authenticity check	Secure communicate. Key distribution

In summary, QKD offers a higher level of security and key distribution flexibility than conventional encryption methods, it's resistant to quantum computer attacks and does not rely on communication channels. It can also be used in combination with other encryption methods for added security (Tables 3 and 4).

Table 2. Comparison of different QKD Protocols (Part 1)

Author/s	Protocol	Proposed Work	Approach	Key Contributions	Advantages	Disadvantages
Hwang et al. [21] (2003)	SSP	Uses six non-orthogonal state to encode the key	Uses decoy states and a single-photon source	The protocol is Secure even under high-loss conditions	More secure than BB84	Moderate Complexity
Scarani, et al. [22] (2004)	SARG04	A simplified version of BB84, and it uses only one non-orthogonal state to encode the key	Uses weak laser pulses and a decoy-state technique	The protocol is secure against photon number splitting attacks	Simple to implement	Less secure than BB84
Samuel L. et al. [23] (2005)	Gaussian-modulated Coherent State QKD	Uses Gaussian modulation to encode the key	Uses the properties of continuous variable systems, such as quadrature amplitude and phase	The approach allows for the implementation of a wide range of quantum information protocols	More efficient than BB84	More complex to implement
Raúl García-Patrón et al. [24] (2006)	Twin-Field QKD	Uses the correlation between two quantum fields to encode the key	Uses mathematical proofs and analysis	The proof provides a theoretical basis for the security of continuous variable QKD	More robust against certain types of attacks	More complex to implement
Inoue, Kyo et al. [25] (2009)	DPS	Uses a decoy-state to detect the presence of eavesdroppers	Uses differential quadrature phase shift and decoy states	The protocol is secure against general attacks	More secure than BB84	Moderate Complexity
Sajeed et al. [10] (2015)	TWIN-QKD	Uses two conjugate quadratures of light field and phase modulation to encode the key	Sagnac interferometer with entangled twin beams	High-dimensional key distribution	Resistant to noise and loss enabling higher key rates over longer distances	Requires complex hardware and may be vulnerable to attacks against the detector

Table 3. Comparison of different QKD Protocols (Part 2)

Author/s	Protocol	Proposed Work	Approach	Key Contributions	Advantages	Disadvantages
Tang et al. [11] (2016)	MDI-QKD	Combines MDI-QKD with MDI to achieve higher security	Phase-encoded coherent states and homodyne detection	Overcome the vulnerabilities of the measurement apparatus and detectors	Resistant to attacks against the detection devices	Requires more complex hardware and may have lower key rates
Sit et al. [12] (2017)	High-dimensional QKD	Uses high-dimensional quantum states (e.g., qubits) to encode the key	Spatial-mode entanglement with homodyne detection	Ability to send more information per photon	Can achieve higher key rate with a smaller number of photons	Requires more complex hardware and may be vulnerable to attacks against the detector
Huang et al. [13] (2018)	CV-MDI-QKD	Combines the security of SARG04 with the practicality of MDI-QKD using CV states and homodyne measurements	Homodyne detection and postselection	Robust against all detector side channels	High security against all types of attacks, including MDI attacks	Requires more complex hardware and may have lower key rates
Hamouda and B.E.H.H [2] (2020)	Various cryptographic algorithms	Comparative study of different cryptographic algorithms	Review and analysis of existing algorithms	Provides insights into the strengths and weaknesses of different cryptographic algorithms	Evaluates the algorithms based on factors like security, performance, efficiency and ease of implementation	Evaluation criteria may not be suitable for all applications
Jouguet et al. [9] (2013)	Coherent one-way QKD	Uses homodyne detection and phase randomization to achieve high key rates	Homodyne detection and reverse reconciliation	Achieved the the longest distance for continuous-variable quantum key distribution	Can achieve high key rates over long distances using standard telecom equipment	Sensitive to channel noise And requires error correction and privacy amplication

Table 4. Comparison of different QKD Protocols (Part 3)

Author/s	Protocol	Proposed Work	Approach	Key Contributions	Advantages	Disadvantages
Wang et al. [6] (2020)	COW-QKD	A new COW-QKD protocol based on non-orthogonal states	Non-orthogonal state encoding and coherent one-way communication	Improved security, higher key rate compared to other COW QKD protocols	Higher secret key rate and better security against certain types of attacks compared to existing COW-QKD protocols	Requires careful alignment of optical elements
Guan et al. [7] (2020)	TF-QKD	A new TF-QKD protocol with improved security and key rate	Introduces additional twin-field mode, improved phase estimation and random basis switching	Improved key rate and security compared to existing twin-field QKD protocols	Better security against certain types of attacks and improved key rates compared to existing TF-QKD protocols	Requires precise timing synchronization between the sender and receiver
Marian Lazro Gheorghies and Emil Simion [15] (2021)	Cryptography key distribution protocols	Comparative study of different cryptographic key distribution protocols	Literature review and analysis	Evaluation and comparison of different cryptographic key distribution algorithms	Helps in selecting an appropriate protocol for specific applications based on the requirements	Limited only to key distribution protocols considered in the study
Bhattachar et al. [8] (2021)	MDI-QKD	A new approach to MDI-QKD using structured coherent states	Encodes information in structured coherent states utilize unambiguous state discrimination technique	Improved security, higher tolerance to channel noise and photon loss	Higher key rates and better security against attacks compared to traditional MDI-QKD protocols	Requires more precise control over the encoding and decoding operations

(continued)

Table 4. (*continued*)

Author/s	Protocol	Proposed Work	Approach	Key Contributions	Advantages	Disadvantages
A.A Abushgra [5] (2022)	Variations of QKD protocols based on a conventional system measurements	Literature review	Review of existing protocols	Analysis of different QKD protocols based on conventional system measurements	Provides a comprehend-sive overview ofvariations of QKD protocols based on conventional system	Limited to variations of QKD protocols based on conventional system measurements
Hoi-Kwong Lo et al. [26] (2012)	MDI-QKD	MDI-QKD eliminates the need for trust in measurement devices	Usesentanglement and a loophole-free Bell test	The protocol is secure against attacks on the measurement devices	More robust against certain types of attacks	More complex to implement

3 Research Challenges and Issues

QKD is a promising technology that offers a high level of security and key distribution flexibility, but it also has some limitations:

- Distance: The distance over which a QKD system can operate is limited by the loss of the quantum signal as it travels through the optical fiber or free space. This limits the range of QKD systems and makes them less practical for long-distance communication.
- Cost: QKD systems can be expensive to build and maintain, especially when compared to conventional encryption systems. This can make them less practical for widespread use.
- Complexity: QKD systems can be complex to set up and operate, requiring specialized equipment and trained personnel. This can make them less accessible to the average user.
- Interception: QKD systems can be intercepted by an attacker who is able to access the quantum channel. However, this can be detected by the legitimate parties.
- Scalability: The scalability of QKD systems is limited and the number of users that can be supported is small in comparison to conventional encryption systems.
- Integration: QKD systems have to be integrated with conventional encryption systems as they are not yet capable of providing end-to-end encryption.
- Noise: QKD is sensitive to noise and errors, which can decrease the secret key generation rate and decrease the overall security of the system.

Despite these limitations, QKD is still considered to be a promising technology and research is ongoing to improve its performance, reduce costs, and make it more accessible.

QKD is a method of securely distributing cryptographic keys using the principles of quantum mechanics. However, there are several challenges that must be overcome in order to make QKD a practical and widely-used technology. One of the primary challenges is the installation of QKD systems into existing infrastructure. This includes the need for specialized hardware, software, and communication networks that can support QKD.

Another challenge is the limited distance that photons can travel before they are absorbed or scattered, which limits the maximum distance over which QKD can be used. Finally, the initial application of QKD is also a challenge, as it requires a significant amount of resources and expertise to implement. Despite these challenges, researchers and industry professionals are working to overcome these limitations and make QKD a viable and secure method of communication. Putting in place a QKD infrastructure that works perfectly is challenging. Although it is theoretically completely secure, security problems arise in real-world applications due to shortcomings in equipment like single photon detectors. Security analysis should always be considered. Modern fibre optic connections often have a maximum distance a photon can travel. Frequently, the range exceeds 100 km. This spectrum for QKD implementation has been widened by some groups and organizations. For instance, the University of Geneva and Corning Inc. collaborated to build a device that, under perfect circumstances, can transport a photon 307 km. With the use of a patent-pending, out-of-band delivery mechanism dubbed Phio Trusted Xchange, Quantum Xchange established Phio, a QKD network in the United States that can transfer quantum keys over an ostensibly limitless distance.

Another difficulty with QKD is that it requires the establishment of a channel of communication with traditional authentication. This indicates that a sufficient level of security was already established because one of the involved users had already exchanged a symmetric key. Without QKD, a system can already be designed to be adequately secure by employing another high-level encryption standard. However, as the use of quantum computers increases, the likelihood that an attacker may utilize quantum computing to break present encryption techniques increases, making QKD increasingly significant.

4 Conclusion

This research paper has provided a thorough analysis of the different QKD protocols, including their strengths and limitations. The reader can gain a comprehensive understanding of these protocols after reviewing this study. Four of the most commonly used QKD protocols: BB84, BB92, E91, and SARG04 have been analyzed here. Each of these protocols has its own strengths and limitations, while BB84 is the most widely used and well-studied protocol. This work has examined their experimental feasibility, and methodology and discussed the future directions and challenges for each of these protocols. Despite their limitations, QKD is a promising technology that offers a high level of security. Ongoing research aims to improve its performance, reduce costs, and make it more accessible. With advancements in technology, QKD is expected to become more widely adopted in the future. Companies are working on making QKD systems more affordable and user-friendly, and scientists are developing new QKD protocols that can operate over longer distances and support more users.

One important direction for future research is the integration of QKD with other technologies such as the internet and cloud computing. As more and more data is stored and transmitted over networks, the need for secure communication methods like QKD becomes increasingly important. Another important area of research is the security proof of QKD in realistic scenarios, where the assumption of idealized conditions is relaxed and the security proof holds in the presence of practical noise, device imperfections, and other side-channel attacks.

Acknowledgements. I would like to acknowledge the support from Vivekananda Global University, India. My special thanks are extended to all professors from Computer Science Department, Vivekananda Global University, Jaipur, India.

References

1. Javed, M., Aziz, K.: A survey of quantum key distribution protocols. In: 7th International Conference on Frontiers of Information Technology (FIT 2009), Abbottabad, 16–18 December 2009. ACM (2009)
2. Hamouda, A., B.E.H.H.: Comparative study of different cryptographic algorithms. J. Inf. Secur. **11**, 138–148 (2020)
3. C.S. et al.: A study and analysis on symmetric cryptography. In: International Conference on Science, Engineering and Management Research (ICSEMR), pp. 978-1-4799-7613-3/14/$31.00. IEEE (2014)
4. Gnatyuk, S.O.: Comparative analysis of quantum key distribution systems. Sci. Based Technol. 78–82 (2013)
5. Abushgra, A.: Variations of QKD protocols based on conventional system measurements: a literature review. Cryptography **6**(1), 1–25 (2022)
6. Wang, C., Huang, Y., Zhang, L., Yu, Z., Guo, J., Liu, Y.: A new coherent one-way QKD protocol based on non-orthogonal states. IEEE Access **8**, 143485–143496 (2020)
7. Guan, J., Zhang, L., Liu, Y., Guo, J.: A new twin-field quantum key distribution protocol with improved security and key rate. Quantum Inf. Process. **19**(3), 1–17 (2020)
8. Bhattacharya, A., Dhar, A., Das, D.: A new approach to measurement device-independent quantum key distribution using structured coherent states. IEEE J. Quantum Electron. **57**(3), 1–10 (2021)
9. Jouguet, P., Kunz-Jacques, S., Leverrier, A., Grangier, P., Diamanti, E.: Experimental demonstration of long-distance continuous-variable quantum key distribution. Nat. Photonics **7**(6), 378–381 (2013)
10. Sajeed, S., Kumar, R., Prakash, G.: Twin-field quantum key distribution. Phys. Rev. **92**(5), 052315 (2015)
11. Tang, D., Qi, B., Lo, H.-K.: Experimental measurement device independent quantum key distribution. Phys. Rev. X **6**(1), 011024 (2016)
12. Sit, A., Fung, C.F., Lo, H.-K.: Measurement device independent quantum key distribution with high-dimensional states. IEEE J. Sel. Top. Quantum Electron. **23**(5), 1–10 (2017)
13. Huang, D., et al.: Continuous-variable measurement device-independent quantum key distribution. Phys. Rev. **98**(3), 032315 (2018)
14. Sharma, A. and Kumar, A.: A survey on quantum key distribution. In 2nd International Conference on Issues and Challenges in Intelligent Computing Techniques (ICICT), (2019)
15. Gheorghies, M.L., Simion, E.: A comparative study of cryptographic key distribution protocols. IACR Crypt. ePrint Arch. **2021**, 31 (2021)
16. Kour, J., Koul, S., Zahid, P.: A survey on quantum key distribution protocols. Int. J. Comput. Sci. Appl. **7**(3) (2017)
17. Al-Shabi, M.A.: A survey on symmetric and asymmetric cryptography algorithms in information security. Int. J. Sci. Res. Publ. **9**(3) (2019)

18. Panhwaret, A., Al Saca, Z.: A study of symmetric and asymmetric cryptographic algorithms. IJCSNS Int. J. Comput. Sci. Netw. Secur. **19**(1), 1–8 (2019)
19. Bharathi, E., Marimuthu, A., Kavitha, A.: Performance analysis of symmetric encryption techniques. Int. J. Comput. Netw. Secur. **5**(1), 1–4 (2013)
20. Nurhadi, A., Syambas, N.: Quantum key distribution (QKD) protocols: a survey. In: 4th International Conference on Wireless and Telematics (ICWT), pp. 1–5. IEEE (2018)
21. Hwang, W.-Y.: Quantum key distribution with high loss: toward global secure communication. Phys. Rev. Lett. **91**, 057901 (2003)
22. Scarani, V., Acin, A., Ribordy, G., Gisin, N.: Quantum cryptography protocols robust against photon number splitting attacks for weak laser pulse implementations. Phys. Rev. Lett. **92**, 057901 (2004)
23. Braunstein, S.L., van Loock, P.: Quantum information with continuous variables. Rev. Mod. Phys. **77**(2), 513–577 (2005)
24. Garcia-Patron, R., Cerf, N.J.: Unconditional optimality of Gaussian attacks against continuous-variable quantum key distribution. Phys. Rev. Lett. **97**, 190503 (2006)
25. Inoue, K., Iwai, Y.: Differential quadrature phase shift quantum key distribution. Phys. Rev. A **79**, 022319 (2009)
26. Lo, H.-K., Curty, M., Qi, B.: Measurement-device-independent quantum key distribution. Phys. Rev. Lett. **108**, 130503 (2012)

Fog Intelligence for Energy Optimized Computation in Industry 4.0

Abhishek Hazra[1]([✉])[iD], Surendra Singh[2][iD], and Lalit Kumar Awasthi[2][iD]

[1] National University of Singapore, Singapore, Singapore
hazra@nus.edu.sg
[2] National Institute of Technology Uttarakhand, Srinagar, India
{surendra,lalit}@nituk.ac.in

Abstract. The field of communication and computation is experiencing ongoing expansion with the emergence of Industry 4.0 technology enabling efficient data transfer among devices. However, this advancement also poses several challenges, particularly in managing the vast amount of data generated by Industrial Internet of Things (IIoT) devices. Despite being widely recognized as an effective solution to these challenges, cloud computing also poses its own challenges. These include high bandwidth usage, latency, security concerns, and energy dissipation. In an effort to mitigate these issues, fog computing has emerged as a more energy-efficient alternative. The primary focus of this paper is the reduction of energy consumption in industrial fog networks. To accomplish this, we propose a novel architecture with the integration of fog networks and Deep Reinforcement Learning (DRL) technique to optimize the overall system reward and reduce energy consumption in industrial applications. The problem of state-action-reward is formulated as a Markov Decision Process (MDP) and optimized using a popular DRL technique. Simulated results indicate that the proposed strategy decreases energy consumption rate by 10% compared to existing offloading strategies by offloading decisions on different computing devices.

Keywords: Edge Computing · Deep Reinforcement Learning
Computation offloading · Energy Optimization · Industry 4.0

1 Introduction

The increasing demand for communication among humans and across devices, such as sensors, actuators, and Industrial Internet of Things (IIoT) devices, has been brought on by rapid advances in mobile technologies and data handling techniques [18]. These communications may include time-sensitive tasks, such as data transfer between smart cars, cameras, laptops, sensors, smartphones, and industrial objects. In order to efficiently process and analyze this data, a fast and reliable method of communication is necessary [4]. However, IIoT devices are often limited in terms of computing capacity, data size, and battery power.

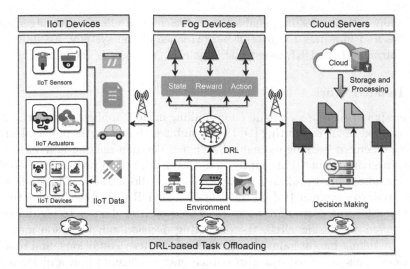

Fig. 1. Illustration of DRL-based task offloading strategy.

In the past, cloud servers were commonly used to offload and process tasks generated by IIoT devices. However, due to the physical distance between cloud servers and IIoT devices, latency issues can arise in data transfer [9]. To address this issue, fog devices, which are geographically close to IIoT devices and possess strong computational capabilities, were introduced. These devices act as a bridge between cloud servers and IIoT devices, enabling faster computation, reducing latency, and decreasing energy consumption [10]. In contrast, traditional fog-cloud hierarchy models make restrictive assumptions and use simplistic mathematical optimization techniques. In real-world Industry 4.0, task offloading and computation are further complicated by dynamic network traffic, user conditions, and mobility [11]. To address these challenges, this work proposes an algorithm utilizing the Markov Decision Process (MDP) and deep reinforcement learning to improve the efficiency of selecting suitable devices for task offloading in industrial IIoT environments.

2   Why DRL is Important in IIoT Offloading?

DRL is important in IIoT computation offloading because it allows for intelligent decision-making in dynamic and uncertain environments. The use of DRL enables the IIoT devices to learn from interactions with their environment and adapt their actions accordingly. This is particularly useful in industrial IIoT applications where tasks are time-sensitive and require immediate processing from computing devices. DRL algorithms can be used to optimize the offloading of tasks to selected computing devices by taking into consideration factors such as task deadlines, computational requirements, and energy consumption. Additionally, DRL can also help to reduce the latency in data transfer by selecting the most appropriate device for offloading tasks. Overall, the use of DRL in

IIoT computation offloading enables the efficient and effective management of resources and improves the overall performance of the system. An illustrative representation of the DRL-based task offloading strategy is shown in Fig. 1.

2.1   Related Work

DRL is a powerful approach to decision-making in IIoT applications, particularly in the context of fog computing [1]. DRL combines the strengths of deep learning and reinforcement learning to enable intelligent decision-making in dynamic and uncertain environments. This approach has been applied to various IIoT applications, including smart grid management, resource allocation, and task offloading in industrial IoT systems [19]. In the field of IIoT, DRL has been applied to optimize the offloading decision of tasks generated by IIoT devices to fog devices or cloud servers. By using DRL, the offloading decision can be made more efficiently by taking into account the network traffic, user conditions, and mobility of devices. This approach can also reduce energy consumption and latency in data transfer. In the field of fog computing, DRL has been applied to optimize resource allocation and management in fog networks. By using DRL, fog nodes can make intelligent decisions about allocating resources to different IIoT devices based on their requirements and constraints. This approach can also improve the scalability and reliability of fog networks.

In recent years, a considerable amount of research has been completed on the topic of fog-cloud hierarchy in the context of the IIoT [12]. The main focus of these studies has been on reducing latency in task offloading and selecting the appropriate device for offloading tasks [20]. In traditional models, only a single user generates tasks, and the optimization of offloading involves selecting the appropriate device, either fog or cloud [5]. However, more recent studies have considered multi-user models, where multiple users generate tasks, and the tasks are offloaded to fog or cloud as needed [1,3]. Additionally, some studies have also introduced deadlines to tasks [7]. Another trend in recent research is the introduction of multiple fog devices that can perform computation in parallel [2,16]. Furthermore, other studies have also investigated the trade-offs between local and remote computation [13,15].

However, simple mathematical and numerical analysis alone were insufficient in addressing the challenges associated with IoT computation offloading in fog computing. To address these challenges, recent research has incorporated machine learning concepts such as Deep Neural Networks (DNN) and DRL. For example, Mukherjee *et al.* have proposed an energy-efficient workload distribution strategy for fog models using distributed deep learning [14]. Additionally, Sarkar *et al.* have designed an Intelligent Service Provisioning model that utilizes DRL to calculate optimal load balancing in SDN-based industrial fog environment [17]. Other studies have explored the use of DRL methodology for minimizing idle remote execution time for fog devices through adaptive resource allocation [8]. Furthermore, Dehury *et al.* have provided a CCEI-IoT framework for IoT applications, security, accuracy, and resource management challenges of devices using edge computing and machine learning [6].

In this work, we aim to address the research gap in the field of task offloading and DRL for IoT devices in industrial fog networks. Our proposed framework introduces an intelligent decision-making system to optimize the energy consumption of the overall system. To achieve this, we utilize the concepts of deep reinforcement learning, which allows IoT devices to make informed decisions in real-time Industry 4.0. The goal of this work is to improve the performance of existing offloading strategies and provide a more efficient solution for industrial IoT applications.

2.2 Contribution

In this study, we design a computation offloading framework utilizing the advancements of DRL for efficient task offloading among IIoT devices. The framework considers multiple wireless devices that generate a set of tasks within a specific time frame. Based on the complexity of the tasks and the amount of computational resources required, decisions are made, and tasks are transferred to a remote device for computation. In addition, the framework takes energy consumption rates and the execution delay of IIoT tasks into account to select the most appropriate device for IIoT task computation. Using the DRL technique, the algorithm is able to select the most appropriate device for task offloading based on intelligent decisions.

- First, we define our optimization objective in terms of a weighted energy-delay minimization problem, subject to satisfying the constraints that are imposed by the problem.
- Decisions are then made and tasks are transferred from the industrial control system to a remote computing server for computation. This is based on the complexity of industrial tasks and the amount of computational resources required for executing those tasks.
- With DRL, the framework selects the most appropriate device for task offloading based on intelligent decisions.
- Furthermore, the results of the simulation indicate that our task offloading strategy is more efficient than the existing baseline algorithms based on a number of performance metrics.

3   Network Model

Consider an industrial fog network with three key components as IIoT devices $\mathcal{I}$, fog devices $\mathcal{F}$ and cloud servers $\mathcal{C}$, defined as $\mathcal{I} = \{1, 2, \ldots, I\}$, $\mathcal{F} = \{1, 2, \ldots, F\}$, and $\mathcal{C} = \{1, 2, \ldots, C\}$. In this network, IIoT devices generate a set of independent tasks $\mathcal{T} = \{1, 2, \ldots, T\}$ defined by two tuples $T = \langle T_i^{in}, T_i^{CPU} \rangle$. Now, the challenge is to execute $x \in T$ tasks to $y \in (\mathcal{I} \cup \mathcal{F} \cup \mathcal{C})$ computing devices. Let $L_x = T_x^{in} \times T_x^{CPU}$, $\forall x \in \mathcal{T}$ be the amount of CPU cycles required to execute a task and $\mathscr{D}(x, y)$ denotes the task offloading decision matrix. The values of $\mathscr{D}(x, y)$ can also be defined by a condition as follows.

$$\mathscr{D}(x, y) = \begin{cases} 0 & \text{if } x^{th} \text{ task executed to } t^{th} \text{ IIoT device} \\ 1 & \text{if } x^{th} \text{ task offloaded to } y^{th} \text{ computing device} \end{cases}$$

Local Computing. Since IIoT devices have limited computational capacity, tasks requiring less processing power are retained on IIoT devices. Then, we can define delay and energy consumption on IIoT devices as follows.

$$\mathbb{T}_{xy}^{local} = \frac{\sum_{x=1}^{T} \mathscr{D}(x, y) L_x}{\mathscr{F}_y}, \quad \forall y \in \mathcal{I}, x \in \mathcal{T} \tag{1}$$

$$\mathbb{E}_{xy}^{local} = \mathbb{T}_{xy}^{local} \times \phi^{local}, \quad \forall y \in \mathcal{I}, x \in \mathcal{T} \tag{2}$$

where ϕ^{local} defines the energy rate of IIoT devices.

Uploading IIoT Tasks. Let B_{xy}^{up} represent the transmission bandwidth and Q_y^{up} denote the transmission power of an IIoT device. Then the uploading data rate for a task x can be defined as $\mathbb{R}_{xy}^{up} = B_{xy}^{up} \log_2\left(1 + \frac{Q_y^{up} w_x^{up}}{\xi_y^2}\right)$.

Where w_x^{up} defines the channel gain and ξ signifies the circuit noise of the device. Thus the uploading time and corresponding energy consumption rate are defined as.

$$\mathbb{T}_{xy}^{up} = \frac{\sum_{x=1}^{T} \mathscr{D}(x, y) T_x^{in}}{\mathbb{R}_{xy}^{up}}, \quad \forall y \in (\mathcal{F} \cup \mathcal{C}), x \in \mathcal{T} \tag{3}$$

$$\mathbb{E}_{xy}^{up} = \mathbb{T}_{xy}^{up} \times \phi^{remote}, \quad \forall y \in (\mathcal{F} \cup \mathcal{C}), x \in \mathcal{T} \tag{4}$$

Processing IIoT Tasks. Devices begin executing tasks as soon as they receive them. Now with the given computation frequency and input data size of the computing devices, we can define task execution delay and execution energy consumption as follows.

$$\mathbb{T}_{xy}^{proc} = \frac{\sum_{x=1}^{T} \mathscr{D}(x, y) L_x}{\mathscr{F}_y}, \quad \forall y \in (\mathcal{F} \cup \mathcal{C}), x \in \mathcal{T} \tag{5}$$

$$\mathbb{E}_{xy}^{proc} = \mathbb{T}_{xy}^{proc} \times \phi^{remote}, \quad \forall y \in (\mathcal{F} \cup \mathcal{C}), x \in \mathcal{T} \tag{6}$$

where ϕ^{local} defines the energy rate of remote devices.

Downloading IIoT Tasks. Similar with B_{xy}^{down} and Q_y^{down} we can define the downloading data rate as $\mathbb{R}_{xy}^{down} = B_{xy}^{down} \log_2 \left(1 + \frac{Q_y^{down} w_x^{down}}{\xi_i^2}\right)$.

Now we can derive the downloading time and corresponding energy consumption as follows.

$$\mathbb{T}_{xy}^{down} = \frac{\sum_{x=1}^{T} \mathscr{D}(x,y) T_x^{in}}{\mathbb{R}_{xy}^{down}}, \quad \forall y \in (\mathcal{F} \cup \mathcal{C}), x \in \mathcal{T} \tag{7}$$

$$\mathbb{E}_{xy}^{down} = \mathbb{T}_{xy}^{down} \times \phi^{remote}, \quad \forall y \in (\mathcal{F} \cup \mathcal{C}), x \in \mathcal{T} \tag{8}$$

Thus overall delay and energy consumption to process x^{th} task on y^{th} computing device can be defined as $\mathbb{TE}_{xy}^{total} = \Im\partial\mathbb{T}_{xy}^{total} + \Re\mathbb{E}_{xy}^{total}$,

where, $\mathbb{E}_{xy}^{total}$ combines the IIoT and remote server execution energy, and $\mathbb{T}_{xy}^{total}$ combines the IIoT and remote server execution delay.

Problem Formulation. This section focuses on formulating the problem of intelligent offloading decisions in industrial IoT-fog-cloud architecture, specifically regarding selecting an appropriate fog device or cloud server for offloading IIoT tasks. Mathematically, the objective functions can be formulated as follows,

$$\text{minimize} \quad \sum_{x=1}^{T} \mathbb{TE}_{xy}^{total} \tag{9a}$$

$$\text{subject to} \quad \mathbb{T}_{xy}^{total} \leq \mathbb{T}_y^{max}, \tag{9b}$$

$$\mathscr{F}_x^{CPU} \leq \mathscr{F}_y^{max}, \tag{9c}$$

$$\sum_{x=1}^{T} \sum_{y=1}^{IFC} \mathscr{D}(x,y) \leq |IFC|, \tag{9d}$$

$$\sum_{x=1}^{|T|} \mathscr{D}(x,y) = 1, \tag{9e}$$

$$\mathscr{D}(x,y) \in 0,1, \tag{9f}$$

$$\mathbb{T}_{xy}^{up} \geq 0 \text{ and } \mathbb{T}_{xy}^{down} \geq 0 \tag{9g}$$

Equation (9) defines our objective function for remote execution on various computing devices. Constraints (9b) establishes the upper bounds of task execution delay. Constraint (9c) identifies the task offloading decision upto the number of devices, and Constraint (9d) limits the selection of computing devices. Similarly, Constraint (9f) identifies the binary task offloading the decision process. Finally, Constraint (9g) specifies a positive transmission value.

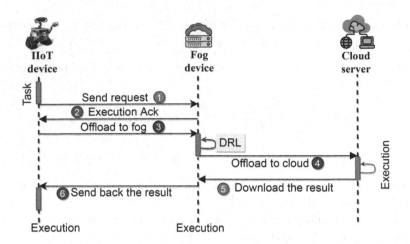

Fig. 2. Workflow of the task offloading strategy.

4   Proposed DRL for Task Offloading

DRL-based computation offloading framework is proposed for IIoT devices in industrial fog networks. The proposed model utilizes the RL approach and a DNN, as shown in Fig. 2. When IIoT devices interact with the environment without an explicit system dynamics model, they act as agents and make optimal decisions. A device part of the IIoT has limited knowledge and experience of its surroundings in the initial phase. As a result, it needs to explore by taking different actions at each offloading state. As the agent gathers more experience through interaction with the environment, it can exploit known information about states while continuing to explore. The state-action function can be updated using the experience tuple of the agent, allowing for continuous improvement in decision-making and energy consumption reduction in the industrial fog network. MDP is a popular technique that can quickly help identify state action and rewards of this environment as follows.

State (S). An overview of the current environment and can be defined by.

$$
\begin{aligned}
\mathsf{S} &= \{s_k = (\mathscr{D}(k),\ \mathscr{F}(k))\} \\
&= \{\mathscr{D}_1(k), \mathscr{D}_2(k), \ldots, \mathscr{D}_x(k), \mathscr{F}_1(k), \mathscr{F}_2(k), \ldots, \mathscr{F}_y(k)\}
\end{aligned} \tag{10}
$$

Action (A). Transitions between states of the environment by the agent and can be defined by.

$$
\mathsf{A} = \{\overline{\mathscr{D}}_k = \{\overline{\mathscr{D}}_1(k), \overline{\mathscr{D}}_2(k), \ldots, \overline{\mathscr{D}}_x(k)\} \mid \overline{\mathscr{D}}_x(k) \in \mathscr{D}^{\max}\} \tag{11}
$$

Algorithm 1: *DRL-Based Task Offloading*

---

1 **INPUT:** $\mathcal{I}$, $\mathcal{F}$, $\mathcal{C}$, T_x^{in}, $\mathscr{D}(x,y)$, Q, w, $\mathscr{F}_y$, and ϕ

2 **OUTPUT:** $\overline{\mathscr{D}^*}$: Best task offloading decision

  1: Initialize state $\mathsf{S} = \{s_k = (\mathscr{D}(k),\ \mathscr{F}(k))\}$;

  2: Initialize action $\mathsf{A} = \{\overline{\mathscr{D}}_k = \{\overline{\mathscr{D}}_1(k), \overline{\mathscr{D}}_2(k), \dots, \overline{\mathscr{D}}_x(k)\}\}$;

  3: Initialize replay memory ζ;

  4: **for** $x = 1$ to $\mathcal{T}$ **do**

  5:    Observe state s_t, execute action a_t, and nest state s_{t+1};

  6:    Calculate reward $\mathsf{R}(s_k, a_k) = \mathcal{J}_{s_k}(\mathscr{D}, \mathscr{F}) - \mathcal{J}_{s_{k+1}}(\mathscr{D}, \mathscr{F})$;

  7:    Store environmental s_k, a_k, r_k, s_{k+1} in replay memory ζ;

  8:    Choose a mini-batch experience from replay memory ζ;

  9:    Calculate $\mathscr{C}^*(s,a) = \max_{\delta} \mathbb{X}\left[r_k + \sum_{b=1}^{\infty} \gamma^b r_{k+b} | s_k = s, a_k = a, \delta\right]$;

10:    Calculate $\mathcal{X}_k = r_k + \gamma \max_{a'} \mathscr{C}\left(s_{k+1}, a'; \partial_{k-1}\right)$;

11:    Update $\mathcal{A}_k(\partial_k) = \mathbb{X}_{s_k, a_k, r_k, s_{k+1} \sim P(.)}\left[\left(\mathcal{X}_k - \mathscr{C}(s_k, a_k; \partial_k)\right)^2\right]$;

12:    Update the environmental experiences;

13:    Make optimal task offloading decision $\overline{\mathscr{D}^*}$;

14: **end for**

Reward (R). Successful states are rewarded by the agent and are defined by. For simple understanding, we use $\mathcal{J}_{s_k}(\mathscr{D}, \mathscr{F}) = \mathcal{J}(\mathscr{D}(k), \mathscr{D}(k)\}$ as the objective function.

$$R(s_k, a_k) = \mathcal{J}_{s_k}(\mathscr{D}, \mathscr{F}) - \mathcal{J}_{s_{k+1}}(\mathscr{D}, \mathscr{F}) \tag{12}$$

To solve the state-action-reward problem, we define the use of a model-free deep learning technique called DRL with a finite amount of replay memory. For a fixed set of action space and state space, we can define the long-term expected environmental reward for the system as follows.

$$\mathscr{C}^*(s, a) = \max_{\delta} \mathbb{X}\left[r_k + \sum_{b=1}^{\infty} \gamma^b r_{k+b} | s_k = s, a_k = a, \delta\right] \tag{13}$$

$$\mathcal{A}_k(\partial_k) = \mathbb{X}_{s_k, a_k, r_k, s_{k+1} \sim P(.)}\left[\left(\mathcal{X}_k - \mathscr{C}(s_k, a_k; \partial_k)\right)^2\right] \tag{14}$$

$$\mathcal{X}_k = r_k + \gamma \max_{a'} \mathscr{C}\left(s_{k+1}, a'; \partial_{k-1}\right) \tag{15}$$

The steps of the proposed DRL strategy are presented in Algorithm 1.

5   Experimental Analysis

This section analyzes the performance of the DRL-based task offloading strategy and compares it to current strategies, such as Random Execution (RE), in which IIoT devices process all sensor-generated tasks randomly without considering remote processing capabilities, heuristic CoT [2] in which IIoT devices assign complete tasks to suitable computing devices for execution, and DRL-based ISD [9] strategies. The proposed framework is compared in terms of energy efficiency and delay minimization in order to find the most efficient solution.

5.1   Simulation Setup

In our simulation, we consider a hybrid IIoT network that consists of a number of industrial fog devices $\mathcal{F}$ and a number of cloud servers $\mathcal{C}$ that run the applications. Depending on the device $y \in (\mathcal{F} \cup \mathcal{C})$, there are varying numbers of IIoT devices $\mathcal{I}$ connected and computing tasks $\mathcal{T}$ to complete. In order to obtain stable results, 1000 timeslots are used during the simulation. Local computation time and computing energy consumption rate are set to their maximum limit for each IIoT device $x \in \mathcal{I}$. A random distribution of 10Kb to 5Mb is assumed for the size of tasks. A 15-second delay threshold is assumed for the local computation workload. This algorithm considers a DNN with an input layer, one output layer and multiple hidden layers.

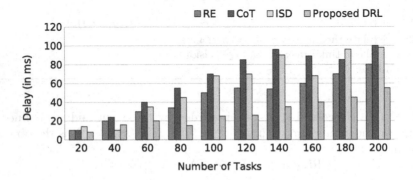

Fig. 3. Delay comparison with existing strategies.

5.2   Delay Analysis

This criterion illustrates the aggregate delay incurred in the execution of all tasks $\mathcal{T}$ to the industrial domain on fog networks. There are several factors contributing to task execution delays. These include $\mathbb{T}_{xy}^{local}$, $\mathscr{D}(x, y)$ and $\mathbb{T}_{xy}^{proc}$. From the formulation of $\mathbb{T}_{xy}^{total}$, it is apparent that the delay calculation is contingent upon the decision variable $\mathscr{D}(x, y)$ and can be optimized by controlling the T_{xy}^{in}. Figure 3 illustrates the delay analysis of our proposed strategy. The objective of this study is to analyze how our proposed strategy compares to existing algorithms in terms of task processing delay. The results indicate that our proposed approach demonstrates superior performance in reducing overall processing delay by 12–14% in comparison to RE, CoT, and ISD strategies. This is attributed to the utilization of a DRL mechanism for the selection of decision-making $\mathscr{D}(x, y)$ and selecting suitable computing devices $\forall y \in (\mathcal{I} \cup \mathcal{F} \cup \mathcal{C})$ with reduced computational complexity, thus resulting in a decrease in overall time complexity.

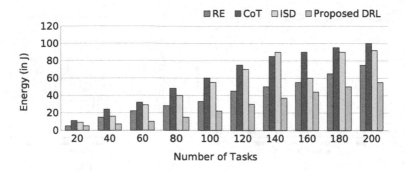

Fig. 4. Energy comparison with existing strategies.

5.3   Energy Analysis

The energy consumption on devices $y \in (\mathcal{I} \cup \mathcal{F} \cup \mathcal{C})$ is a crucial parameter that significantly impacts network performance. Energy consumption is dependent on various factors, such as T_x^{in}, B_{xy}^{up}, B_{xy}^{down}, $\mathcal{D}(xy)$ and $\mathcal{F}_y$, $\forall x \in \mathcal{T}, y \in (\mathcal{I} \cup \mathcal{F} \cup \mathcal{C})$ of the processing devices y. In our proposed strategy, we take into account the energy consumption during $\mathbb{T}_{xy}^{proc}$, $\mathbb{T}_{xy}^{up}$, and $\mathbb{T}_{xy}^{down}$ in successive stages of execution. By controlling $\mathbb{R}_{xy}^{up}$, $\mathbb{R}_{xy}^{down}$, and increasing $\mathcal{F}_{xy}$ of the processing devices x, $\forall x \in \mathcal{T}, y \in (\mathcal{I} \cup \mathcal{F} \cup \mathcal{C})$ it is possible to optimize industrial energy consumption rates. We have demonstrated the energy performance improvement achieved by our DRL-based task offloading strategy through the results presented in Fig. 4. Our proposed task offloading technique has achieved a performance improvement of 13–18% compared to standard RE, CoT, and ISD strategies while satisfying multiple network- and system-based constraints.

5.4   Performance Analysis of DRL

To evaluate the performance and show the effectiveness of the DRL-based task offloading method, we employ mini-batch samples with a batch size of 16, a learning rate of 0.0001, and a replay memory ζ of 5000. In order to obtain a stable output, the model undergoes 1000 training epochs discounted by γ. The scratch-made DNN architecture employed in this study includes 1 output layer, 3 hidden layers, and 1 input layer for estimating the next most appropriate action. We utilize dense layers without dropout in DNN and ReLU activation to extract features and capture dynamic network parameters. Additionally, we apply gradient descent-based optimizers, the Boltzmann Q-policy, and the optimal set of hyperparameters from the available parameter set. To train the industrial fog network, initial network weights are randomly picked, and high targets are set. The performance of the DRL model and subsequent decision and long-term rewards are shown in Fig. 5. Our proposed strategy maintains high rewards over time using a finite number of state and action spaces.

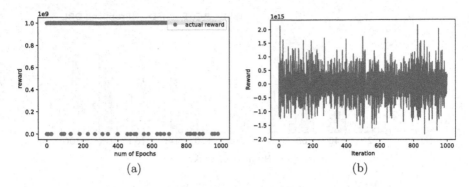

Fig. 5. Performance analysis of DNN, (a) Actual reward while making decisions. (b) Cumulative reward while training the network.

6   Conclusion

In this study, we present an intelligent computation offloading framework for IoT devices in industrial fog networks. Our focus is on providing a decision-making system for time-sensitive and critical industrial applications. We propose a hierarchical computation offloading framework that includes a set of fog devices and cloud servers. The framework is based on a constraint-oriented optimization problem, where the objective function is the time average constraint among computing devices. To solve this problem, we propose a DRL-aware computation offloading strategy for IoT devices to make intelligent decisions in the industrial environment. Additionally, we use replay memory to enhance the decision-making system. Simulation results on various network parameters confirm the efficacy of our proposed computation offloading strategy in comparison to greedy-based algorithms.

References

1. Aazam, M., Harras, K.A., Zeadally, S.: Fog computing for 5G tactile industrial internet of things: QoE-aware resource allocation model. IEEE Trans. Industr. Inf. **15**(5), 3085–3092 (2019)
2. Aazam, M., Islam, S.U., Lone, S.T., Abbas, A.: Cloud of things (CoT): cloud-fog-IoT task offloading for sustainable internet of things. IEEE Trans. Sustain. Comput. **7**(1), 87–98 (2020)
3. Aazam, M., Zeadally, S., Harras, K.A.: Deploying fog computing in industrial internet of things and industry 4.0. IEEE Trans. Industr. Inform. **14**(10), 4674–4682 (2018)
4. Adhikari, M., Hazra, A., Menon, V.G., Chaurasia, B.K., Mumtaz, S.: A roadmap of next-generation wireless technology for 6G-enabled vehicular networks. IEEE Internet Things Mag. **4**(4), 79–85 (2021). https://doi.org/10.1109/IOTM.001.2100075
5. Dehury, C., Srirama, S.N., Donta, P.K., Dustdar, S.: Securing clustered edge intelligence with blockchain. IEEE Consum. Electron. Mag. (2022)

6. Dehury, C.K., Donta, P.K., Dustdar, S., Srirama, S.N.: CCEI-IoT: clustered and cohesive edge intelligence in internet of things. In: 2022 IEEE International Conference on Edge Computing and Communications (EDGE), pp. 33–40. IEEE (2022)

7. Guo, M., Mukherjee, M., Liang, G., Zhang, J.: Computation offloading for machine learning in industrial environments. In: IECON 2020 The 46th Annual Conference of the IEEE Industrial Electronics Society, pp. 4465–4470. IEEE (2020)

8. Hazra, A., Adhikari, M., Amgoth, T.: Dynamic service deployment strategy using reinforcement learning in edge networks. In: 2022 International Conference on Computing, Communication, Security and Intelligent Systems (IC3SIS), pp. 1–6. IEEE (2022)

9. Hazra, A., Adhikari, M., Amgoth, T., Srirama, S.N.: Intelligent service deployment policy for next-generation industrial edge networks. IEEE Trans. Netw. Sci. Eng. **9**(5), 3057–3066 (2022). https://doi.org/10.1109/TNSE.2021.3122178

10. Hazra, A., Adhikari, M., Nandy, S., Doulani, K., Menon, V.G.: Federated-learning-aided next-generation edge networks for intelligent services. IEEE Network **36**(3), 56–64 (2022). https://doi.org/10.1109/MNET.007.2100549

11. Misra, S., Roy, C., Sauter, T., Mukherjee, A., Maiti, J.: Industrial internet of things for safety management applications: a survey. IEEE Access **10**, 83415–83439 (2022)

12. Misra, S., Tiwari, M., Ojha, T., Raj, Y.: PANDA: preference-based bandwidth allocation in fog-enabled internet of underground-mine things. IEEE Syst. J. **15**(4), 5144–5151 (2021). https://doi.org/10.1109/JSYST.2021.3086150

13. Mukherjee, M., et al.: Latency-driven parallel task data offloading in fog computing networks for industrial applications. IEEE Trans. Industr. Inf. **16**(9), 6050–6058 (2019)

14. Mukherjee, M., Kumar, V., Lat, A., Guo, M., Matam, R., Lv, Y.: Distributed deep learning-based task offloading for UAV-enabled mobile edge computing. In: IEEE INFOCOM 2020-IEEE Conference on Computer Communications Workshops (INFOCOM WKSHPS), pp. 1208–1212. IEEE (2020)

15. Rathee, G., Ahmad, F., Iqbal, R., Mukherjee, M.: Cognitive automation for smart decision-making in industrial internet of things. IEEE Trans. Industr. Inf. **17**(3), 2152–2159 (2020)

16. Sarkar, I., Adhikari, M., Kumar, N., Kumar, S.: Dynamic task placement for deadline-aware IoT applications in federated fog networks. IEEE Internet Things J. **9**(2), 1469–1478 (2022). https://doi.org/10.1109/JIOT.2021.3088227

17. Sarkar, I., Adhikari, M., Kumar, S., Menon, V.G.: Deep reinforcement learning for intelligent service provisioning in software-defined industrial fog networks. IEEE Internet Things J. **9**(18), 16953–16961 (2022). https://doi.org/10.1109/JIOT.2022.3142079

18. Singh, S., Pal, S.: SDTS: security driven task scheduling algorithm for real-time applications using fog computing. IETE J. Res. 1–20 (2021)

19. Singh, S., Tripathi, S.: A security-driven scheduling model for delay-sensitive tasks in fog networks. In: Nicopolitidis, P., Misra, S., Yang, L.T., Zeigler, B., Ning, Z. (eds.) Advances in Computing, Informatics, Networking and Cybersecurity. LNNS, vol. 289, pp. 781–807. Springer, Cham (2022). https://doi.org/10.1007/978-3-030-87049-2_29

20. Srirama, S.N., Vemuri, D.: CANTO: an actor model-based distributed fog framework supporting neural networks training in IoT applications. Comput. Commun. **199**, 1–9 (2023)

Security Enhancement of Content in Fog Environment

Ayushi$^{(\boxtimes)}$ and Manisha Agarwal

Banasthali Vidyapith, Banasthali, Radha Kishnpura, Rajasthan, India
`Ayushi.hce@gmail.com`

Abstract. Research has discovered that typical security measures only provide a limited level of protection for cloud-based data. However, neither dependability nor security can be guaranteed at this time. This study gives a review of previous research that has been done. In addition, a more secure strategy has been proposed in this study. In the future, research activities may divide data as it is being sent. Because of this, there will be a lower likelihood of any security being breached. A soon-to-be-developed encryption technique could have the ability to prevent data from being lost or stolen. Not only would this give security at the application layer, but it would also provide security at the session level. It is not possible to send all of the data on a single channel. The data would be broken up into two parts: one for transmission through the fog, and another via the cloud. In the light of this, forthcoming work will make use of an integrated fog-based strategy for improving the security of the cloud. Because existing security mechanisms suffer from some drawbacks, there is a pressing need for such a system.

Keywords: Cloud Computing · Fog Computing · Security Enhancement · Data Splitting · Encryption

1 Introduction

In this section Cloud Computing (CC) and its application areas are discussed. Cloud computing service model Software as services. Cloud Integrators could play an important part in deciding the right cloud way for every company. They choose private, public, and hybrid clouds according to the requirement of the organization. However, Cloud computing is providing the benefit of better storage and economical solution.

2 Cloud Computing

The use of distant servers and apps accessed over the internet to manage a user's data processing needs is considered "cloud computing," which is self-explanatory given the name of the concept. There is a wide variety of services available as examples of cloud computing. Cloud computing may be segmented into the following three basic categories: SaaS, PaaS, and IaaS.

R. K. Challa et al. (Eds.): ICAIoT 2023, CCIS 1929, pp. 176–188, 2024.
https://doi.org/10.1007/978-3-031-48774-3_12

SaaS. To use a SaaS, you just go to the service's website instead of downloading and installing software on your computer. Some examples include (Fig. 1):

- Applications developed by Google, such as Drive or Calendar
- Slack, empowers its users to collaborate and communicate with one another
- Square, is an e-commerce company that handles payments.

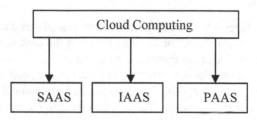

Fig. 1. Services of Cloud Computing

IaaS or Infrastructure as a Service: The cloud, servers, storage, networking, and security are all part of the IaaS offering. Some examples include:

- Dropbox is an online service for sharing and syncing large files and working together in real-time.
- Hosting, backup, and disaster recovery are just a few of the many services that can be found on Microsoft Azure, a cloud computing platform.
- Rackspace is a company that provides services in the areas of data, security, and infrastructure.

PaaS: Computing platforms such as web servers are all provided by PaaS. To cite a few instances:

- GAE and Heroku, which are platforms that enable developers to build and host mobile applications

Serverless Computing: Sometimes shortened to "Server less" this term refers to the use of a cloud-based server without the need for any additional management or maintenance on the part of the user. When compared to housing servers in-house, these options provide more flexibility, less frequent maintenance, and typically lower costs.

3   Challenges of Cloud Computing

One may argue that the benefits of CC are masking the drawbacks. As a result, the model is growing rapidly. The most often encountered problems are as follows:

1. Information safety.
2. Information revival and accessibility.
3. Administration capability
4. Malicious Insiders
5. Data Loss Leakage
6. Taking Control of Traffic
7. Profile of Unknown Risk Unknown

3.1 Fog Computing

Fog computing is a dispersed computing architecture that places data, processing, storage, and applications in the "fog," or the area between the data source and the cloud. Both fog computing and edge computing bring some of the cloud computing benefits and capabilities to locations closer to where data is being generated and used. The terms "fog computing" and "edge computing" are sometimes interchanged because of the parallels that exist between the two ideas. This is often done to increase output, but it may also be done to ensure safety and conformity with regulations.

Like fog, which tends to gather near the system's periphery, the word "fog" refers to a low-lying cloud in the atmosphere. Many people identify the phrase with Cisco, and legend has it that Ginny Nichols, Cisco's product line manager, came up with the name. The term "Cisco Fog Computing" is a trademark, although the concept of "fog computing" is available to everyone.

3.2 Benefits of Fog Computing

Fog computing, like any other technology, offers both advantages and disadvantages. As an example of the benefits that Fog Computing offers:

- **Bandwidth Conservation**. The quantity of data that is uploaded to the cloud is reduced while using fog computing, which results in cost savings related to bandwidth expenses.
- **Improved Response Time.** The first processing of the data takes place close to the data itself, which reduces latency and boosts responsiveness. The intended reaction time is on the order of milliseconds, which enables the processing of data in a timeframe that is very close to real-time.
- **Network-Agnostic**. Even though fog computing typically deploys computational resources at the LAN level, one could argue that the network is still an essential component of the design of fog computing. This is because edge computing deploys computational resources at the device level. On other hand, fog computing doesn't care if the underlying network is wired, wireless, or even 5G; it simply wants to finish the task at hand as quickly as possible.

3.3 Disadvantages of Fog Computing

Naturally, there are drawbacks to fog computing as well.

- **Physical location.** Fog computing negates some of the "anytime/anywhere" advantages of CC due to its reliance on a specific physical location.

- **Potential security issues.** In certain contexts, the use of fog computing might be compromised by security flaws such as IP address spoofing and MitM attacks.
- **Startup costs.** When implementing a solution like fog computing, which draws from both the edge and the cloud, you'll need to budget for the necessary hardware.
- **Ambiguous concept.** The term "fog computing" has been in use for a while, but its precise meaning remains unclear since different vendors define it in different ways.

3.4 Cloud vs Fog and Edge Computing

The most important difference between cloud computing, fog computing, and edge computing is at which place, at which time, and how the information collected from endpoint devices is processed and stored. When compared to traditional storage media, cloud storage is distinguished by its remote location from user devices. This explains why there is so much lag, how expensive bandwidth is, and how demanding the network needs to be. In contrast, the cloud is a robust worldwide solution that can grow efficiently by enlisting more computer resources and server space to deal with ever-increasing data loads.

Fog is a layer between the cloud and the edge that offers services from both. It is dependent on and interacts directly with, the cloud to distribute data that does not need processing locally. Meanwhile, fog is moved to the perimeter. To do analytics in real-time and react rapidly to events, it may tap into local processing and storage facilities if required.

Edge has the shortest possible latency and the quickest reaction time since it is the node closest to the end devices. Data processing and storage may be handled locally at the device, application, or edge gateway level using this method. It features a decentralized structure with nodes at the edges processing data separately. This is the main distinction between fog at the edge and fog inside a network.

4 Literature Review

There have been many different kinds of research conducted in connection with fog computing.

Abbasi et al. demonstrated problems with security, solutions to those problems, and robust practices for fog computing. The services offered by cloud computing have been expanded by Fog Computing. It takes on some of the qualities of cloud computing as a result of its inheritance. According to what they said, fog computing also has several distinguishing characteristics [1].

Abdulqadir et al. explained the transition from the cloud to the edge. High latency, restricted capacity, and network failure are just a few of the issues that the rapid expansion of IoT technology presents to the traditional centralized cloud computing model. By bringing the cloud closer to IoT devices, cloud and fog computing aim to solve these issues. In contrast to transmitting data to the cloud, IoT devices may be processed and stored locally with the help of cloud and fog computing. When used in tandem with the cloud, Cloud and Fog allow for faster responses and increased productivity. The IoT needs to provide dependable and stable resources to a wide variety of clients, and cloud and fog computing should be seen as the most secure method to achieve this goal.

This essay highlights the benefits and complications of deployment while discussing the newest developments in cloud and Fog computing and their confluence with the Internet of Things [2].

Abubaker et al. discussed a model for privacy-preserving fog computing for computing on the cloud on a short-term basis. In these modern times, fog computing has shown to be of great assistance. The problems inherent in the paradigm of CC have led to the development of a solution known as FC. It is useful to overcome a tremendous quantity of traffic, as many people have discovered. The majority of the time, this traffic is brought on by an abundance of IOT systems. Regularly, these gadgets are fastened to the network [3].

Bhavani et al. researched the cloud computing system's provisioning method for resource allocation. The concept of cloud computing has been taken into consideration as a paradigm. It makes possible a network that may be accessed on demand and conveniently. The purpose of this is to get access to the centralized pool of programmable computing devices. Networks, servers, apps, and services might be used as an example [4].

D. Bermbach et al. focused on the analysis of fog computing from a research standpoint. Modern apps were frequently hosted on the cloud because of their seemingly endless computing power, elasticity, and straightforward pricing structure. This has very high access latency for end users but was highly handy for developers. However, low latency access is essential for future application areas like the Internet of Things, autonomous driving, and future 5G mobile applications. This is often done by relocating computing to the network's edge. While Fog Computing has shown promise as a deployment platform, it has yet to achieve mainstream acceptance. They believe that a more consistent application of the concept of service-oriented computing to the services provided by fog infrastructure might assist bring about this transition. This study provides a thorough explanation of Fog Computing based on this impetus, addresses the primary barriers to Fog Computing acceptance and generates new areas for research [5].

Dolui et al. explores the differences between various edge computing implementations such as cloudlet computing, mobile edge computing, and fog computing [6].

Firdhous et al. predicted that it will become the standard for cloud computing in the future. Along with the use and scope of a cloud-based system, the notion of fog was presented in this study [7].

Georgescu et al. have published research on why businesses need cloud computing. They proved that cloud computing is becoming one of the most important business trends overall, not just for IT companies [8].

Gorelik researched Cloud Computing Models. Information Technology is a basic weak point of organizations in terms of cost and management and has experienced dramatic changes within the past decades. The agile IT processes have permuted the progress of innovative technology as well as business models. In the present time, cloud computing provides companies with more choices [9].

Guan et al. (2018) provided information on the risks associated with fog computing. They stressed the risks of using a fog-based technology in a cloud setting [10].

Malkowski et al. outlined both the challenges and opportunities presented by consolidation. This results in significant use of resources. It is connected to non-monotonic changes in response time, which occurs when dealing with n-tier applications. It has been suggested that an in-depth investigative study of the efficiency of consolidated n-tier applications under high usage be conducted. It has overcome the difficulty by using procedures that are replicable [11].

Masarweh et al. introduced the broker management system for real time congestion control of the fog, cloud, and IoT Environment. Fog computing is one approach suggested to address these issues; it brings the cloud closer to IoT devices. Because of the large number of requests to the cloud coming from the fog broker layer, the proposed system can fulfill the IoT's Quality of Service (QoS) criteria as specified by the service-level agreement. They also provide a means for delivering requests from fog brokers and cloud users to the most appropriate cloud resources, which are used by the cloud service broker. This proposed method takes its cue from Cisco's Weighted Fair Queuing (WFQ) mechanism and tries to make it easier to handle congestion from the cloud service broker's perspective. After testing the system with Fog Sim and Cloud Sim, the authors found that it increased IoT QoS compliance and prevented cloud SLA breaches [12].

Pazowski et al. performed an analysis of the most forward-thinking IS/IT presenting ideals in cloud computing. The investigation's overarching objective is to illustrate the notion of cloud computing. It also defines the word, as well as its primary service and presentation types. Comparing the traditional ways of organizing and presenting information systems and information technology in projects with cloud computing is the goal of the writers [13].

Sareen discussed the types of cloud computing, it's architecture, applications, and concerns. Computing in the cloud is a relatively new idea in the field of information technology that enables businesses or individual users to spread many services flawlessly and cost-effectively. In addition to this, it examines the similarities and differences between cloud computing and grid computing applications [14].

Shenoy et al. elucidated the concept of fog computing. Cloud service providers are likely to face the most significant challenge in the shape of information security threats posed by users. This is because the amount of attempts to steal information has been steadily rising over recent years [15].

Unnisa et al. provided\ an introduction to cloud and fog computing. This article provides a broad overview of computing in the cloud and computing in the fog, comparing and contrasting the two. This enables us to select which computer platform gives the largest analysis of service delivery most expediently and straightforwardly possible. Fog computing, on the other hand, which employs a new set of protocols and standards, is more stable and less likely to fail than cloud computing, which is reliant on the internet and hence more prone to failure. Because of this, the level of security offered by fog computing is far higher than that of cloud computing in the context of this situation. In this study, we investigate differences between cloud, & edge computing from some angles. A computer system is that functions at the "edge" of a network to deliver faster responses to requests by reducing the amount of latency that is experienced. To speed up responses and reduce data transfer costs, "edge computing" moves data processing and storage closer to the point of use [16].

Yashpal et al. about how to beef up security to combat both brute force and timing attacks in networked environments. It strengthens the security mechanism so that it can better defend against assaults on the application layer [17].

Verma et al. discussed the structural design of a load-balancing device with fog computing. In this study, the advantages of load balancing in cloud environments for achieving high availability and achieving zero downtime were investigated [18].

Zhou et al. linked to Fog Computing, introduced Hierarchic Secure Cloud Storage Scheme, as the approach of cloud computing is expanding rapidly. The use of cloud computing provided evidence for a claim. The reality is that it has been regarded as the principal controller all along. There are some different approaches to cloud computing. That is something that results from using cloud computing [19].

5 Problem Statement

The cloud's security can't be guaranteed by using the same tried-and-true methods that have always been used on other networks.

- Transmission of data often occurs via the protocol-specific port, such as 21 for FTP or 80 for HTTP. The success rate of an assault grows as the number of preset ports grows in usage.
- The conventional encryption methods failed to prevent data loss.
- Only the application layer was protected by the Tradition system. It has been disregarded that the session layer has any security at all.
- All information is sent over a single channel. It may have been dangerous if the data were decrypted by an unauthorized party. As a result, cryptanalysis can decipher it with relative ease.

6 Proposed Model

Here, an IP filter has been deployed to prevent packets from being sent between the server and the client without proper authentication. If the packet is correct, then the improved AES ENCRYPTION will function. A potential data transmission process flow is shown below (Fig. 2).

6.1 At Sending End

1) Consider the text
2) Concern Round_Key and set as count = 1
3) if the count is less than i-1 (where i is iteration)
   (a) Procedure sub_byte
   (b) Make Shift_row
   (c) Merge col
   (d) count = count + 1;
4) otherwise
   a) procedure sub_byte
   b) shift_row
   c) concern round_key
5) Cipher_text could be generated

6.2 At the Receiver End

1) Consider Cipher_text
2) Concern Round_Key and set count = 1
3) if the count is less than i-1
   a) procedure contrary shift_row.work
   b) make inverse sub_byte
   c) contrary merge col
   d) count = count + 1;
4) otherwise
   a) contrary shift rows
   b) contrary sub byte
   c) concern round key
5) text could be generated

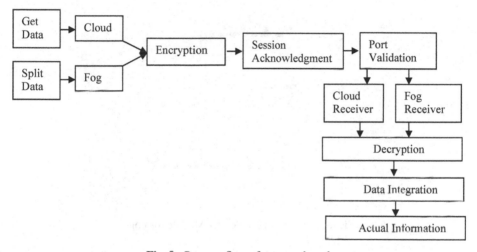

Fig. 2. Process flow of proposed work

7 Result and Discussion

The process and outcome of the planned work have been outlined. Data would be divided into two files, one for the cloud and one for the fog, with the help of the FILE splitter. The file with the provided name and authentication code has been split in two and scattered throughout the cloud and fog. Because of this, communication is safer and more dependable. Data is sent to the server using a file-sending interface. Include the security token and AES CODE here, as well as the user ID, password, port, IP address, and path to the file. A graphical user interface for a server would upload information to the server. Enter the port number, AES code, and path to the file that will contain the security token here. During transmission, the file contents are encoded as cipher text.

Incomprehensible cipher text has been used. Someone trying to hack such data wouldn't be able to make sense of it. The information would be sent from the cloud to the user through the transmitter module. Using an authentication code, the information would be encrypted. The recipient's port and the sender's port would be identical. Here, you would enter the final user's IP address. The following diagram illustrates the blueprint for the cloud data transmitter module. There are three separate text boxes. The sender's data would be encrypted using the port number entered in the first input box, the server's IP address entered in the second, and the authentication code entered in the third.

7.1 Implementation of Fog

Data from Fog would be sent to the user through the transmitter module. Using an authentication code, the information would be encrypted. The recipient's port and the sender's port would be identical. Here, you would enter the final user's IP address. Design of the data sender module for Fog has been represented in Fig. 3.

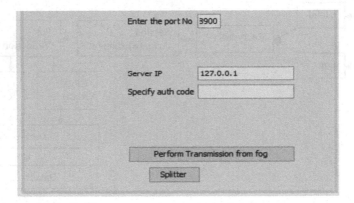

Fig. 3. Design of Data Sender Module for Fog

End user module has been divided into three parts as follows (Fig. 4).

- **Ready to Receive from the Fog.** The fog may now enter, and its data can be collected in this part. Each end should have access to the same port.
- **Ready to Receive from Cloud.** This part unlocks the cloud and accepts data collection and transmission from the cloud. Each end should have access to the same port.
- **Merge and Decode.** The information received would be combined here and decoded using the authentication key.

Table 1 is presenting the packet dropping in case of different number of packets for traditional and proposed work. It has been observed that proposed work is providing less packet dropping as compared to traditional work (Fig. 5).

Table 2 is presenting the accuracy in case of different number of packets for traditional and proposed work. It has been observed that proposed work is providing more than 93% accuracy whereas traditional work has provided accuracy below 91% (Fig. 6).

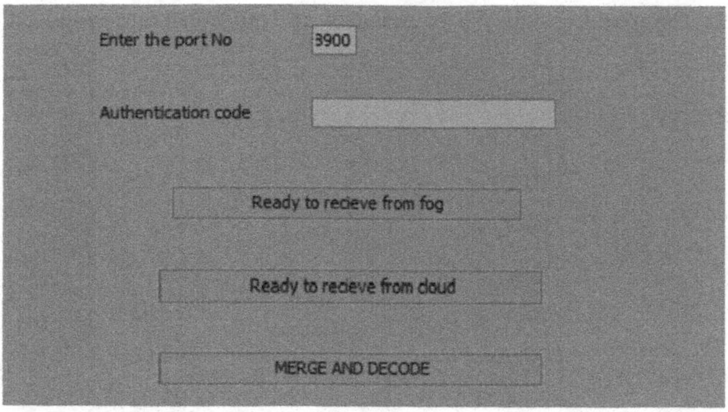

Enter the port No 3900

Authentication code

Ready to recieve from fog

Ready to recieve from cloud

MERGE AND DECODE

Fig. 4. End User Module

Table 1. Comparative Analysis of Packet Dropping In Traditional & Proposed Work

Packets	Traditional	Proposed
100	4	2
200	8	2
300	9	5
400	11	6
500	13	6
600	17	7
700	29	10
800	38	19

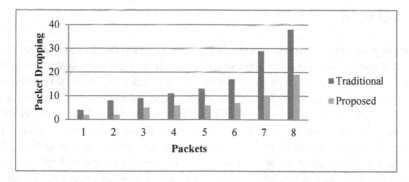

Fig. 5. Comparative analysis of packet dropping in traditional and proposed work

Table 2. Comparative Analysis of Accuracy in Traditional and Proposed Work

Packets	Traditional	Proposed
100	90.54%	93.48%
200	90.20%	93.55%
300	90.67%	93.11%
400	90.80%	93.76%
500	90.37%	93.43%
600	90.90%	93.60%
700	90.43%	93.21%
800	90.27%	93.80%

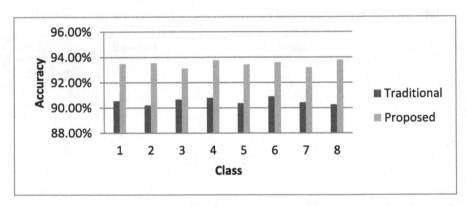

Fig. 6. Comparative analysis of Accuracy in traditional and proposed work

8   Conclusion

The rate of progress in digital transformations has quickened thanks to cloud computing and related computing paradigms. Industries, smart cities, healthcare systems, academic institutions, factories, and even governments are all benefiting from this technology. Some different computer paradigms have been suggested by researchers. This article provides a brief introduction to Cloud and Fog computing, two of the most recent and promising developments in the field of computing. Meanwhile, we've given a brief introduction to EC, dew computing, mobile CC, and mobile edge computing. When considering a wide variety of applications, fog-assisted cloud computing emerges as the most practical and dependable option. One major distinction between edge and FC is the location of network nodes. Research on cloud computing aided by fog is just getting started. Thus, it is essential to do a great deal of research in this field. Some of the challenges of Fog-assisted cloud computing are discussed in this research. Data at the application layer is protected from both active and passive attacks due to the suggested solution. Studies have compared the suggested approach to the current security paradigm. It has been shown that the likelihood of packet loss is lower in the suggested work compared to the conventional method. Traditional security measures have been found wanting. Data security has been guaranteed in the proposed system by the use of a sophisticated cryptographic process that splits and encrypts data. There are lesser chances that packets will be lost or backed up under this arrangement. In this study, we investigate both active and passive attacks to secure the network on several levels. The suggested technique provides security for packets by breaking them up into smaller pieces. As a result of the shortcomings of these older security methods, a new security system is needed. The simulation came to the conclusion that the suggested work results in roughly half the amount of packet losing compared to the standard work. In addition, the suggested job provides an accuracy of more than 93%, while the accuracy supplied by the old work was less than 91%.

9 Future Scope

In earlier research work, there was simply protection for the data at the application layer. The proposed work includes a provision for the packet's security. It became necessary to create an entirely new security mechanism as a result of the limitations of the ones that were already in place. Because of this, the likelihood of decryption occurring without authentication is significantly decreased. To protect against an attacker coming from a separate network, it is necessary to put security measures in place. The data has been broken up into various portions to establish a trustworthy transmission method with the help of the suggested work. Because of this technique, the security system would be impervious to any assaults carried out by hackers or crackers.

References

1. Abbasi, B.Z., Shah, M.A.: Fog computing: security issues, solutions and robust practice. In: 23rd International Conference on Automation and Computing (ICAC) (2017). https://doi.org/10.23919/iconac.2017.8082079
2. Abdulqadir, H.R., et al.: A study of moving from cloud computing to fog computing. Qubahan Acad. J. 1(2), 60–70 (2021). https://doi.org/10.48161/qaj.v1n2a49
3. Abubaker, N., Dervishi, L., Ayday, E.: Privacy-preserving fog computing paradigm. In: IEEE Conference on Communications and Network Security (CNS) (2017). https://doi.org/10.1109/cns.2017.8228709
4. Nagesh, B.: Resource provisioning techniques in cloud computing environment-a survey, IJRCCT 3(3), 395–401 (2014)
5. Bermbach, D., et al.: A research perspective on fog computing. In: Braubach, L., Murillo, J.M., Kaviani, N., Lama, M., Burgueño, L., Moha, N., Oriol, M. (eds.) Service-Oriented Computing – ICSOC 2017 Workshops: ASOCA, ISyCC, WESOACS, and Satellite Events, Málaga, Spain, November 13–16, 2017, Revised Selected Papers, pp. 198–210. Springer, Cham (2018). https://doi.org/10.1007/978-3-319-91764-1_16
6. Dolui, K., Datta, S.K.: Comparison of edge computing implementations: fog computing, cloudlet and mobile edge computing. In: 2017 Global Internet of Things Summit (GIoTS) (2017)

7. Firdhous, M., Ghazali, O., Hassan, S.: Fog computing: will it be the future of cloud computing? In: 3rd International Conference on Informatics & Applications (ICIA2014), October, 8–15 (2017)
8. Georgescu, M., Matei, M.: The value of cloud computing in the business environment. The USV Annals of Economics and Public Administration. Stefan cel Mare University of Suceava. Romania. Faculty of Economics and Public Administration, pp 222–228 (2013)
9. Gorelik, E.: Cloud computing models, M. Sc. Thesis, pp 34–38 (2013)
10. Guan, Y., Shao, J., Wei, G., Xie, M.: Data security and privacy in fog computing. IEEE Network **32**(5), 106–111 (2018). https://doi.org/10.1109/MNET.2018.1700250
11. Malkowski, S., et al.: Challenges and opportunities in consolidation at high resource utilization: non-monotonic response time variations in n-tier applications. In: 2012 IEEE Fifth International Conference on Cloud Computing (CLOUD). IEEE (2012). https://doi.org/10.1109/cloud.2012.99
12. Masarweh, M.A., Alwada'n, T., Afandi, W.: Fog computing, cloud computing and IoT environment: advanced broker management system. J. Sens. Actuator Networks **11**(4), 84 (2022). https://doi.org/10.3390/jsan11040084
13. Pazowski, P.: Cloud computing – a case study for the new ideal of the IS/IT implementation. In: International Conference on Management, Knowledge, and Learning, Zadar, Croatia, pp. 855–862 (2013)
14. Sareen, P.: Cloud computing: types, architecture, applications, concerns, virtualization and role of it governance in cloud. Int. J. Adv. Res. Comput. Sci. Software Eng. **3**(3), 2277 (2013). www.ijarcsse.com
15. Shenoy, K., Bhokare, P., Pai, U.: FOG computing future of cloud computing. Int. J. Sci. Res. **4**(6), 55–56 (2015)
16. Unnisa, A.: A study on review and analysis of cloud. Fog Edge Comput. Platforms. **5**(11), 325–331 (2020)
17. Stiawan, D., Idris, M.Y., Malik, R.F., Nurmaini, S., Alsharif, N., Budiarto, R.: Investigating brute force attack patterns in IoT network. J. Electr. Comput. Eng. **2019**, 1–13 (2019). https://doi.org/10.1155/2019/4568368
18. Verma, M., Bhardawaj, N., Yadav, A.K.: An architecture for load balancing techniques for fog computing environment. Int. J. Comput. Sci. Commun. **6**(2), 269–274 (2015). https://doi.org/10.090592/IJCSC.2015.627
19. Zhou, J., Wang, T., Bhuiyan, M.Z.A., Liu, A.: A hierarchic secure cloud storage scheme based on fog computing. In: 2017 IEEE 15th International Conference on Dependable, Autonomic and Secure Computing (2017).https://doi.org/10.1109/dasc-picom-datacom-cyberscitec.2017.90

Local Database Connectivity and UI Design for the Smart Automated Cooker

Varsha Goyal$^{(\boxtimes)}$, Kavita Sharma⬤, and S. R. N. Reddy

Compute Science and Engineering, Indira Gandhi Delhi Technical University for Women, Opposite to James Church, Kashmere Gate, New Delhi 110006, India
goyalvarsha107@gmail.com, srnreddy@igdtuw.ac.in

Abstract. In today's technologically advanced world with modern wireless systems, everyone is in the need of smart and intelligent tools. As a result, smart kitchen appliances are necessary for comfortable and easy lifestyle. For this purpose, IGDTUW has designed and developed a smart, automated cooker in the stimulated environment. To develop the stimulated automated cooker in real environment, there is a requirement of GUI and local database creation for the manual control of the system. Thus, the objective of this paper is to design and develop the GUI with local database connectivity for the system. This paper also discuss, the use of PyQt5 in designing, developing, and connecting the user login credentials to MySQL including storage of data. The well known Python wrapper PyQt5 is used for GUI design and development. It create modern, user friendly, cross-platform GUI for the system while retaining python flexibility. Highly reliable MySQL is used to store user authentication and recipe data locally on the system. Using PyQt5 and Mysql, Graphical User Interface (GUI) is successfully developed which is used to record user credentials via the login framework.

Keywords: smart cooker · PyQt5 · MySQL · smart kitchen

1 Introduction

In the new age of technology, IoT devices, applications can respond intelligently. With the aid of embedded technology, all big or small activity in the society are expanded with the use of Internet of Things to increase human comfort levels in healthcare, agriculture, communication, manufacturing, transportation, and smart homes. The use of scientific knowledge to affect and manipulate the human environment is a practical goal of the human life. Various applications are there for different purposes, i.e. the application [1] is intended to keep track of product expiration dates and do a cost analysis. It necessitate GUI design for framework. PyQt5 is a Python wrapper for the well-known Qt desktop application framework. PyQt5 is also used in [2] for user interface design for their Tele-operated Vehicles. PyQt5 [3] is used to design the framework for the Smart Tab device.

K. Sharma and S. R. N. Reddy—Contributed equally to this work.

© The Author(s), under exclusive license to Springer Nature Switzerland AG 2024
R. K. Challa et al. (Eds.): ICAIoT 2023, CCIS 1929, pp. 189–200, 2024.
https://doi.org/10.1007/978-3-031-48774-3_13

Today, the number of IoT-connected devices in smart homes is rapidly increasing. The house now has smart appliances such as lighting systems and a smart kitchen as a result of the smart devices installed inside. It takes a long time to create a dish by hand and go through all the steps. Due to everyone in the family being so busy at work and in the office, there are relatively few hours in the day for the kitchen in the modern era. The elderly and little children who were left at home are now also anticipating someone's arrival for meal preparation. The hectic schedule of modern society, with less time to spend in the kitchen, inspired the creation of smart automated cooker. Hence, in IGDTUW Lab, a smart compact cooker is designed to serve the purpose of smart cooking system. It is a more sophisticated and intelligent cooking appliance called a "smart compact cooker", to advance the smart kitchen.

Consequently, this study's objective is to design and develop a user interface login page and recipe details framework using PyQt5 technology and local database connection with the MySQL database on raspberry pi for manual control of smart compact cooker. This paper's key contribution is:

- To design and develop GUI for login form, inserting recipe details using PyQt5 framework.
- To integrate the GUI with Mysql local database in order to save the user's credentials for future use.

The rest of the paper is organized into various sections. Section 1 is introduction, Sect. 2 is the literature survey of tools for user interface design and database creation. In Sect. 3, GUI Design and database connectivity is discussed while Sect. 4 is results and discussion while Sect. 5 discuss the challenges faced in GUI design and database creation. Future scope is discussed in Sect. 6 and Sect. 7 conclusion the overall paper.

2   Literature Survey

2.1   PyQt5

Python provides numerous options for developing GUI applications i.e. PyQt, Thinkercad. PyQt5 [3–5] cross-platform, GUI toolkit for python developers. Python bindings are included in the cross-platform GUI toolkit PyQt5 for Qt v5 framework. Because of the tools and simplicity provided by this library, it is possible to create an interactive desktop application with great ease. In [1] and [3], authors have used PyQt5 for UI framework of their application because of the minimalistic nature and simplicity of Qt designer provided by PyQt5 framework. Table 1 highlights on the analysis of PyQt5 that exists in scientific literature. There are two components that make up a GUI application:

- The front end.
- The back end.

2.2 MySqL Database

The most widely used language for accessing and maintaining database records is Structured Query Language (SQL), and MySQL [6] is a relational database management system built on SQL. After the survey as shown in Table 1, MySQL is being used as local database because of it's reliability, speed efficiency and security. This paper has MySQL as local database to store users information submitted through registration form. MySQL database connection is established with the form for information storage using python command in cmd. The author [1] focuses on Shopwell, a retail management system (RMS) created with Android Studio, Python and MySql to track product expiry dates as well as total expenditure analysis. It notifies them when a product's expiration date is approaching, whether they are shopping online or offline. The suggested system is a mobile, tablet, and laptop app that can be connected to any retail establishment.

Teleoperated system [2] is created using Python's Socket Programming to send and receive control signals between the vehicle and the base station as well as the "PyQt" App Development Framework for UI application development. This provides in-depth knowledge on using PyQt5 for user interface design. Each package comes with its own controls and widgets. The author [3] concentrates on creating a SmartTab, a prototype Smart Tablet that combines tablet and laptop features. The tablet that runs a desktop operating system (Raspbian), which sets it apart from other tablets that run a mobile operating system. Smart tab efficiently implements PyQt5 for UI framework. It explains the process, implementations and requirements in detail. The time needed for field information labelling prior to the competition is what this study intends to shorten.

A semi-automatic labelling system [7] combines manual and semi-automatic labelling techniques for robot contests. The client interface is created by PyQt5, which works together with the SiamRPN algorithm at the back end to provide a semi-automatic labelling function. The framework is used to connect the front and back-end calling interfaces. The article [8] covers the creation of a licence plate recognition programme utilising neural network technologies. Recognizing the licence plate number can be accomplished using a variety of methods, including internet services. The application of neural network technologies is the most promising strategy for overcoming this issue.

The tools presented in [9] were developed or modified to allow for seamless integration of future PyQt GUI development with existing Java-oriented processes and the controls environment. The Beam Instrumentation (BI) group at CERN has looked into alternatives as Java GUI toolkits grow obsolete and has chosen PyQt as one of the viable technologies for future GUIs. The Food Classifier and Nutrition Interpreter (FCNI), a user-friendly tool that categorises various food kinds with multiple graphical representations of food nutrients values in terms of calorie estimation and a multimedia auditory response, has been proposed by [10].

A GUI-Based Penetration Testing Tool for Scanning and Enumeration called EAGLE is presented in [11]. It was created to help novice penetration testers. Features from many scanning and enumeration programmes, including

Nmap, Gobuster, Hydra, Nikto, Enum4Linux, and Whatweb, are integrated into EAGLE. In [12], the Galaxy Detection and Classification Tool (GalaDC) is developed, which can efficiently and accurately detect and classify galaxies using a trained neural network and a number of computer vision methods. GalaDC is easy to use, allows batch processing, and can handle photos that contain many galaxies as well as do statistical analysis.

After analysis of different papers, it concluded that most of them has used PyQt5 for GUI design and development. Thus, in this GUI, QT Designer provided by PyQt5 is used to develop and design front-end for user login form. Qt designer provides .ui files from QT designer, which has converted to .py file for further integration with database.

3   GUI Design and Database Connectivity

When it comes to Python packages, to achieve our objective, a number of programmes can be employed. Numerous Development Tools for the GUI are widely accessible in Python versions. Two of these packages that are most widely used are PyQt and Tkinter. Each package comes with its own set of controls and widgets. PyQt varies from the other two in that it supports micro-controllers and allows the development of multi-threaded programmes.

The framework's design and implementation primarily address the range of duties involved in developing the entire application. Following the creation of the application's User Interface, the underlying reasoning is put into place in accordance with the project's requirements and desired functionality. This is a two-phase development process that begins with designing the User Interface for the learning programmes and ends with construction of the backend local database storage system.

3.1   UI Designing

A variety of issues are addressed in UI design. These include installing the PyQt environment and the designer tools discussed in Sects. 3.1.1 and 3.1.2

3.1.1   Setup Environment

Any operating system can use the well-liked User Interface design framework PyQt, including Linux, Windows, and iOS. It is also available in both open-source and proprietary forms. The Windows version of the PyQt framework can be downloaded directly from the internet, subject to system hardware and Python version compatibility. The iOS version, or PyQt framework, is available on the web as separate files that can be downloaded using the Homebrew installer with this command:

brew install pyqt

The Linux version of the PyQt framework can be downloaded directly via a terminal and the supplied commands:

Table 1. Study and analysis for Database and Problem Statement

S. No	Ref	Problem	Application	Database
1	[1]	This paper design and develop a retail management system (RMS) "Shopwell" with the help of Python, MySQL, and Android Studio to track product expiration dates and perform total cost analysis	Shopping application	Mysql
2	[2]	In this study, a teleoperated vehicle control system is proposed, as well as a methodology for developing a Python-based desktop application	UI application	–
3	[3]	This essay emphasizes the design and creation of a SmartTab, a prototype for a Smart Tablet that combines tablet and laptop features. One of the tablets that utilises a desktop operating system (Raspbian) is this one	Smart Tab	–
4	[13]	Behind the scenes, the GUI application communicates with a Java-based service. This article discusses the generic architecture adopted for the project, our development process, also the difficulties and lessons discovered when integrating Python with Qt	Timing Control application	–
5	[14]	In this paper, they design and implement an open source verification GUI with openness, sharing, and freedom characteristics	Open Source	–
6	[15]	In this study, using the most well-liked Python clients, the performance of several MySQL and not solely Structured Query Language (NoSQL) DBs (Redis, Cassandra, mongoDB) is compared	–	Mysql, No sql, Mongo DB
7	[16]	In this article, they build a straightforward sine wave generator and measure it using the NI USB-6361 multipurpose I/O device. They compare Matlab, LabVIEW and Python in the context of data gathering. To see if there are any variations, they will compare how quickly LabVIEW, MATLAB, and Python apps execute	–	Mysql

sudo apt-get install python-qt4 OR sudo apt-get install pyqt5-dev-tools

The Qt Designer, the primary PyQt framework utility, is a highly effective tool for dragging and dropping Graphical User Interfaces into place. A number of multipurpose widgets in Qt Designer can be properly chosen to build an interactive and efficient GUI. This technology not only streamlines the design process but also cuts down on time. Unlike other designer programmes like Android Studio, which can also include code, it can just give User Interface design.

Qt Designer has produced a UI file with the .ui extension. Using the following cmd command, you could easily convert this UI file into a Python file with a .py extension.

pyuic4 -x sample.ui -o sample.py

When you run the Python file after conversion, the UI of the application should look like the one created Fig. 1.

Step 1: First GUI is to Enter your credentials in form. In this, user will enter their credentials as a part of user authentication as shown in Fig. 2, and Fig. 3.

Step 2: After entering credentials, users will successfully login to system to make changes and add recipe information as shown in Fig. 4.

Step 3: After entering login credentials, the GUI made using PyQt5 to enter recipe details for cooking comprises of all parameters shown in Fig. 5.

Fig. 1. Login page GUI

Fig. 2. Adding credentials to login page

Fig. 3. Logged in successfully

Fig. 4. Recipe details GUI

3.1.2 Database Connection

For connecting designed user interface to local database for storing credentials, a python code is discussed in this section that establishes connection between user interface and MySQL database. Here is the python code for database connection:

```
def login(self):
        try:
            email = self.lineEditEmail.text()
            password = self.lineEditPassword.text()
```

Fig. 5. How to enter data in parameters for recipe details

```
mydb = mc.connect(
    host="localhost",
    user="root",
    password="smartcooker",
    database="users"

)

mycursor = mydb.cursor()
mycursor.execute("SELECT email,password from users where email
like '"+email + "'and password like '"+password+"'")
result = mycursor.fetchone()

if result == None:
    self.labelResult.setText("Incorrect Email & Password")

else:
    self.labelResult.setText("You are logged in")
    mydialog = QDialog()
    mydialog.setModal(True)
    mydialog.exec()


except mc.Error as e:
    self.labelResult.setText("Error")
```

As you can see from the above code, all the information will be stored in the database in the form table using this code after connecting the GUI with the local database.

After establishing connection, a class function is used with sql queries to enter data to the database table whenever the user submit the form with information.

4 Results and Discussion

The smart automated cooker is designed and developed in the lab environment of Indira Gandhi Delhi Technical University for Women using simulated environment. It is smart as it measures, wash and cook the food as per user requirement. It accepts cooking instructions and cooking time from the user and automatically start cooking at user set time following the user instructions without human intervention.

The designed cooker accepts commands locally from the user to control the smart cooker operations locally. The user friendly user database and recipe database, to control the smart cooker operations and authenticate user locally, GUI is designed and developed using Raspberry pi3 development board, Linux operating system and PyQt5 environment in this paper. The system keep record of user information to authenticate user for controlling systems operations and recipe record, cooked earlier recipes by user or new recipe record as per user input details as shown in Fig. 6. The stored user information helps in synchronization of system operations and recipe record recommend the recipe to user in future.

Fig. 6. User and recipe details in database

5 Challenges

Python is a strong, adaptable, straightforward, and easy-to-learn programming language. Less code is needed to accomplish the same purpose than in other popular languages like C++ or Java, resulting in higher productivity. These benefits also apply to GUI programming. PyQt is one of the most popular Python GUI frameworks, so it has a large community support. There are numerous learning materials for PyQt and Qt, but it was discovered that it was not comprehensive and there wasn't a current PyQt beginner's tutorial including basic and advanced code examples. PyQt5 installations and importing python faces some issues initially because of less resources. On several occasions, segmentation faults were encountered. These were initially difficult to interpret, but with time, all concepts were clear, and the target source was quickly identified.

Developing the GUI also presented some challenges, such as using QT designer widgets, converting the a .ui file to a .py file and integrating the GUI with a database to store data. It is concluded that a .ui file can be converted using the mentioned command:

pyuic5 -x yourfile.ui -o yourfile.py (use yourfile name as mainPage).

Database connection was also accomplished using python code in PyQt5.

6   Future Scope

The primary goal of this work was to design, create an easy-to-use online login form, which would enable users to enter their credentials for login processes and recipe details form for cooking. The intended paper successfully developed to serve the manual control purpose based on the basic requirements and specifications. Finally, the PyQt5 technology and MySQL database were used to meet the basic requirements outlined in the specifications. Finally, the designers successfully debugged, implemented and tested the paper to meet the necessary criteria. The basic functionalities of the Login system and recipe details form have been implemented by the Qt designer and integrated with local database. However, greater sophistication can be applied to improve the system's features and facilities, such as:

- adding a button to update their password and recipe details.
- adding new pages to see previous stored recipe to choose.
- update any data without recurrence of previous data.

7   Conclusion

The GUI design and development using PyQt5, local database creation using MySQL were discussed in this article. The GUI is capable of manually controlling smart cooker operations, storing recipe records for future user recommendations and storing user authentication information to synchronize the integrated device operations. This resulted in the conclusion that PyQt5 is an excellent choice for GUI design because it is compatible with micro-controllers and may be used to create multi-threaded programs. Qt designer feature in PyQt5 made development part very easy and accessible as it is very user-friendly. To store data, this GUI is integrated with MySQL local database. The task will be completed with smart cooker hardware, raspberry pi integration so that it can work in real time.

References

1. Shah, S., Patel, Y., Panchal, K., Gandhi, P., Patel, P., Desai, A.: Python and MySQL based smart digital retail management system. In: 2021 6th International Conference for Convergence in Technology (I2CT), pp. 1–6. IEEE, April 2021
2. Cynthia, J., Mohankumar, T., Arjun, T., Naveenkumar, C.: Development of Python based UI application for tele-operated vehicles. In: 2021 IEEE 6th International Conference on Computing, Communication and Automation (ICCCA) (2021)
3. Shrunkhla, I., Tripathi, B.S., Reddy, S.R.N.: SmartTab: a design & implementation of tablet for learning purposes based on PyQT framework. In: 2019 IEEE International Conference on Electrical, Computer and Communication Technologies (ICECCT), Coimbatore, India, pp. 1–7 (2019). https://doi.org/10.1109/ICECCT. 2019.8869403
4. Python: Introduction to PyQt5. GeeksforGeeks (2022). GeeksforGeeks. https:// www.geeksforgeeks.org/python-introduction-to-pyqt5/. Accessed 29 Apr 2023
5. "PyQt5 Tutorial - Setup and a Basic GUI Application." YouTube, 3 July 2019. www.youtube.com/watch?v=Vde5SH8e1OQ&list=PLzMcBGfZo4-lB8MZfHPLTEHO9zJDDLpYj
6. "Python MySQL Create Database." Python MySQL Create Database. www.w3schools.com/python/python_mysql_create_db.asp. Accessed 29 Jan 2023
7. Renyi, L., Yingzi, T.: Semi-automatic marking system for robot competition based on PyQT5. In: 2021 International Conference on Intelligent Computing, Automation and Systems (ICICAS), Chongqing, China, pp. 251–254 (2021). https://doi.org/10.1109/ICICAS53977.2021.00058
8. Varkentin, V., Schukin, M.: Development of an application for car license plates recognition using neural network technologies. In: 2019 International Conference "Quality Management, Transport and Information Security, Information Technologies" (IT&QM&IS), Sochi, Russia, pp. 203–208 (2019). https://doi.org/10.1109/ITQMIS.2019.8928373.
9. Zanzottera, S., Jensen, S., Jackson, S.: Adopting PyQt for beam instrumentation GUI development at CERN. In: JACoW ICALEPCS, 2021, pp. 899–903 (2022)
10. Sundarramurthi, M., Nihar, M., Giridharan, A.: Personalised food classifier and nutrition interpreter multimedia tool using deep learning. In: 2020 IEEE Region 10 Conference (TENCON), Osaka, Japan, pp. 881–884 (2020). https://doi.org/10.1109/TENCON50793.2020.9293908
11. Singh, A.S.B., Yusof, Y., Nathan, Y.: EAGLE: GUI-based penetration testing tool for scanning and enumeration. In: 2021 14th International Conference on Developments in eSystems Engineering (DeSE), pp. 97–101. IEEE, December 2021
12. Cai, E.: GalaDC: galaxy detection and classification tool. In: 2020 IEEE 5th International Conference on Image, Vision and Computing (ICIVC), pp. 261–266. IEEE, July 2020
13. Kovari, Z., Kruk, G.: New timing sequencer application in Python with Qt-development workflow and lessons learnt. In: JACoW ICALEPCS 2021, pp. 904–907 (2022)
14. Peiming, G., et al.: A PyQt5-based GUI for operational verification of wave forecasting system. In: 2020 International Conference on Information Science, Parallel and Distributed Systems (ISPDS), Xi'an, China, pp. 204–211 (2020). https://doi.org/10.1109/ISPDS51347.2020.00049

15. Reichardt, M., Gundall, M., Schotten, H.D.: Benchmarking the operation times of NoSQL and MySQL databases for Python clients. In: IECON 2021–47th Annual Conference of the IEEE Industrial Electronics Society, Toronto, ON, Canada, pp. 1–8 (2021). https://doi.org/10.1109/IECON48115.2021.9589382
16. Csokmai, L.S., Novac, C.M., Novac, O.C., Bujdosó, G., Oproescu, M., Codrean, M.: Comparative study about data speed acquisition and recording in a MySQL database of LabVIEW, MATLAB and Python programming languages. In: 2021 13th International Conference on Electronics, Computers and Artificial Intelligence (ECAI), Pitesti, Romania, pp. 1–4 (2021). https://doi.org/10.1109/ECAI52376.2021.9515034

A Detection Approach for IoT Traffic-Based DDoS Attacks

Praveen Shukla[1]($\boxtimes$) (iD), C. Rama Krishna[1] (iD), and Nilesh Vishwasrao Patil[2] (iD)

[1] Computer Science and Engineering, NITTTR, Sector-26, Chandigarh, U.T.
Chandigarh 160019, India
praveenshukla805@gmail.com
[2] Computer Engineering, Government Polytechnic, Burudgaon Road, Ahmednagar,
Maharashtra 414001, India

Abstract. The Internet of Things (IoT) is a rapidly growing technology that significantly changed the human life by automating everything around us. It enables us to manage IoT devices 24/7 from anywhere. However, it also brings several cyber security risks, such as Distributed Denial of Service (DDoS) attacks. A large-scale DDoS attack immediately overwhelms the victim's network or system with a massive volume of unwanted traffic from the pool of compromised IoT devices. As a result, it prevents legitimate users from accessing the victim's services or applications. Further, protecting Internet-based applications and networks from large-scale IoT traffic-based DDoS attacks is a challenging task. In the literature, several techniques are available to protect networks and services from IoT traffic-based DDoS attacks. However, the occurrence and sophistication of IoT traffic-based DDoS attacks are expanding every year. In this article, we propose a comprehensive approach for identifying IoT traffic-based DDoS attacks. The experimental results demonstrate that the proposed XGB-based model achieves significant accuracy of 99.89%.

Keywords: IoT devices · Internet of Things · DDoS attacks · Machine learning techniques · Bot-IoT Dataset

1 Introduction

Today, we live in a world surrounded by numerous IoT applications that significantly transform our life. IoT technology has profoundly impacted each industry and society. It increases efficiency and productivity with the use of intelligent machines. Several factors have contributed to the quick adoption of IoT technology in recent years. One of the biggest reasons is that consumers and businesses have realized its benefits. According to recent statistics, there will be 31 billion devices connected worldwide by 2025. It is shown in Fig. 1. Further, the IoT devices market is predicted to grow up to 1.6 trillion US dollars by 2025 [1, 2].

Intelligent devices are incorporated with various significant sectors, such as smart houses, smart cities, agriculture, industries, healthcare, logistics, transportation, defense,

R. K. Challa et al. (Eds.): ICAIoT 2023, CCIS 1929, pp. 201–214, 2024.
https://doi.org/10.1007/978-3-031-48774-3_14

etc., to enhance the smartness in day-to-day activities [3–5]. Some of society's fundamental problems can be solved using IoT applications. For example, an intelligent park-
ing system prevents traffic congestion. A smart water and energy management system
enable authorities to save water and efficiently distribute the power. However, with all
these significant benefits, IoT technology comes with a number of security concerns. As
a result, cybercriminals take advantage of this chance to compromise more devices in
order to launch large-scale DDoS attacks.

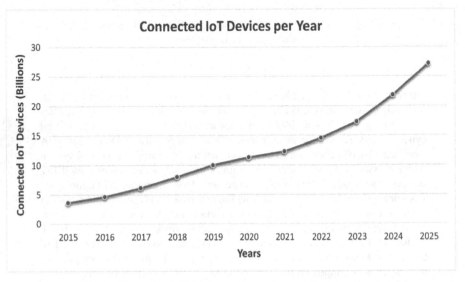

Fig. 1. Year-wise IoT devices connected to public network [1].

1.1 IoT Platform

The IoT network refers to a giant group of "interrelated or Internet-connected" thousands of objects/devices that can accumulate and communicate information via a wireless
medium without human interference. These devices may include computing machines,
mechanical machines, and digital devices that work together, sense/gather, process, and
analyze the data from heterogeneous smart IoT gadgets [6]. The main idea behind designing IoT devices is to accomplish a particular task based on valuable information or trigger points without explicit programming. Typically, in smart homes, most appliances
are connected to the Internet, such as smart watches, smart cars, air conditioners, smart
TVs, IP cameras, etc. These appliances make decisions based on triggers, environmental
conditions, etc.

1.2 DDoS Attack

DDoS attacks are among the most severe cyberattacks executed on Internet enabled applications and IoT systems/networks. It immediately crashes the victim's server, system,

or network [7, 8]. An attacker overwhelms the victim or its essential Internet services by flooding a large number of irrelevant packets from each compromised IoT device. Therefore, it denies access to legitimate users. An attacker is referred to as a "bot herder" and compromised devices are "bots". A typical setup for performing IoT traffic-based large-scale DDoS attacks is shown in Fig. 2. A botnet is typically made up of a collection of malicious devices infected by malware and controlled by an attacker.

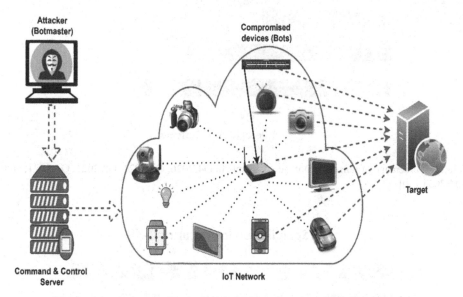

Fig. 2. A typical setup for performing IoT devices-based DDoS attack [9].

1.3 Recent Statistical Information of DDoS Attacks

According to the current work culture, numerous activities are being conducted online, such as education, healthcare, gaming, shopping, work, etc. Therefore, the demand for Internet-connected IoT devices has increased even more. However, less-secure IoT devices are prone to several critical security issues due to the lack of storage, processing capacity, and not the availability of security features. As per the report from Symantec, on average, an IoT device is breached every two minutes [12]. As per the recent statistical report of Cloudflare, ransom DDoS attacks increased by 75% in Quarter-4 2021 compared to Quarter-3 2021. Figure 3 shows that there is always year-by-year growth in attacks [10]. In another study, cybersecurity ventures predict that the damage caused by cybercrime will be approximately $10.5 trillion annually by the year 2025 [13].

Therefore, in 2021, it has been observed that significant global spending on IoT security, which is approximately 3.5 billion USD and 21% higher than in 2020 [11]. It is shown in Fig. 4.

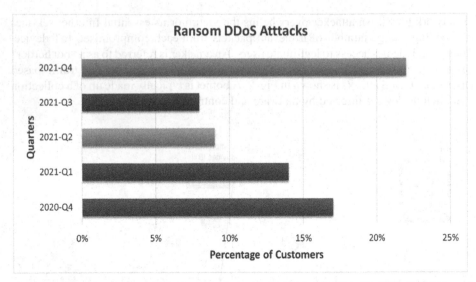

Fig. 3. Percentage of ransom DDoS attack events in Q4:2020, Q1:2021, Q2:2021, Q3:2021, and Q4:2021 [10].

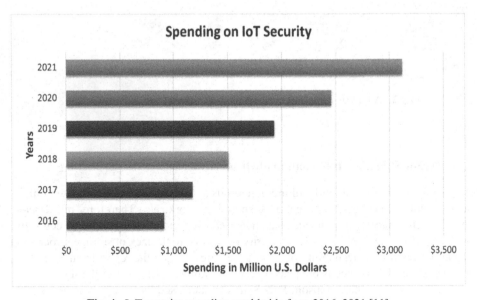

Fig. 4. IoT security spending worldwide from 2016–2021 [11].

1.4 Contributions

The primary contributions of this article are listed as follows:

i. A novel detection approach has been proposed for identifying IoT traffic-based DDoS attacks.

ii. Design and implement a detection mechanism using different machine learning algorithms.
iii. Determine the most significant features using Chi-2, ANOVA, and mutual information gain techniques.
iv. Evaluate and compare the proposed mechanism's performance with or without feature selection scenarios.

1.5 Structure of the Article

The rest of the article is arranged as follows: Sect. 2 is a summarized description of existing IoT traffic-based classification approaches. Section 3 discusses the complete working of the presented classification technique for detecting DDoS attacks. Section 4 provides detailed information about the practical development environment and software tools utilized to implement the proposed classification technique and subsequently, Sect. 5 demonstrates results and analysis. Eventually, Sect. 6 concludes the suggested work of this article.

2 Related Work

In the literature, several strategies have been published by researchers to defend Internet-enabled services from different category of destructive DDoS attacks. The severity and frequency of DDoS attacks based on IoT traffic, on the other hand, are increasing day-by-day. The recent IoT traffic-based DDoS attack detection approaches have been summarized in Table 1.

Kumari et al. [14] have implemented an ensemble-based model using two ML techniques: J48 DT and SVM. They employed particle swarm optimization for extracting the most significant features from the sample KDD Cup dataset. This approach classified network traces into two classes: legitimate and attack. It classified network traces with an accuracy of 90%.

In [15], the DDoSNet model has been presented as an intrusion detection system for large-scale DDoS attacks. They employed a Recurrent Neural Network (RNN) and an autoencoder. The proposed model evaluated using the CICDDoS2019 dataset. This dataset includes different types of DDoS attack traces and legitimate flows. The results indicate that the proposed approach provides an accuracy of 98%.

Shurman et al. [16] presented two approaches for identifying DDoS attacks based on reflection in IoT environments. Firstly, they employed hybrid IDS to identify IoT traffic-based DoS attacks, and secondly, they employed Long Short-Term Memory (LSTM) based trained model. Therefore, they deployed collaboratively both signature and anomaly-based approaches in hybrid IDS. This method is evaluated using the recently published dataset CICDDoS2019. The results demonstrate that it detects cyberattacks more precisely (99.19%).

Pokharel et al. [24] suggested a supervised learning-based IDS for network anomaly detection. This approach is designed using hybrid ML strategies, such as the Naive bayes (NB) and Support vector machine (SVM) classifier. They pre-processed and normalized the real-time historical log dataset for experimentation. The proposed hybrid IDS demonstrates accuracy (92%) and precision (95%) and minimizes false positives.

Table 1. Summary of existing DDoS attack detection approaches.

Authors	Detection method	Dataset	Functionality	Limitations
Tang et al. [17]	Deep Neural Network	NSL-KDD	Proposed IDS system in a	Results obtained with
			Software-Defined Network (SDN)	the unbalanced dataset
				suffered from overfitting
Kim et al. [18]	Deep Neural Network	KDD Cup 99	The suggested method was trained	No feature selection
			with only 10% of the data, achie-	method applied
			ving accuracy with full features	
Meidan et al. [19]	SVM Autoencoder	N-BaIoT	Applied auto-encoder to detect	This approach focused
			botnet attacks using anomaly in	only on the Mirai and
			IoT network	Bashlite attacks
Feng et al. [20]	CNN and LSTM	KDD-CUP 99	Deep learning detection approaches	Class wise prediction
			are used for detecting XSS and	is not provided, only
			SQL attacks	focus on DoS attack
Yang et al. [21]	ICVAE-DNN	NSL-KDD,	ICVAE-DNN has the potential to	Low detection rate for
		UNSW-NB15	correctly predict minority attacks	U2R and R2L while
			and not defined attacks	using NSL-KDD
Gaur et al. [22]	KNN, RF, DT,	CICDDoS	The proposed system employed	The model may fail to ana-
	XGBoost	2019	a hybrid feature selection method	lyze real IoT traffic beca-

(*continued*)

Table 1. (*continued*)

Authors	Detection method	Dataset	Functionality	Limitations
			that integrates chi-square, Extra	use it is designed with a
			tree, and Anova techniques for	non-IoT dataset, and it
			DDoS attack detection	does not consider all
				types of DDoS attack
Parra et al. [23]	LSTM RNN	N-BaIoT	Distributed deep learning	The proposed approach
			cloud-based framework to counter	validated with low variance
			phishing and botnet attacks	attack traffic

Krishnaveni et al. [25] proposed an efficient anomaly detection system (ADS) to be used in the domain of cloud computing. They applied an information gain technique to extract an optimal feature set from the NSL-KDD dataset. They employed SVM ML techniques for anomaly detection. The results indicate that the RBF kernel provides a detection accuracy of 96.24%. The study illustrates that SVM offers substantial advantages for developing IDS methods for cloud computing environments.

In [26], authors proposed an ensemble learning approach that combines the advantages of each classification algorithm. They used three classifiers: gradient boosting (GB), logistic regression (LR), and decision trees (DT) for implementing the ensemble learning model in a stacking manner. The Chi square or correlation coefficient techniques are applied to determine the most appropriate 23 features. The proposed approach is evaluated on the publicly available dataset CICIDS2018. The proposed model provided 98.9% detection accuracy.

In the literature, several approaches have been proposed for protecting Internet-based systems from IoT traffic-based cyberattacks. However, many approaches have been developed and validated with outdated/traditional or imbalanced datasets. Further, several existing approaches [7, 8, 27–30] detect various types of DDoS attacks in a distributed manner by deploying their systems on distributed frameworks, including Hadoop, Spark, Kafka, etc. However, they failed to provide a solution against IoT network traffic-based DDoS attacks.

In the subsequent section, the detailed functioning of the presented IoT traffic-based DDoS attack detection approach, as demonstrated.

3 Proposed Methodology

In this section, the working of the proposed detection approach for IoT traffic-based DDoS attacks is presented. The logical workflow of the suggested attack detection approach is shown in Fig. 6.

To provide an efficient solution, this classification approach is designed using the recently released and publicly available realistic Bot-IoT dataset. The Bot-IoT dataset incorporates a massive volume of IoT traffic flows, including benign and different DDoS attack types. The Bot-IoT dataset [31] comprises 74 publicly available CSV files that contain 72 million records and address the problem of existing datasets. In the detection process of this approach, several steps are involved, which are discussed one-by-one as follows:

The preprocessing step includes: cleaning the data (removal of incorrect or incomplete records), removing outliers, and transforming categorical variables into numerical values. In the next step, the data values of features will be standardized on an identical scale (lie in the range of 0 and 1). After that, the data is normalized, and the network traces are labeled before performing the data balancing process. The up-sampling technique is employed to balance the dataset by injecting synthetically generated records into the minority class. Because an imbalanced dataset adversely affects the accuracy of models and is biased toward a particular class.

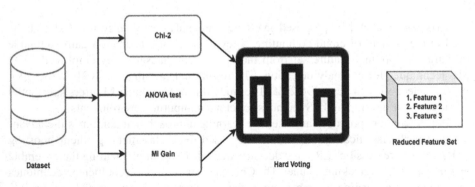

Fig. 5. Conceptual flowchart of the feature selection process.

In the process of building an ML model, feature selection is a significant step. It plays a critical role in improving the performance of the model. In addition to dimensionality reduction, feature selection reduces the computation costs associated with data mining and avoids the overfitting problem. Chi-2, ANOVA, and mutual information gain methods are employed to rank the top 15 key features, followed by hard voting to narrow down the feature set. This is shown in Fig. 5. After obtaining the optimal features, the samples were divided into training and testing samples and fed into the ML algorithms for developing models.

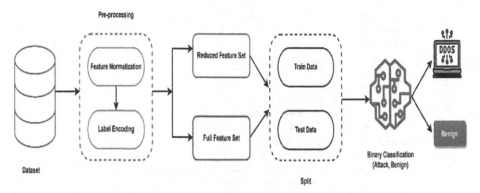

Fig. 6. Framework of the proposed attack detection approach.

3.1 Training Process

In this training process, the prepared data samples are divided into a training sample and a testing sample. Further, the 80/20 rule is used in which 80% data samples are fixed for training and remaining 20% for testing. The performance of the ML model depends on the training data. Hence the stratified technique for splitting the data samples into two parts is employed. The proposed approach is developed in two ways: (i) Designing model using extracted feature set and (ii) Designing model using a complete feature set as shown in Fig. 6. The performance of these two scenarios are evaluated.

4   Experimental Environment

This section illustrates the experimental environment utilized for developing and validating the proposed classification approach. A powerful server running Ubuntu 18.04.5 LTS, a 64-bit operating system (AMD64) with 64GB RAM is utilized. The processor is an Intel(R) Xeon(R) Processor E5–2420 v2 with a clock speed of 2.20GHz, and the graphics card is llvmpipe (LLVM 6.0 256 bits), while the storage capacity is 984GB. For developing and evaluating the proposed classification approach, the recently published Bot-IoT dataset is utilized as a source of network traffic.

5   Results and Analysis

This section illustrates the performance evaluation and analysis of the proposed attack detection method. The proposed classification approach classifies IoT network traffic into two target classes: attack and benign. The classification accuracy and other standard metrics (precision, sensitivity, and F-measure) were utilized to evaluate the model's performance for demonstrating its effectiveness.

After preprocessing step, ranked the leading 15 features with the greatest gain values in descending order using various feature selection methods as depicted in Fig. 7. Further, selected the ten best features by applying the hard voting technique. In the next step, fed these most relevant features into ML models to classify incoming traffic.

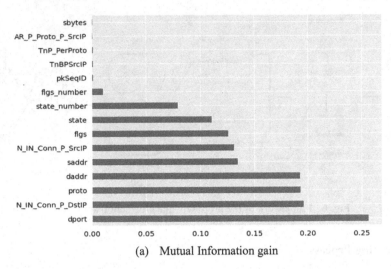

(a) Mutual Information gain

Out[30]:

	Features	Score
0	pkSeqID	6.376447e+08
18	sbytes	3.047612e+04
25	TnP_PerProto	2.283741e+04
9	pkts	4.488947e+03
22	TnBPSrcIP	4.471625e+03
20	srate	1.600373e+03
16	sum	4.037729e+02
23	TnP_PSrcIP	3.978651e+02
24	TnP_PDstIP	3.007522e+02
28	N_IN_Conn_P_DstIP	2.979059e+02
13	dur	2.700528e+02
6	sport	2.650689e+02
8	dport	2.276476e+02
1	stime	2.151520e+02
29	N_IN_Conn_P_SrcIP	1.897153e+02

(b) Chi-2 test

Fig. 7. Ranked top 15 features using feature selection methods.

The performance evaluation of the five ML models is summarized in Table 2. In this table, researchers compared the performance (classification accuracy) of each model w.r.t. various selected feature scenarios. These results are obtained by implementing the five ML models with features selected through feature selection techniques such as chi-2 test, Anova test, MIG, ten best features and all features. The Table 2, exhibits that the XGB model achieves higher average classification accuracy 99.50% than other techniques, such as RF 98.89%, NB 93.78%, DT 98.25%, and KNN 98.06%. Therefore, it has been inferred that the XGB model outperforms other models in terms of average classification accuracy. Another pertinent observation is that the accuracy of models

Table 2. Performance evaluation of ML models with different selected feature scenarios.

Features	Accuracy of the ML Models				
	XGB [32]	RF [32]	NB [32]	KNN [32]	DT [32]
Chi-2	99.54	98.54	93.73	97.97	97.99
Anova	99.09	97.78	92.38	96.83	97.29
MIG	99.12	98.52	94.18	98.85	98.48
All	99.85	99.78	93.73	97.89	98.57
10-Best	99.89	99.83	94.88	98.78	98.94
Average	99.50	98.89	93.78	98.06	98.25

Table 3. Represents class wise performance analysis of the proposed detection approach with respect to various metrics.

Classifier	Classes	Metrics			
		Precision	Recall	F1-Score	Accuracy
XGB [32]	Attack	01.00	01.00	01.00	99.89
	Benign	00.99	01.00	01.00	
RF [32]	Attack	01.00	01.00	01.00	99.83
	Benign	00.99	00.99	01.00	
NB [32]	Attack	00.97	01.00	00.98	94.88
	Benign	00.99	00.96	00.96	
KNN [32]	Attack	01.00	01.00	00.99	98.78
	Benign	00.99	00.98	00.98	
DT [32]	Attack	01.00	01.00	00.99	98.94
	Benign	00.99	00.99	01.00	

(XGB 99.54%, RF 98.54%, NB 93.73%, KNN 97.97%, and DT 97.99%) with the ten best features is higher than with a full set of features due to the overfitting problem, as presented in Fig. 8.

It is always important to measure the performance (effectiveness) of a classification model using multiple metrics, including precision, recall, F-score, and accuracy, to get a complete picture of its performance. Table 3, represents the class wise performance analysis of the proposed model with respect to various metrics (precision, recall, F1-score, and classification accuracy). In Table 3, the accuracy column presents classification accuracy of different ML-based detection models with ten best features XGB 99.89%, RF 99.83%, NB 94.88%, KNN 98.78%, and DT 98.94%. It has been observed that the XGB-based detection model has given better accuracy than others.

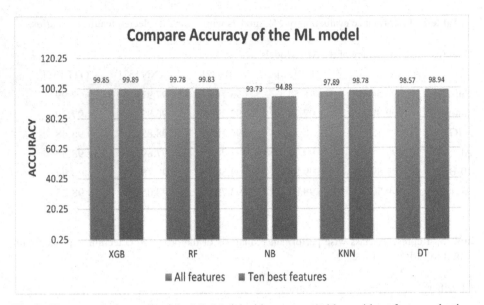

Fig. 8. Compare the accuracy of the ML Model with respect to with or without feature selection.

6   Conclusions

With the massive expansion of modern technologies and IoT networks, attackers are getting more opportunities for compromising a large number of devices to perform different types of DDoS attacks on the public networks. Therefore, this paper proposed an IoT network traffic-based DDoS attacks detection approach for protecting public networks and devices. This detection approach is designed and validated using a publically available Bot-IoT dataset. Further, various feature extraction techniques, such as Chi-2, ANOVA, and mutual information gain techniques are employed. With these feature extraction techniques, ranked the top 15 most significant features and further reduced them by applying a hard voting technique. The XGB model has given significant detection accuracy (99.89%) with low computational energy, low resource usage, and low false alarm rates. In future work, intend to apply ensemble learning and deep learning to make IoT environments more secure. Additionally, consider the Flash Events scenarios while distinguishing IoT network traffic-based DDoS attacks and legitimate traces.

References

1. Sinha, S.: Number of connected IoT devices 2021 (2021). https://iot-analytics.12.com/number-connected-iot-devices/
2. Vailshery, L.S.: Global IoT end-user spending worldwide 2017–2025 (2021). https://www.statista.com/statistics/976313/global-iot-market-size/
3. Jan, S.U., Ahmed, S., Shakhov, V., Koo, I.: Toward a lightweight intrusion detection system for the internet of things. IEEE Access **7**, 42450–42471 (2019)

4. Nivaashini, M., Thangaraj, P.: A framework of novel feature set extraction-based intrusion detection system for internet of things using hybrid machine learning algorithms. In: 2018 International Conference on Computing, Power and Communication Technologies (GUCON), pp. 44–49 (2018)
5. Thaseen, I.S., Poorva, B., Ushasree, P.S.: Network intrusion detection using machine learning techniques. In: 2020 International Conference on Emerging Trends in Information Technology and Engineering (ICETITE), pp. 1–7 (2020)
6. Ashton, K.: That 'internet of things' thing. RFID J. **22**(7), 97–114 (2009)
7. Patil, N.V., Rama Krishna, C., Kumar, K.: Apache hadoop based distributed denial of service detection framework. In: Gani, A.B., Das, P.K., Kharb, L., Chahal, D. (eds.) ICICCT 2019. CCIS, vol. 1025, pp. 25–35. Springer, Singapore (2019). https://doi.org/10.1007/978-981-15-1384-8_3
8. Patil, N.V., Rama Krishna, C., Kumar, K.: S-DDoS: apache spark based real-time DDoS detection system. Journal of Intelligent & Fuzzy Systems **38**, 1–9 (2020)
9. Vishwakarma, R., Jain, A.K.: A survey of DDoS attacking techniques and defence mechanisms in the IoT network. Telecommun. Syst. **73**(1), 3–25 (2020)
10. Yoachimik O, Ganti V.: DDoS attack trends for Q4 2021 (2022). https://blog.cloudflare.com/ddos-attack-trends-for-2021-q4/
11. Alsop, T.: Global internet of things security spending 2016–2021 (2020). https://www.statista.com/statistics/543089/iot-security-spending-worldwide/
12. Symantec: symantec internet security threat report 2019 (2019). https://docs.broadcom.com/doc/istr-24-2019-en
13. Steve Morgan, C. Sausalito: cybercrime to cost the world 10.5 trillion annually by 2025 (2020). https://cybersecurityventures.com/cybercrime-damages-6-trillion-by-2021/
14. Kumari, A., Mehta, A.K.: A hybrid intrusion detection system based on decision tree and support vector machine. In: 2020 IEEE 5th International Conference on Computing Communication and Automation (ICCCA), pp. 396–400 (2020)
15. Elsayed, M.S., Le-Khac, N.-A., Dev, S., Jurcut, A.D.: DDoSNet: a deeplearning model for detecting network attacks. In: 2020 IEEE 21st International Symposium On "A World of Wireless, Mobile and Multimedia Networks" (WoWMoM), pp. 391–396 (2020)
16. Shurman, M.M., Khrais, R.M., Yateem, A.A., et al.: Dos and DDoS attack detection using deep learning and ids. Int. Arab J. Inf. Technol. **17**(4A), 655–661 (2020)
17. Tang, T.A., Mhamdi, L., McLernon, D., Zaidi, S.A R , Ghogho, M.. Deep learning approach for network intrusion detection in software defined networking. In: 2016 International Conference on Wireless Networks and Mobile Communications (WINCOM), pp. 258–263 (2016)
18. Kim, J., Shin, N., Jo, S.Y., Kim, S.H.: Method of intrusion detection using deep neural network. In: 2017 IEEE International Conference on Big Data and Smart Computing (BigComp), pp. 313–316 (2017)
19. Meidan, Y., et al.: N-baiot—network-based detection of IoT botnet attacks using deep autoencoders. IEEE Pervasive Comput. **17**(3), 12–22 (2018)
20. Feng, F., Liu, X., Yong, B., Zhou, R., Zhou, Q.: Anomaly detection in ad-hoc networks based on deep learning model: a plug and play device. Ad Hoc Netw. **84**, 82–89 (2019)
21. Yang, Y., Zheng, K., Wu, C., Yang, Y.: Improving the classification effectiveness of intrusion detection by using improved conditional variational autoencoder and deep neural network. Sensors **19**(11), 2528 (2019)
22. Gaur, V., Kumar, R.: Analysis of machine learning classifiers for early detection of DDoS attacks on IoT devices. Arab. J. Sci. Eng. **47**(2), 1353–1374 (2022)
23. De La Torre, G., Parra, P.R., Choo, K.-K.R., Beebe, N.: Detecting internet of things attacks using distributed deep learning. J. Network Comput. Appl. **163**, 102662 (2020). https://doi.org/10.1016/j.jnca.2020.102662

24. Pokharel, P., Pokhrel, R., Sigdel, S.: Intrusion detection system based on hybrid classifier and user profile enhancement techniques. In: 2020 International Workshop on Big Data and Information Security (IWBIS), pp. 137–144 (2020)

25. Krishnaveni, S., Palani Vigneshwar, S., Kishore, B.J., Sivamohan, S.: Anomaly-based intrusion detection system using support vector machine. In: Dash, S.S., Lakshmi, C., Das, S., Panigrahi, B.K. (eds.) Artificial Intelligence and Evolutionary Computations in Engineering Systems. AISC, vol. 1056, pp. 723–731. Springer, Singapore (2020). https://doi.org/10.1007/978-981-15-0199-9_62

26. Fitni, Q.R.S., Ramli, K.: Implementation of ensemble learning and feature selection for performance improvements in anomaly-based intrusion detection systems. In: 2020 IEEE International Conference on Industry 4.0, Artificial Intelligence, and Communications Technology (IAICT), pp. 118–124 (2020)

27. Patil, N.V., Krishna, C.R., Kumar, K., Behal, S.: E-had: A distributed and collaborative detection framework for early detection of DDoS attacks. J. King Saud Univ. Comput. Inf. Sci. **34**, 1373–1387 (2019)

28. Patil, N.V., Rama Krishna, C., Kumar, K.: KS-DDoS: kafka streams-based classification approach for DDoS attacks. J. Supercomputing **78**(6), 8946–8976 (2021). https://doi.org/10.1007/s11227-021-04241-1

29. Patil, N.V., Rama Krishna, C., Kumar, K.: SSK-DDoS: distributed stream processing framework based classification system for DDoS attacks. Cluster Comput. **25**(2), 1355–1372 (2022). https://doi.org/10.1007/s10586-022-03538-x

30. Patil, N.V., Rama Krishna, C., Kumar, K.: SS-DDOS: In: Kumar, K., Behal, S., Bhandari, A., Bhatia, S. (eds.) Security and Resilience of Cyber Physical Systems, pp. 81–90. Chapman and Hall/CRC, Boca Raton (2022). https://doi.org/10.1201/9781003185543-7

31. Koroniotis, N., Moustafa, N., Sitnikova, E., Turnbull, B.: Towards the development of realistic botnet dataset in the internet of things for network forensic analytics: Bot-IoT dataset. Futur. Gener. Comput. Syst. **100**, 779–796 (2019)

32. Shukla, P., Krishna, C.R., Patil, N.V.: EIoT-DDoS: embedded classification approach for IoT traffic-based DDoS attacks. Cluster Comput (2023). https://doi.org/10.1007/s10586-023-04027-5

Comparison between Performance of Constraint Solver for Prediction Model in Symbolic Execution

Meenakshi Tripathi[1]([✉]) [iD], Shadab Khan[1], and Sangharatna Godboley[2] [iD]

[1] Department of CSE, MNIT, Jaipur, India
{mtripathi.cse,2020pcp5577}@mnit.ac.in
[2] Department of CSE, NIT, Warangal, India
sanghu@nitw.ac.in

Abstract. Nowadays, automated software testing is too popular, and symbolic execution takes too much attention for software testing automation. Constraint solvers are used to solve the constraint set produced during the Symbolic Execution process. Various popular constraint solvers (Z3, mathsat, CVC4, Yices, boolector) are available. On average, constraint solver takes 90–98% of the total time symbolic execution. Constraint solver might get stuck in solving complex problems hence, degrades the overall performance of the symbolic execution. In literature timeout threshold-based solutions are given but they are not so efficient. One solution to this problem is to have a time prediction model which can predict the time required for solving a constraint model and checks whether to continue the ongoing solving process or not in symbolic execution. In this paper, we present this model in which we have used the Mathsat Constraint Solver to collect the data by running the GNU Coreutils program and Busybox utilities. The performance comparison of the different solvers was done over three constraint datasets.

Keywords: Symbolic Execution · Constraint Solver · prediction model · SMT format

1 Introduction

Software testing and dynamic program analysis often use the automated test case generation method known as symbolic execution [1]. It is a white-box method that builds execution path conditions into a constraint model. By using a constraint solver to solve the constraint model, the symbolic execution tool may be able to identify the test input for initiating specific paths and errors. Nevertheless, Symbolic Execution tools (e.g., KLEE [2], Angr [3]) often have clear limitations. For example, some code pieces restrict complex theories, such as floating-point arithmetic and arrays. Furthermore, Symbolic Execution (SE) struggles to scale when dealing with extensive programs since the number and complexity of paths grow exponentially with the height of the Control Flow Graph (CFG) [4]. Specific paths necessarily contain challenging constraint models that are time-consuming to resolve. It could take a few days to fully explore all paths for

© The Author(s), under exclusive license to Springer Nature Switzerland AG 2024
R. K. Challa et al. (Eds.): ICAIoT 2023, CCIS 1929, pp. 215–226, 2024.
https://doi.org/10.1007/978-3-031-48774-3_15

a problematic program, and Constraint Solvers might significantly drain the resources allotted for Symbolic Execution. In other words, constraint solving costs constitute a significant barrier to symbolic execution being more helpful in debugging real-world systems. Most symbolic execution technologies currently use a timeout threshold to stop the constraint-solving process automatically and prevent uncontrollable time growth [4]. For example, angr's the standard threshold for constraint solving is five minutes. If a solution procedure takes longer than the allotted time, angr will stop examining the path and choose a different path for analysis. As a result, angr would ultimately waste a lot of time on the paths that would eventually give up. The threshold and the number of constraint models that go beyond the time limit significantly impact how well can solve the problem.

One possible solution to improve this problem is to create a prediction model that will predict the completion time of a constraint set by learning the data collected from a constraint solver [5]. In this method, one important point is the selection of a constraint solver. In the literature, authors have used the prediction model based on the data collected from the Z3 constraint solver only. In this paper, two constraint solvers, Z3 and Mathsat [6], are compared for the prediction model, and it is found that the performance has increased by 1.25x to 3x for the constraint solver Mathsat, depending on the distribution of the hardness of their constraint model.

2 Symbolic Execution Environment

This section introduces the background on symbolic execution and constraint solvers.

2.1 Symbolic Execution (SE)

The main idea behind symbolic execution is to represent the values of program variables as symbolic expressions and to use symbolic values rather than actual data as input values. As a result, the values computed by a program are stated as a function of the symbolic values supplied. The symbolic values of program variables, a path condition (PC), and a program counter make up the state of a symbolically run program. The path condition is a (quantifier-free) Boolean formula over the symbolic inputs called constraint set or constraint module. It builds up constraints that inputs must meet for an execution to take the specifically related path. Then the constraint module sends into the Constraint Solver. Constraint Solver finds the constraint model's satisfiability and gives the corresponding input to follow this path as shown in Fig. 1.

2.2 Constraint Solver

Constraint Solvers are gaining attention now due to their success in various application fields. In addition to program analysis, theorem proving, automatic test creation, and many other contexts, they serve as the workhorses in many formal verification systems. One significant factor in this success story, aside from the solvers' inherent efficacy and efficiency, is thanks to the universally supported standard input language. A global and simple-to-parse everyday input language is defined by the SMT-LIB [7, 8] effort. The

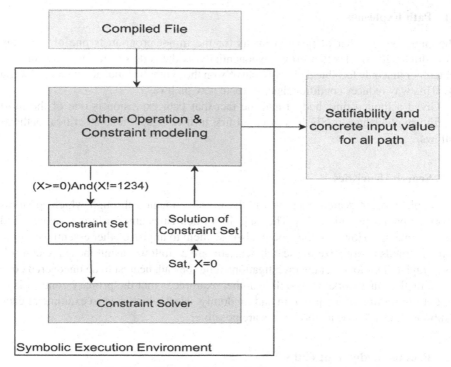

Fig. 1. Block diagram of Symbolic Execution Environment

SMT-LIB format performs well when a problem can be prepared in advance and then provided to the solver. PySMT [9] is a well-known library that includes a variety of Constraint Solvers. PySMT is a library for solver-independent formulas manipulation, so you may construct an expression, make changes to it, and carry out fundamental operations like simplification, replacement, visualization, and file-based importation. All of this is possible without the installation of an SMT solver. Modulo the UFLIRA(and all of its pieces) and BV theories, PySMT provides the representation and satisfiability verification of quantified logic. MathSAT, Z3, CVC4 [10], and Yices are a few of the solvers that are now integrated through their native APIs.

3   Limitation of Symbolic Execution

SE can find any unsafe input that might cause the assert to fail. This is accomplished by thoroughly investigating all potential execution states. Theoretically, exhaustive SE gives every decidable analysis a solid and thorough technique. Since all potentially unsafe inputs are guaranteed to be identified, soundness avoids false negatives, and completeness prevents FP (false positives), which happens when input values that are thought unsafe are safe [11]. Exhaustive Symbolic Execution is unlikely to scale beyond small applications, as described later. As a result, in reality, less challenging objectives are chosen frequently, such as trading performance for soundness. Several challenges in SE and their solutions are discussed in this section.

3.1 Path Explosion

The enormous number of program paths for the minor program is one of the main difficulties with SE. It often grows exponentially as the code's static branches do, but remember that only feasible pathways that rely on the symbolic input are explored during SE. This way reduces conditionals that create new pathways.

Given a limited time budget and the fact that path explosion is one of the most significant challenges to SE, it is essential first to investigate the most critical paths as follows.

3.2 Search Heuristics

The application of search heuristics is the primary method through which SE tools prioritize path exploration [11]. The majority of heuristics aim for high statement and branch coverage. However, they might also use them to improve other essential criteria. Some methods to optimize the search heuristic are as follows: using the static control-flow graph (CFG) to direct the investigation along the path nearest to an uncovered code is one method that works very well. Another example is that the primary concept is to start at the starting of the program and randomly select which side to examine at each symbolic branch for which both sides are possible.

3.3 Pruning Redundant Paths

Another approach to automatically prune duplicate pathways during exploration is to prevent repeatedly exploring the same lines of code repeatedly [11]. The key in sight in this technique is that a program path can be discarded if it arrives at the same program point with the same symbolic constraints as a previously investigated path since it will execute exactly the same way. This method is improved by a significant optimization, which eliminates constraints that depend on variables that won't be afterward read by the program when comparing the constraints on the two running pathways.

3.4 Lazy Test Generation

The counterexample-guided refinement methodology from static software testing is analogous to lazy test generation. By substituting each called function with public input, the approach initially investigates an abstraction of the process being tested via concolic execution [11]. Second, it makes an effort to expand each (possibly fictitious) trace produced by this abstraction to a concretely realizable execution by recursively expanding the called functions and discovering SE in the called functions that can be combined with the original trace to create a full program execution. Thus, in the exploration phase, symbolic reasoning about intraprocedural pathways is reduced to symbolic logic about intraprocedural paths and a confined and constrained find by functions (in the concretization section).

3.5 Constraint Solving

Despite significant recent developments in constraint-solving technologies, which first made SE practicable. In Symbolic Execution, where it frequently takes up the majority of run-time constraint sets, Constraint solving remains one of the major bottlenecks. Some programs' code produces queries that are blowing up the solver, which is one of the critical reasons SE fails to scale on some systems. We study two types of optimizations of constraint solving as follows:

3.6 Redundant Constraint Elimination

In SE, many queries are sent to determine if adopting a specific branch side is feasible. To establish if the program may take the opposite side of the branch, which corresponds to the negated constraint, one condition branch predicate of an existing path constraint is negated. The new constraint set is then examined for satisfiability. The fact that a program branch often only depends on a minimal number of program variables and, thus, limited constraints from the path condition is a significant finding. Therefore, removing constraints from the path condition that do not affect how the current branch will turn out is an efficient optimization [11].

Example let path condition (PC) Current execution (PC) $= (A + B > 5) \cap (C > 2)$ $\cap (B < 6) \cap (C-A = 0)$ and by the Symbolic Execution algorithm generate a new input by solving $(A + B > 5) \cap (C > 2) \cap \sim (B < 6)$ where $\sim (B < 6)$ is the negation branch condition whose feasibility trying to establish. So it's feasible to eliminate the constraint on B because the $B < 6$ branch cannot be affected by this constraint.

3.7 Prioritization of Constraint Set

To find the new path that is not found able in a limited period because Constraint Solver take generally lots of time to solve the hard constraint set, so if possible that we can do the prioritization of constraint set so we can first solver easy constraint set that high priority and solve hard constraint set those given low priority as we can improve the performance of SE.

Different decisions and suppositions are made depending on the particular situation in which Symbolic Execution is employed to answer the problems raised above. A partial investigation of the set of potential execution states may be adequate in some cases to accomplish the goal (for example, detecting a crashing input for an application) within a constrained time frame, even if these decisions often influence soundness or completeness.

4 Literature Review

To address the various limitations of SE and enhance its performance, there are many methods, such as:-

4.1 By Automatically Learning Search Heuristics

A fundamental challenge in dynamic SE is how to effectively investigate the program's execution in SE paths to acquire high code coverage on a tight budget. Symbolic Execution utilises a search method that favours looking at particular types of paths during symbolic execution, which is most likely to boost the total coverage to get around this problem [4]. However, it may be challenging and frequently produces uneven and poor results when proper search criteria are manually created.

SE can circumvent this issue by automatically detecting search strategies. They provide a method that, with the aid of defining a category of search heuristics called a parametric search heuristic, successfully identifies the optimal search heuristic for each topic program. Their method successfully produces search heuristics that significantly exceed current manually crafted heuristics in branch coverage and bug-finding, according to experimental results with industrial-strength SE tools (such as KLEE). So as we can overcome the search heuristic search problem to path exploration.

4.2 Improving Symbolic Execution through solver Selection Based on Machine Learning [12]

Based on the data received, authors attempt to demonstrate that no solution can consistently beat the others and that a primary classification based on constraint logic is adequate. They used SMT-LIB as the common format to express constraints. They choose five popular constraint solvers from SMT-COMP and identify some results to prove the statement that no solver can consistently outperform the others.

They collect SMT-LIB format data from SE Tools Klee and Angr. And run into various popular constraint solvers, such as Z3, MathSAT, and CVC4. An analysis of specific findings shows that Yices needs the least time to solve all the data. In contrast, Boolector is the most excellent solver for the SE dataset. They concluded that the boolvector may have trouble answering the specific restrictions, which lengthens the overall solution time. However, Yices is the quickest solution for restrictions that range from 5% to 37%. It is stable and uses the least total solving time. Therefore, it's essential to boost performance overall through careful solver selection.

They developed a new SE component called a path constraint classifier (PCC) to address this issue. PCC will use machine learning techniques to estimate how well each solution will perform while resolving a specific path constraint. The solver with the best expected performance will be chosen.

4.3 SMT Time Prediction Model [15]

The use of SE is necessary for automated test case technology. The technique, however, typically cannot handle massive programs. The difficulty of constraint-resolving problems in SE is a critical factor. To solve such a complex problem, the SE process may be stuck on these problems. SE tools typically use a timeout threshold to end solving to address this problem. The tool's effectiveness depends on knowing how to set a timeout correctly.

Therefore, estimating the amount of time needed to solve a constraint model provides one potential solution to this problem. The SE engine can use the prediction to decide whether to continue solving the problem. They demonstrate that such a predictor can perform well with various machine learning models and datasets. The predictor can get an F1 score on these datasets ranging from 0.743 to 0.800 by using an adaptive approach. Based on the distribution of the hardness of their constraint models, the results show that the efficiency of the constraint solution for SE can be increased by $1.25 \times$ to $3 \times$.

5 Proposed Approach

In literature timeout threshold-based techniques are proposed to improve the efficiency of SE but in these methods calculation of threshold is quite difficult. So, prediction-based methods are preferred which can predict the time required for solving a constraint model and based upon this prediction SE engine can take the decision whether to continue with the current solving process or not [15].

In real world, many SMT Constraint Solvers are available and trendy. Hence, this paper has checked the effect of two popular constraint solvers i.e. Z3 and MathSat on prediction model.

5.1 Methodology

- First, the time prediction model was installed, that predicts the solving timing of SMT-lib format data in the particular constraint solver was installed. The constraint sets are collected and then, their solving time was noted from Mathsat by running GNU Coreutils and Busybox utilities programs in angr and KLEE Symbolic Execution tools.
- The collected data was again given as input to Mathsat with the help of PySMT library and updated timing for constraint solving was recorded. This step is needed because to get the ideal solving time of the constraint solver.
- The pre-proposed prediction model was set to corresponding constraint solver data and the results were collected for various parameters.

5.2 Framework

To prevent SE's ineffective constraint solving process SMTimer [7] was built, to predict the amount of time required for constraint solving. Figure 2 shows how the system uses the expected time and interacts with an active SE tool. The constraint solvability from a constraint model that was generated by an ongoing SE program is output by their prediction model. After that, they either return to the SMT solver or help process the path via the SE tool. The unique design of SMTimer includes a conditional trigger for the prediction process. The reason is that, given the massive volume of inquiries, adding unmanageable overhead if they predicted each time it questioned, even for a brief constraint solver cost, would be impractical. Only 0.2% of queries would take more than one second. Thus, they may first solve the constraint model quickly to exclude simple constraint models. They can avoid the majority of predictions by adding a little more

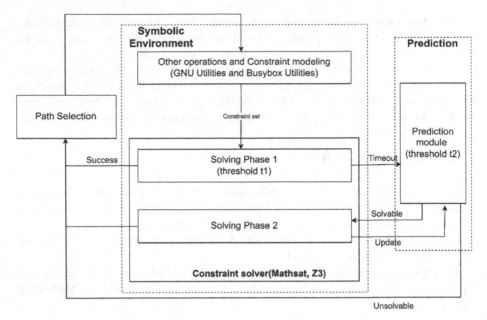

Fig. 2. System Framework

solution time. Additionally, this approach guarantees that programs that have fast-solving queries will not experience any negative effects from their prediction module.

We go into further depth about the crucial phases of the new constraint solution method as shown Fig. 2.

- Solving Phase 1: The method begins by attempting to solve the constraint model in less than one second, much like the traditional constraint solving method. There is no difference between it and traditional SE if it can produce a solution. The prediction is triggered if the constraint model cannot be solved in one second.
- Prediction: In this phase the Mathsat constraint solver is used. The system predicts the solution time using a predictor and compares it to the time threshold 2. If the predicted time is larger than 2, It would skip the following solution stage and provide the SE tool with an unknown result. Otherwise, for the second step of solution, it would return to constraint solving.
- Solving Phase 2: The constraint set is solved by the system within the allotted time. The adaptive model would then be updated using the constraint set and the ground truth to improve the prediction.
- This process is used in our simulation experiment to test the accuracy of our prediction and demonstrate how the time optimization problem, which is being defined in the following section, has improved for constraint solver Mathsat.

5.3 Experimental Setup

The evaluation of the proposed approach was done by comparing a complete state-of-the-art SMT solver, namely Z3 and Mathsat in conjunction with the SE tools such as Klee and Angr. The parameters used for the implementation as mentioned in the Table-1.

6 Result and Analysis

Experimental results are evaluated along three dimensions: F1 score, timeout query, and original solving time. The first dimension sums up the predictive performance of the model by combining two otherwise competing metrics: precision and recall. The second dimension focuses on the number of timeout queries, i.e., queries that have to be cancelled once a certain amount of time has passed since they were started. This helps protect the constraint solver from running infinitely. The third dimension keeps track of original solving time of a constraint solver.

6.1 Comparison by F1 Score

The precision, recall, and F1-score measures are used to assess the classification outcome. A successful prediction model should be able to retain high precision and strong recall. True positive means we get the constraint model that exceeds the time threshold, and true negative means we get the constraint model under the time threshold. To elaborate, the false negative cases mean that we miss the constraint set that exceeds the time threshold we set. On the other hand, the false positive (FP) cases indicate that the predictor predicts that the constraint set within the allotted time limit is complex but easy. As a result, you may delay the paths' exploration by giving them a low priority and scheduling their SE after other paths. Overall, we want to find more true-positive cases while avoiding false-positive ones. The evaluation of various experiments conducted on Mathsat constraint solver and multiple results were collected as given in Fig. 3.

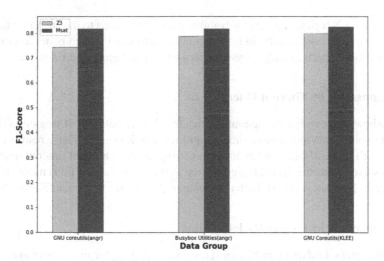

Fig. 3. Comparison By F1 Score.

A comparison of the Mathsat constraint solver was also done with Z3 solver as shown in Table 2. It was observed that Mathsat solver gets a better outcome as compared to Z3.

Table 1. System configuration

Component	Name
OS	Ubuntu 18.04.6 LTS
OS type	64 *bit*
RAM	32 GB
Processor	Intel@ Xeon(R) CPU E5–2620 v2 @ 2.10 GHz x 12
Used Symbolic Execution Tools	Klee and Agnr

Table 2. Comparison by F1 Score

Data Set	Model	Precision		Recall		F1-Score	
		Z3	Msat	Z3	Msat	Z3	Msat
GNU (Angr)	KNN	0.552	0.765	0.422	0.506	0.478	0.609
	I-KNN^	0.737	0.812	0.750	0.824	0.743	0.818
GNU (Klee)	KNN	0.198	0.337	0.244	0.446	0.219	0.385
	I-KNN	0.769	0.820	0.808	0.813	0.788	0.817
Busybox (Angr)	KNN	0.770	0.793	0.747	0.754	0.758	0.773
	I-KNN	0.768	0.785	0.834	0.870	0.800	0.826

The naive KNN cannot predict solvability better compared to the Incremental KNN (I-KNN) due to its adaptability. In both models, Mathsat performed better as compared to Z3 for all three performance measures, i.e., precision, recall, and F1-score.

6.2 Comparison by Timeout Query

Table 3 shows the number of timeout queries in all three datasets. It was observed that MathSAT solver data found more timeout queries than Z3 solver. These results confirm the higher F1 score of Mathsat because it's getting more significant timeout queries, so Mathsat solver trains the data better compared to the Z3 solver and finds more similarity data. Hence, Mathsat performs better as compared to Z3 for the prediction model.

6.3 Comparison by Original Solving Time

For Mathsat data, I-KNN identifies most timeout situations for most programs with the fewest errors. The prediction outcome is consistent with the earlier classification results, demonstrating that when the model adds more negative examples, the vector similarity in KNN is unaffected by our data selection. According to Table 4, the adaptive KNN often accelerates the solution by 1.25x(19347/14589) to 3x(128031/39890) times.

Table 3. Comparison by Timeout query

Distract	Program	Timeout Query		Test Query
		Z3	Mathsat	
GNU(Angr)	Sha1sum	315	362	1366
	tail	97	109	1115
GNU(klee)	Stat	156	163	1925
	cp	189	198	1706
Busybox(Angr)	ifconfig	18	25	433
	Klogd	16	19	408

Table 4. Comparison original time for particular programs with Symbolic Execution with prediction model for Mathsat data

Program	Original Solving Time	I-KNN
sha1sum	128031	39890
Klogd	19347	14589

7 Conclusion and Future work

This paper compares the Z3 constraint solver data on prediction with calculated data from the Mathsat solver. Based on the experimental results, it was observed that the Mathsat constraint solver data is better for the prediction model than the Z3 solver data, depending on the distribution of the hardness of their constraint models. Experimental results show that the adaptive approach for Mathsat constraint solver data can achieve a F1 score of 0.818, which is higher than Z3 solver data and better than the adaptive non-approach model. In addition, a simulation experiment showed that the adaptive technique might boost constraint solution effectiveness by 1.25x to 3x. In conclusion, with the help of a Mathsat solver with a prediction model, one can speed up constraint solving effectively and support symbolic execution by quickly analysing more paths. In the future, after updating the SMT-LIB format data of the GNU Coreutils and Busybox utilities with other constraint solvers, a comparison of the prediction model can be made. Another possible future work can be the deployment of the prediction model to real-time symbolic execution tools and checking the performance.

References

1. Li, J., Zhao, B., Zhang, C.: Fuzzing: a survey. Cybersecurity **1**(1), 1–13 (2018)
2. Cadar, C., Dunbar, D., Engler, D.R., et al.: Klee: unassisted and automatic generation of high-coverage tests for complex systems programs. OSDI **8**, 209–224 (2008)
3. Godefroid, P., Levin, M.Y., Molnar, D.A., et al.: Automated whitebox fuzz testing. NDSS **8**, 151–166 (2008)

4. Cha, S., Hong, S., Bak, J., Kim, J., Lee, J., Oh, H.: Enhancing dynamic Symbolic Execution by automatically learning search heuristics. IEEE Trans. Softw. Eng. (2021)

5. Moura, L.d., Bjørner, N.: Z3: an efficient smt solver. In: International conference on Tools and Algorithms for the Construction and Analysis of Systems, pp. 337–340, Springer (2008)

6. Cimatti, A., Griggio, A., Schaafsma, B.J., Sebastiani, R.: The mathsat5 smt solver. In: Piterman, N., Smolka, S.A. (eds.) ETAPS 2013, pp. 93–107. Springer, Berlin, (2013). https://doi.org/10.1007/978-3-642-36742-7_7

7. Barrett, C., Stump, A., Tinelli, C., et al.: The smt-lib standard: Version 2.0. In: Proceedings of the 8th international workshop on satisfiability modulo theories (Edinburgh, UK), vol. 13, p. 14 (2010)

8. Chen, Z.: Synthesize solving strategy for symbolic execution. In: Proceedings of the 30th ACM SIGSOFT International Symposium on Software Testing and Analysis, pp. 348–360 (2021)

9. Gario, M., Micheli, A.: Pysmt: a solver-agnostic library for fast prototyping of smt-based algorithms. In: SMT workshop, vol. 2015 (2015), Zhang, T., Wang, P., Guo, X.: A survey of Symbolic Execution and its tool klee. Procedia Computer Science **166**, 330–334 (2020)

10. Barrett, C., Conway, C.L., Deters, M., Hadarean, L., Jovanović, D., King, T., Reynolds, A., Tinelli, C.: Cvc4. In: Gopalakrishnan, G., Qadeer, S. (eds.) Computer Aided Verification, pp. 171–177. Springer, Berlin (2011). https://doi.org/10.1007/978-3-642-22110-1_14

11. Baldoni, R., Coppa, E., D'elia, D.C., Demetrescu, C., Finocchi, I.: A survey of Symbolic Execution techniques. ACM Comput. Surv. (CSUR) **51**(3), 1–39 (2018)

12. Cadar, C., Sen, K.: Symbolic Execution for software testing: three decades later. Commun. ACM **56**(2), 82–90 (2013)

13. Wen, S.-H., Mow, W.-L., Chen, W.-N., Wang, C.-Y., Hsiao, H.-C.: Enhancing Symbolic Execution by machine learning based solver selection. In: Proceedings of the NDSS Workshop on Binary Analysis Research (2019)

14. Barrett, C., Tinelli, C.: Satisfiability modulo theories. In: Clarke, E.M., Henzinger, T.A., Veith, H., Bloem, R. (eds.) Handbook of model checking, pp. 305–343. Springer International Publishing, Cham (2018). https://doi.org/10.1007/978-3-319-10575-8_11

15. Luo, S., Xu, H., Bi, Y., Wang, X., Zhou, Y.: Boosting Symbolic Execution via constraint solving time prediction (experience paper). In: Proceedings of the 30th ACMSIGSOFT International Symposium on Software Testing and Analysis, pp. 336–347 (2021)

Cost-Deadline Constrained Robust Scheduling of Workflows Using Hybrid Instances in IaaS Cloud

Urvashi Nag[1](✉) ⓘ, Amrendra Sharan[1] ⓘ, and Mala Kalra[2] ⓘ

[1] CSIR-Central Scientific Instruments Organisation (CSIO), Chandigarh, India
`urvashinag07@gmail.com`
[2] National Institute of Technical Teachers Training and Research, Chandigarh, India

Abstract. Cloud Computing has gained popularity due to the on-demand resource allocation in the distributed computing environments and provides resources that are dynamically scalable on the "pay as you go" model. Over the past years, Amazon has started providing a new service called EC2 Spot Instances which provides their idle machines on rent in the spot market at a lower cost. Spot instances are the unused virtual machines which are accessible at almost 75% lower price than their on-demand price to perform compute-intensive tasks. Spot instances will end up till the current spot price is less than the user's bid price. In this paper for the cloud environment, the IaaS Cloud Partial Critical Paths(IC-PCP) algorithm is expanded with the aim of reducing execution costs while still meeting user-defined deadlines. ICPCP schedules the task by finding a computation service which can execute complete critical path before its latest finish time. The work presented in this paper proposes a workflow scheduling algorithm, IaaS Cloud Partial Critical Paths with robustness(IC-PCPR) uses both on-demand and spot instances to reduce the cost of workflow execution while satisfying a user-defined deadline and making system robust that runs on heterogeneous resources. The proposed work is simulated in MATLAB framework and the experimental results based on three scientific workflows show that the ICPCPR performs much better than ICPCP algorithm.

Keywords: Fault tolerance · Workflow scheduling · Execution Cost · Deadline · Spot Instances

1 Introduction

Cloud Computing is an emerging trend which offers computation or storage service on subscription-basis. These resources are dynamically available, accessible, and distributed without the need for manual intervention. [1]. It basically focuses on reusing the IT resources. Software as a Service (SaaS), Platform as a Service (PaaS), and Infrastructure as a Service(IaaS) are the various service types offered by cloud providers as a service. In software as a service, consumers can access software using web browsers after it has been deployed by the service provider. In PaaS, the service provider provides

© The Author(s), under exclusive license to Springer Nature Switzerland AG 2024
R. K. Challa et al. (Eds.): ICAIoT 2023, CCIS 1929, pp. 227–240, 2024.
https://doi.org/10.1007/978-3-031-48774-3_16

and manages libraries, programming frameworks and languages for the creation and deployment of applications. IaaS offers access to infrastructure such as storage, network and computing servers [2].

In cloud environment, one of the most difficult challenges to address is workflow scheduling since it requires both expensive calculation and communication. The basic motive of scheduling is to assign the appropriate resources to the tasks in order to achieve one or more objectives. It is a NP-Hard problem as it takes a long time to discover the best solution for a large solution space[3]. There are no such algorithms which may give an optimal solution within polynomial time. Therefore many heuristic and metaheuristic approaches have been introduced in past many years to achieve a near optimal solution in reasonably less time.

Despite of these opportunities in the cloud computing, there is a chance of failures which could unfavorably affect the applications which are executing in the cloud. The failure here means the unsuccessful execution of applications and unable to achieve set goals due to some fault in the system. The failures can be majorly categorized as a permanent and transient failure. The permanent failures are unrecoverable and the processor suffering from permanent failure will not be able to work for the remaining execution while in case of transient failure, the current task is revoked and will be recovered after a span of time and then it will be re-executed [4].

Task failures can occur due to missing input and system errors. Virtual Machine failures can occur due to hardware failure or overloading. Workflow level failure can take place due to server failure and cloud outages [5].

Fault tolerance is an approach in which a system reacts rapidly to an unexpected computation or system error. To overcome these failures there are different fault tolerant techniques such as checkpointing, replication, retry, VM migration and resubmission [6]. The fault-tolerance strategies are classified into two types:–reactive and proactive. Reactive fault tolerance techniques diminish the consequence of failures on the system when failures really occurred. Proactive fault tolerance approaches predict failures and recover the suspected components using other operating components [7].

Before 2009, cloud service providers were providing the on-demand pricing model in which the instances were provided on the demand of the user's request as per the declared prices based on the datacenter regions. Since then, a new pricing model has been proposed using spot instances. Amazon has started providing the idle or unused resources in the auction like market. The providers decide the cost of the spot instance (which varies with time, instance type, regions, and zones) and users will be provided the resource whenever the bid price exceeds the spot price. The user could retain the resources until the current spot price becomes greater than the user's bid price. The cost is reduced up to 70% in the spot market but the quality of service is also degraded. Hence the spot instances are not consistent and exposed to out-of-bid failures. These failures occur in spot market when the spot instance shut down due to increase in spot price which is greater than the user's bid price [8].

The customer has to pay the same spot price for that period irrespective of whether their maximum bid price was higher. The user has not to pay for the partial hours if a failure occurs but has to pay when he himself terminates the instance. The partial hours are converted into full hours for both the models (e.g. 2.3 h is converted into 3 h) [8].

Spot Instance Requests can be categorized into one-time or persistent requests. A one-time request will only be satisfied once but a persistent request will continue submitting the request after each instance termination until the user cancel it. These requests can be helpful if the user has to do continuous work that can be stopped and resumed, such as data processing or video rendering [9].

Fault-Tolerant in Spot-based pricing model means scheduling of tasks on the unused resources provided by the service provider in such a way that it tolerates the out-of-bid failures. Fault tolerance techniques help the system to run without halting when an error occurs.

In 2013, Saeid et al. [10], proposed a deadline-based algorithm IaaS Cloud Partial Critical Paths(IC-PCP) that uses the partial critical path to find the most cost-effective solution for performing business while meeting the specified deadline. The IC-PCP algorithm then attempts to schedule tasks allocated to the appropriate cloud service instances before the deadline. Although the system is able to reduce the costs, it still has room for improvement.

The previous workflow scheduling algorithms focused on minimizing the cost while keeping the makespan within a user-defined deadline. In an attempt to further decrease the execution cost, the trend is to use spot instances in addition to on-demand instances, but spot instances are volatile. Spot Instances are complementary to On-Demand Instances and Reserved Instances, the users will be provided with the resource when the bid price is equal to or greater than the spot price. A workflow scheduling algorithm, IaaS Cloud Partial Critical Paths with robustness(IC-PCPR) using both on-demand and spot instances is proposed which goals to diminish the cost of process execution while satisfying user-defined deadline and making system robust.

The rest of the paper is organised as follows. Section 2 shows the related work and application and resource model are described in Sect. 3. Section 4 covers the details of the proposed work. The performance evaluation can be seen in Sect. 5. Finally, conclusion and future work are put forward in Sect. 6.

2 Related Work

Significant work has been done in the area of robust workflow scheduling algorithms which minimize cost within deadline constraints using spot and on-demand instances. Chandrashekar [5] has proposed different resource allocation strategies and an algorithm which provides robustness. The tasks are scheduled using both the spot and on-demand instances in order to meet the deadline and reduce the cost. The proposed algorithm is robust against prior completion of tasks and performance variations of resources. They have used checkpointing at regular intervals and rollbacking is done in case of failure. It also minimizes the total elapsed time and cost. Bala et al. [7] have also proposed an algorithm that schedules the tasks and reduces the makespan and increases the resource utilisation. They have used proactive fault tolerance approach by handling the failures by predicting them along with the scheduling. If over-utilization of VM occur and the tasks fail due to it than by using VM migration approach, it automatically migrates the overloaded VM to another running host. But if resource overutilisation does not occur, then the task would be executed on the assigned machines.

Poola et al. [8] have proposed two workflow scheduling approaches which uses both on-demand and spot instances out of which one provides fault tolerance using replication when the deadline is short whilst optimizing cost and time. Another is retry or resubmission as a fault tolerant strategy when the deadline is loose. It is cost effective but makespan is increased. When the deadline is indulgent, replication is used to mitigate failure and maximize the resource utilisation. They have used intelligent bidding strategy for bidding the spot instances. The two metrics used to provide fault tolerance is failure probability and replication factor.

Voorsluys et al. [11] achieves reliability by saving the state of VM on a global file system in the same or different datacenter and when out-of-bid failures occur then the VM is relocated to some appropriate host machine. They have used different resource policies for optimizing the cost as a deadline constraint and has used hourly-based checkpointing, where the state of an application is saved at hourly basis so that in case of failures the partial hours should not be charged. They have proposed the fault tolerance technique in which a replica of the task which is running for more than 1 h is taken and submitted to the other datacenter with the same scheduling policy. They have used mean, minimum, on-demand, maximum and current spot price as bidding strategies.

The Fault and Intrusion-tolerant Workflow Scheduling method (FITSW), developed by Farid et al.[12], uses several virtual machines to perform workflow activities and boost workflow dependability. FITSW repeats each sub-task three times to produce sub-deadlines, then employs an intermediate data decision-making procedure and a deadline segmentation technique. The suggested method creates or reuses task executors, maintains a tidy workflow, and enhance productivity. According to the findings, FITSW outperforms intrusion-tolerant scientific workflow (ITSW) systems in terms of success rate by approximately 12%, task completion rate by 6.2%, and completion time by about 15.6%.

Liu et. al. [13] has used Amazon EC2 dataset and predicted the spot price by creating a k-Nearest Neighbors (kNN) regression model based on numerical explanation of spot instance price prediction problem and then compared the model with Linear Regression, Support Vector, gcForest, Multi-layer Perception and Random Forest. The results shows that the k-NN approach gives the best performance with Mean Absolute Percentage Error (MAPE5%) is around 94% in 1-day-ahead prediction and 94.06% in 1-week-ahead, while RF and gcForest having 85.61% and 83.86% on average and other even has less then it. Wang et. al. [14] used an ant colony algorithm-based approach to host data in multi-cloud setups, which helps in optimizing objectives like cost and availability.

Scheduling with on-demand instances can reduce the cost of doing business by around 90%. The paper presents a method that uses neural network techniques to predict spot prices. The algorithm is able to perform well with the help of recurrent neural network.

Yi et al. [15] have used checkpointing strategy for overcoming out-of-bid failures and reducing the cost of execution. They have proposed two policies for deciding the frequency of checkpointing in a running application. One is to take checkpoints on hourly basis and another is when the spot price increases for a given VM. Poola et al. [16] have proposed an algorithm that takes checkpoints at different intervals of 5, 15 and 30 min. They have used different bidding strategies (intelligent, naïve and on-demand) for bidding the price so that the bid price is always higher than the spot price. The scheduling algorithm calculates the slack which is the variation between the critical path time and deadline. If the slack period is greater than the failure probability, it runs spot instances otherwise on-demand instances are executed. By using checkpointing the cost is saved up to 14%.

3 Application and Resource Model

A workflow application is represented in the form of Directed Acyclic Graph (DAG), with a tuple $G = (T, E)$, where T depicts the set of tasks (vertices) and E represents set of the edges that specify the dependencies among the tasks. The precedence relationship is defined as $E = (t_p, t_i)$ where t_p is the parent task and t_i is the child task and $t_p, t_i \in T$, $t_p \neq t_i$. Once the parent tasks have completed their execution, the child task begins. A task without a parent is referred to as the start or entry task while a job without a child is referred to as the exit task or end task. One single and one exit node is required, dummy input and output node t_{in} and t_{out} respectively is added. Both the execution time and the weight requirements for the real entry and exit tasks are absent from these dummy nodes.

The resource model is based on the IaaS cloud which provides a wide range of virtual machines for the computation services. There are different machines available for the computation $M = \{m_1, m_2, \ldots, m_n\}$ having different processing speeds and costs. A virtual machine is primarily described by its processing speed, $PS(rj)$, measured in million instructions per second (MIPS) and cost per unit of time $C(r_j)$.

Basic Terminology.

Running time ($RT(t_i, m_j)$) - It is defined as the completion time of a tasks t_i on any machine m_j.

Minimum Running Time ($MRT(t_i)$) - It is the minimum running time of task t_i on a machine m_j.

$MRT(t_{in}) - 0$

$MRT(t_{out}) = 0$

Data Transfer Time ($DTT(e_{p,i})$) - This time depends on the amount of data to be transferred between the corresponding tasks and does not depend on the machine that is executing that tasks. $DTT(e_{p,i}) = 0$, when both the parent and child tasks are executing on the same instance.

Primary Start Time ($PST(t_i)$) - The Primary Start Time is the start time of the task t_i which is calculated before scheduling the workflow. $PST(t_i)$, is defined as follows:

$PST(t_{in}) = 0$

$PST(ti) = \max\{PST(t_p) + MRT(t_p) + DTT(e_{p,i})\}$

Real Start Time($RST(t_i)$) - This is the real execution time of the task on the selected machine which is determined after the execution of that task.

Estimated Finish Time($EFT(t_i)$) - This is the estimated time of the task by which the task should complete its execution.

$EFT(t_i) = PST(t_i) + MRT(t_i)$.

Maximum Finish Time($MFT(t_i)$) - It is defined as the maximum time at which task t_i should finish its computation.

MFT $(t_{out}) = D$

MFT$(t_i)= min(MFT(t_i)-MRT(t c)-DTT(e p,i))$

Partial Critical Path (P) - It is the longest path between the entry and exit node of the given workflow.

Selected Machine (SM(t_i) = $M_{k,l}$)   It is the selected machine for computation of partial critical path where $M_{k,l}$ is the kth instance o**Time for Spot (TS)** - TS is the time to switch to spot instances in order to lower the cost as shown in Eq. (1). It is the mean of the difference between the MFT and EFT. It is basically used to determine the right instance for speeding up and slowing down the processing by choosing the correct pricing model.

Timeleft= MFT$(t i)-$EFT$(t i)$, where timeleft is the variation between the maximum and estimated finish time of the task t_i, where t_i belongs to tasks in P.

$$TS = \frac{Timeleft}{No.of\ tasks\ in\ P} \tag{1}$$

When the spot instance is shut down due to increase in spot price and becomes greater than the user's bid price, it leads to out-of-bid failures. To mitigate such failures workflow scheduling algorithm should incorporate some kind of fault-tolerant and bidding strategy.

Bidding Strategies. These strategies are used to obtain the bid price in the scheduling algorithm. So, an intelligent bidding strategy should be proposed such that the bidding price should remain greater than the spot price so that the out-of-bid failure doesn't occur. The bidding strategy used here to acquire the bid price for obtaining the spot instance is described below.

Pricing Model.
This algorithm uses both an on-demand and a spot pricing strategy. When using a spot instance, users must place a bid on it in an auction-style market, and the instance is only made accessible when the bid price is more than or equal to the spot price. In the on-demand pricing model, customers must pay by the hour based on the type of instance they are using. Different geographies, availability zones, times, and instance kinds have different spot prices.

Table 1. List of Abbreviations

Abbreviation	Description
EFT	*Earliest Finish Time*
MFT	*Maximum Finish Time*
PST	*Primary Start Time*
RST	*Real Start Time*
TS	*Time for Spot Instances*

4   Proposed Work

The previous workflow scheduling algorithms focused on minimizing cost while keeping the makespan within a user-defined deadline. In order to further reduce the execution cost, the trend is to use spot instances in addition to on-demand instances, but spot instances are volatile. The proposed strategy ICPCPR consists of three algorithms: InitializeWorkFlow(), AssignInstance() and SchedulePath() which are discussed below:

Algorithm 1: Initialize Workflow Algorithm.
Algorithm 1 shows the pseudo-code of ICPCPR algorithm for scheduling a workflow. Firstly, all available machines for computation are determined. Line 2 computes the different times. $RST(t_{out})$ is assigned the deadline and $RST(t_{in})$ is set to 0 and mark t_{in} and t_{out} as executed. Finally, AssignInstance is called for t_{out} which schedules the unexecuted tasks.

```
ALGORITHM 1: InitializeWorkFlow(G(T,E),D)
Input: Graph G, Deadline D
1.    determine available Virtual Machines M
2.    compute PST(tᵢ), EFT(tᵢ) and MFT(tᵢ) for all tasks.
3.    RST(tᵢₙ) ← 0, RST(tₒᵤₜ) ←D
4.    mark tᵢₙ and tₒᵤₜ as executed
5.    AssignInstance(tₒᵤₜ)
```

Algorithm 2: Assign Instance Algorithm.
The pseudo code for AssignInstance is illustrated in Algorithm 2. When given tasks as input, this method schedules all of the incomplete tasks before the deadline. Line 3–7 shows the creation of the partial critical path P of the tasks given as input. It starts from t_{out} and goes back to critical parents until it reaches t_{in}. It also calculates timeleft of all the tasks by calculating the difference between MFT and EFT. Line 8 computes the time for a spot by calculating the mean of the timeleft. Lines 8–15 are executed only when the TS calculated is higher than the threshold value. Firstly, the bid price is predicted using bidding strategy and if it is greater than spot price then SchedulePath algorithm is called for spot model else it is called for on-demand model. Line 16 is executed when the deadline is lenient. In the next section of the code IST and EFT of all predecessors and MFT for all successors are updated. After that, the algorithm recursively called AssignInstance for the remaining unexecuted tasks.

```
ALGORITHM 2: AssignInstance (t)
Input: Tasks t
1.      while (t is not mapped to any resource) do
2.              P←NULL, tᵢ←t, status←false
3.              while(there is unexecuted parents of task tᵢ) do //
Creation of partial critical path
4.                      add critical parent tₚ of tᵢ to start of P
5.                          tᵢ ← tₚ
6.                          Timeleft+= MFT(tᵢ) - EFT(tᵢ)
7.              end while
8.              TS← (Timeleft)/(No. of tasks in P)
9.              if (TS > threshold) then
10.             bᵢ_d←PredictBidPrice(tᵢ, instance type)
11.                     if (bᵢ_d > p_spot) and (bᵢ_d < p_on-demand)  then
12.                         call SchedulePath(P,Spot)
13.                     else
14.                         call SchedulePath(P,On-demand)
15.                     end if
16.             else    call SchedulePath(P,On-demand)// cheapest machine M
which can execute the  tasks before its MFT
17.             end if
18.             for all (tᵢ ∈ P) do:
19.                     update PST and EFT for all successors of tᵢ
20.                     update MFT for all predecessors of tᵢ
21.                     call AssignInstance(tᵢ)
22.             end for
23.     end while
```

Algorithm 3: Schedule Path Algorithm

Algorithm 3 demonstrates the pseudo code of SchedulePath algorithm. The SchedulePath algorithm takes partial critical path P and the pricing model PM as inputs and schedules all the tasks in P on the single instance of the virtual machine which can execute them early than its MFT and with the minimal cost. The time of data transfer between the tasks is zero as they are running on the same instance. Firstly, a low-cost running machine is determined for the pricing model passed as a parameter in the Algorithm 2, which can run the tasks in P before its MFT. Secondly, if the running instance is not available then new instance is launched which can finish tasks before MFT. After finding the appropriate machine, P is executed on that machine and values of $RST(t_i)$ and $SM(t_i)$ are set. Mark all the tasks in the P as executed.

```
ALGORITHM 3: SchedulePath (P, PM)
input: Partial Critical Path P, Price Model PM
1.      Mᵢ,ⱼ←the low-cost running machine which can finish tasks before MFT
2.              if (Mᵢ,ⱼ == NULL) then
3.                  launch new instance which can finish tasks before MFT
4.              end if
5.      schedule P on Mᵢ,ⱼ and set RST(tᵢ), SM(tᵢ ) for each tᵢ ∈ P
6.      set all tasks of P as executed
```

Fault-Tolerant Strategy

The strategy used for fault-tolerant is checkpointing. It is an adequate fault tolerant procedure in which it takes snapshots repeatedly and if a task fails, it restarts the execution from the recently checked point. The limitation of this technique is that it requires fault tolerance in cloud scheduling. For failure recovery, checkpointing is used in case of any kind of failure occurs. For a faster machine, checkpointing can be done at frequent interval of time and for slower machine checkpointing can be done at a larger interval in order to overcome the overhead indulge due to checkpointing.

5   Performance Evaluation

The experiment was performed using the MATLAB [18] framework to assess how well the suggested technique performed. In this paper, three workflows[17]: Montage, Cyber-shake, SIPHT of 100 tasks from different scientific areas, are considered. They have their different data and computational requirements. They have different structures including pipeline, data aggregation, data distribution, and data redistribution.

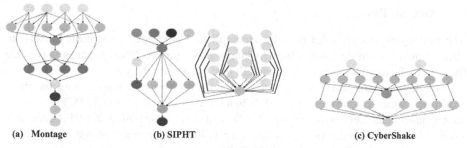

(a)   Montage   (b) SIPHT   (c) CyberShake

Fig. 1. Structure of three scientific workflows used in the proposed work [18]

Bidding Strategy. Table 2 shows the on-demand pricing of the machine available for the US East Region (North Virginia) as per the Amazon EC2 On-Demand Pricing Model [19]. In order to evaluate the bid price, artificial neural network is used. Figure 2 shows the generation of bid value through the ANN for different values of TS times.

Table 2. VM Cost for US East Region (North Virginia)

Name	EC2 Units	Memory (GB)	Cost per hour (On-demand Instances hourly rate)
m3.large	6.5	7.5	$0.133
m3.xlarge	13	15	$0.266
m3.2xlarge	26	30	$0.532

Deadline is an important parameter in workflow scheduling. In this experiment, eight different deadlines are taken. The deadlines are calculated so that the value lies between the slowest and fastest runtimes. Each workflow is executed 10 times for the different deadlines.

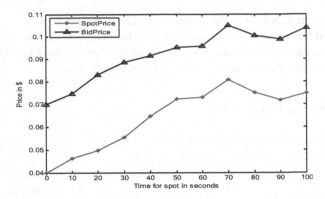

Fig. 2. Creation of Bid Price using Artificial Neural Network (ANN)

5.1 Analysis Based on Cost

The first significant goal that must be accomplished is the execution cost. In order to reduce costs, both on-demand and spot instances are employed. Spot instances can reduce the cost by up to 40% when there is enough free time to do the job.

For the Montage workflow, the mean execution cost is the least among the three workflows which can be seen in Fig. 3. When the deadline is relaxed, ICPCPR uses the spot instances and reduces the cost by 52.87% w.r.t. ICPCP algorithm. ICPCPR generates savings of 34.63% under tight deadline w.r.t ICPCP algorithm. For the Cybershake, the cost obtained is shown in Fig. 4. ICPCPR reduces the execution cost by 44.81% w.r.t. ICPCP under strict deadline. ICPCPR shows significantly better performance w.r.t ICPCP when deadline is relaxed and reduces the execution cost by 51.89%. The result of SIPHT shows better performance among the three workflows under the relaxed deadline it reduces the cost by 57.45% w.r.t. ICPCP which can be seen in Fig. 5. Since ANN predicts the bid price based on history prices, it achieves good results and the proposed algorithm shows better performance. ICPCPR reduces the execution cost 35.05% in the SIPHT workflow under tight deadline.

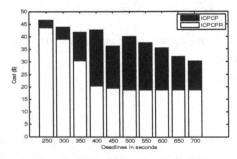

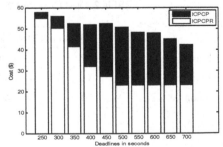

Fig. 3. Mean Cost of Montage with   **Fig. 4.** Mean Cost of with Cybershake

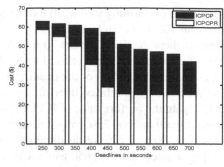

Fig. 5. Mean Cost of SIPHT with

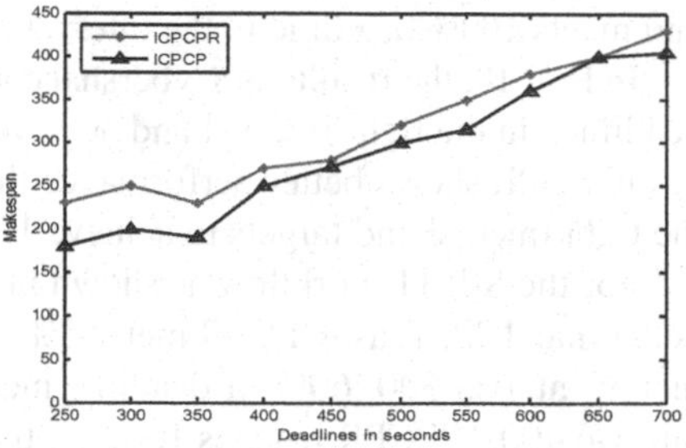

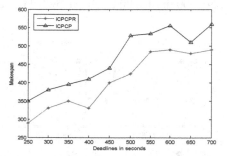

Fig. 6. Mean Makespan of Montage with

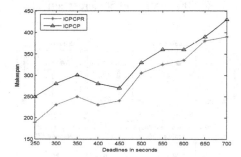

Fig. 7. Mean Makespan of Cybershake with

Fig. 8. Mean Makespan of SIPHT with

5.2 Analysis Based on Makespan

The proposed work in comparison with the base algorithm ICPCP lowers the makespan. Mean makespan of Montage, Cybershake and SIPHT is demonstrated in Fig. 6, Fig. 7 and Fig. 8 respectively. In spite of the failure, the makespan of ICPCPR is lower than the ICPCP. In the Montage workflow, ICPCP and ICPCPR have almost same makespan. Montage has a lesser makespan as compared to SIPHT as it has serial structure than Cybershake and SIPHT.

5.3 Analysis Based on Fault Tolerance

Fault tolerance is another main objective of our algorithm. The main metric of fault tolerance is failure probability. Failure probability P(f) is the execution of workflow without meeting the deadline which is shown as in Eq. (2).

$$P(f) = \frac{AbsoluteRun - UnsuccessfulRun}{AbsoluteRun} \tag{2}$$

The graphs generated displaying the percentage of deadlines met for each workflow and deadline intervals in order to assess the algorithm's failure probability. For Montage workflow in Fig. 9, ICPCP shows significantly worse performance than ICPCPR as it fails to meet the all the deadlines in the tight interval (250 to 400) and achieves 35% of the deadline in the relaxed interval whereas ICPCPR is performing well in all the cases and meets 100% deadline in the interval ranges from 450 to 600 (relaxed interval).

In Fig. 10, the results of Cybershake application show that ICPCP meet 20% of the deadlines in the tight interval and achieves 40% of the deadline in the relaxed interval but ICPCPR shows better performance than ICPCP as it meets 70% of the deadline in the tight interval and targets to achieve 100% of the deadline in relaxed interval.

For the SIPHT workflow as shown in Fig. 11, again the performance of ICPCPR is better than ICPCP, as ICPCPR mets 55% of the deadline at interval 250, 65% of deadline met at interval 300, 60% of deadline met at interval 350 and 75% of deadline met at interval 400. ICPCPR reaches 100% of the deadline at interval 450 to 650 while ICPCP reaches 20% of the deadline at interval 250, 30% of the deadline at 300 interval, 45% of the deadline at interval 350, 35% of the deadline at 400. In the relaxed interval 450, 500, 550, 600 the deadline achieve by ICPCP are 50%, 60%, 55% and 50% respectively.

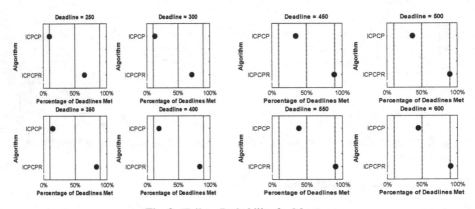

Fig. 9. Failure Probability for Montage

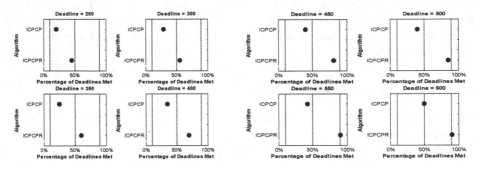

Fig. 10. Failure Probability for Cybershake

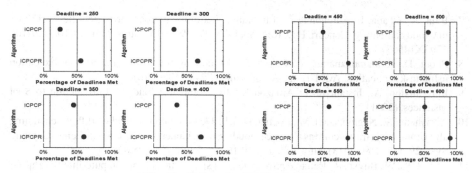

Fig. 11. Failure Probability for SIPHT

6 Conclusion and Future Scope

This paper presented the ICPCPR algorithm for deadline constrained workflow scheduling which uses both spot and on-demand instances. The algorithm seeks to minimize to reduce the execution cost of the given workflow within a user-defined deadline. It provides robust scheduling over out-of-bid failures using checkpointing. The bid price is anticipated using an artificial neural network, which produces extremely effective results because the forecast bid price is typically greater than the spot price. The simulation results prove that ICPCPR performs better than ICPCP in terms of execution time as well as execution cost. The failure probability is measured in terms of percentage of deadlines met and in most of the cases, ICPCPR is able to behave better than ICPCP.

In future work, the algorithm on real infrastructure can be implemented. Other failures can also be consider like VM failure, server failure etc. for making the algorithm fault tolerant with regard to these failures. In order to increase robustness and minimization of cost, task duplication can be used.

References

1. Zhang, Q., Cheng, L., Boutaba, R.: Cloud computing: State-of-the-art and research challenges. J. Internet Serv. Appl. **1**, 7–18 (2010). https://doi.org/10.1007/s13174-010-0007-6
2. Amin, Z., Sethi, N., Singh, H.: Review on fault tolerance techniques in cloud computing. Int. J. Comput. Appl.Comput. Appl. **116**, 11–17 (2015). https://doi.org/10.5120/20435-2768
3. Keshk, A.E., El-Sisi, A.B., Tawfeek, M.A.: Cloud task scheduling for load balancing based on intelligent strategy. Int. J. Intell. Syst. Appl. **6**, 25–36 (2014). https://doi.org/10.5815/ijisa.2014.05.02
4. Mei, J., Li, K., Zhou, X., Li, K.: Fault-tolerant dynamic rescheduling for heterogeneous computing systems. J. Grid Comput. **13**, 507–525 (2015). https://doi.org/10.1007/s10723-015-9331-1
5. Chandrashekar, D.P.: Robust and fault-tolerant scheduling for scientific workflows in cloud computing environments. In: IEEE 28th International Conference on Advanced Information Networking and Applications, pp. 858–865 (2015)
6. Bala, A., Chana, I.: Fault tolerance- challenges, techniques and implementation in cloud computing. Int. J. Comput. Sci. Softw. Issues. **9**, 288–293 (2012)

7. Bala, A., Chana, I.: Autonomic fault tolerant scheduling approach for scientific workflows in Cloud computing. Concurr. Eng. Res. Appl. **23**, 27–39 (2015). https://doi.org/10.1177/106 3293X14567783

8. Poola, D., Ramamohanarao, K., Buyya, R.: Enhancing reliability of workflow Execution using task replication. ACM Trans. Auton. Adapt. Syst. **10** (2016)

9. Introduction, A., Instances, S.: Amazon Elastic Compute Cloud An Introduction to Spot Instances. (2011)

10. Abrishami, S., Naghibzadeh, M., Epema, D.H.J.: Deadline-constrained workflow scheduling algorithms for infrastructure as a service clouds. Futur. Gener. Comput. Syst.. Gener. Comput. Syst. **29**, 158–169 (2013). https://doi.org/10.1016/j.future.2012.05.004

11. Voorsluys, W., Buyya, R.: Reliable provisioning of spot instances for compute-intensive applications. In: Proceedings of IEEE 26th International Conference on Advanced Information Networking and Applications, pp. 542–549 (2011)

12. Farid, M., Latip, R., Hussin, M., Hamid, N.A.W.A.: A fault-intrusion-tolerant system and deadline-aware algorithm for scheduling scientific workflow in the cloud. PeerJ Comput. Sci. **7**, 1–21 (2021). https://doi.org/10.7717/PEERJ-CS.747

13. Liu, W., Wang, P., Meng, Y., Zhao, C., Zhang, Z.: Cloud spot instance price prediction using kNN regression. https://doi.org/10.1186/s13673-020-00239-5

14. Wang, P., Zhao, C., Zhang, Z.: An ant colony algorithm-based approach for cost-effective data hosting with high availability in multi-cloud environments. In: ICNSC 2018 - 15th IEEE International Conference on Networking, Sensing and Control, pp.1–6 (2018). https://doi.org/10.1109/ICNSC.2018.8361288

15. Yi, S., Kondo, D., Andrzejak, A.: Reducing costs of spot instances via checkpointing in the Amazon elastic compute cloud. In: Proceedings of IEEE 3rd International Conference of Cloud Computing, CLOUD, pp. 236–243 (2010). https://doi.org/10.1109/CLOUD.2010.35

16. Poola, D., Ramamohanarao, K., Buyya, R.: Fault-Tolerant workflow scheduling using spot instances on clouds. Procedia Comput. Sci. **29**, 523–533 (2014). https://doi.org/10.1016/j.procs.2014.05.047

17. Bharathi, S., Chervenak, A., Deelman, E., Mehta, G., Su, M.H., Vahi, K.: Characterization of scientific workflows. Work. Support Large-Scale Sci. Work (2017). https://doi.org/10.1109/WORKS.2008.4723958

18. Ma, X., Gao, H., Xu, H., Bian, M.: An IoT-based task scheduling optimization scheme considering the deadline and cost- aware scientific workflow for cloud computing. EURASIP J. Wirel. Commun. Netw. (2019). https://doi.org/10.1186/s13638-019-1557-3

19. Amazon EC2 On-Demand Pricing, https://aws.amazon.com/ec2/pricing/on-demand/ (last visited on August 2017)

AI and IoT for Smart Healthcare

Comparative Performance Analysis of Machine Learning Algorithms for COVID-19 Cases in India

Apoorva Sharma[1]($\boxtimes$) , Maitreyee Dutta[1] , and Ravi Prakash[2]

[1] National Institute of Technical Teachers Training & Research (NITTTR), Chandigarh, India
{apoorva.cse20,maitreyee}@nitttrchd.ac.in
[2] Digital University Kerala (formerly IIITM-K), Thiruvananthapuram, India

Abstract. A novel corona virus is the cause of the viral infection recognized as COVID-19 (initially named as SARC-CoV-2). Since the pandemic emerged in the Wuhan province of China in November 2019, it has been recognized as a global threat. However, over the next two years, it has been witnessed that the novel corona virus tends to evolve rapidly. In this paper, we leverage our time-series data collected since the initial spread of COVID-19, mainly in India, to better understand the growth of this pandemic in different regions throughout the country. The research is based on cases reported in India in chronological order. In addition to numerous previous works, we have tried to come up with the most appropriate solution to estimate and predict the newly reported COVID-19 cases in the upcoming days, with the least possible error through machine learning. This study also aims to compare multiple machine learning algorithms on various factors and their trade-off for prediction. The experimental results indicate that Orthogonal Matching Pursuit is the best algorithm for this problem. We make our dataset available for further research.

Keywords: COVID-19 · Corona · Machine Learning · Regression · Predictive Model

1 Introduction

The novel corona virus is known to emerge in China in November 2019. It was a viral infarction that rapidly spread from bats to humans, and within a few months, it reached almost every other country worldwide. COVID-19 virus was found to affect the human respiratory system severely. Unfortunately, even after detecting the virus and its behavior in the early days, it took over six months to come up with proper vaccination. COVID-19 was declared a global pandemic due to its very sharp growth rate. Moreover, the virus mutated rapidly. It made

The link to access the final code and dataset used for essential data preparation and testing of the model is: https://github.com/apoorva46/COVID-19-Project-2023.

© The Author(s), under exclusive license to Springer Nature Switzerland AG 2024
R. K. Challa et al. (Eds.): ICAIoT 2023, CCIS 1929, pp. 243–257, 2024.
https://doi.org/10.1007/978-3-031-48774-3_17

the situation even more challenging as the virus was also found to evolve over time. Moreover, it was found that different waves of COVID-19 proved fatal for people of different age groups.

Even after a lot of effort regarding its prevention, COVID-19 is still found to be frequently spreading. In such a situation, the estimation of a possible number of new cases seems very difficult. Based on the several approaches involved in predictive data analysis, people have tried to estimate the count of new cases and the rate of its growth [10, 23]. This paper brings in a more detailed comparison of different machine learning algorithms. It proposes methodologies to accurately predict the counts of new cases in the following few days based on the analysis. This study describes the disease's state in terms of total confirmed cases till a specific date. Here, the time-series data has been cumulatively stored on a daily basis, along with the counts of death cases and recovered cases every day. In this work, we are predicting the total number of cases in the upcoming days with the help of different supervised ML algorithms and comparing their performance based on the five evaluation metrics MAE, MSE, $RMSE$, $MAPE$ & R^2 [20]. Here, we have also analyzed the trends in the number of COVID-19 instances in different states of India.

The organisation of this paper is - Introduction to the problem is discussed in Sect. 1. It is followed by the Literature Review in Sect. 2, which covers a detailed description of relevant previous work and the research gaps. Section 3 provides the complete methodology. The outcomes of our work are given in Sect. 4. Conclusion and Future Scopes of this work is given in Sect. 5.

2 Literature Review

Our motivation behind this paper is to develop a solution to estimate the possible count of newly reported COVID-19 cases for the near future using some regression-based ML model.

After going through recently published papers, we found that Zhong *et al.* [26] developed a mathematical model to forecast the spread using epidemiological data in March 2020. Over time, it was discovered that it had no effect. In April 2020, Benitez Pena *et al.* [5] used RF and SVM for the analysis of the disease. They leveraged Gurobi to solve the problem.

Chakraborty *et al.* [8] used ML approach for solving two problems, i.e., forecasting short-term future COVID-19 cases and risk assessment based on fatality rate. Here authors explained that hybrid time-series models using ARIMA & wavelet-based forecasting techniques for predicting cases in five different countries could be the best approach for the problem. As the pandemic was dynamically spreading throughout the world, all of these researchers discovered that there were kinds of dissimilarity in the rate of its growth.

In April 2020, Vashisht *et al.* [23] explained the growth rate of a novel corona virus in China based on the regression models. Kanagarathinam *et al.* [13] proposed the SEIRS model and used data of a month for prediction. Sujath et. al. [21] performed Linear Regression along with a Neural Network based method,

Table 1. Abbreviations

Abbreviation	Full Forms
ARIMA	AutoRegressive Integrated Moving Average
DT	Decision Tree
ES	Exponential Smoothing
GB	Gradient Boosting
KNN	K-Nearest Neighbors
LR	Linear Regression
MAE	Mean Absolute Error
MAPE	Mean Absolute Percentage Error
ML	Machine Learning
MSE	Mean Square Error
NB	Naive Bayes
NN	Neural Network
PR	Polynomial Regression
RF	Random Forest
RL	Reinforcement Learning
RMSE	Root Mean Square Error
SVM	Support Vector Machine
VAR	Vector AutoRegression

by applying Multi-Layer Perceptron and a multivariate solution using Vector AutoRegression on kaggle data having 80 instances of COVID-19 (Table 1).

Rustam *et al.* [19] used Linear Regression, Least Absolute Shrinkage and Selection Operator, Support Vector Machine, and Exponential Smoothing and concluded that ES, LASSO & LR performed better than SVM. Nabi [17] employed Trust Region Reflective algorithm for tentative predictions of the epidemic peak. Goswami *et al.* [11] used the Verhulst Logistic Population Model and also used the Generalized Additive Model of regression to examine the impact of various meteorological parameters on the prediction of COVID-19 instances. In order to estimate the trend of the outbreak, Wang *et al.* [24] incorporated epidemiological data collected before June 16, 2020, into the logistic model.

Burdick *et al.* [6] assessed an ML algorithm's effectiveness for predicting invasive mechanical ventilation in COVID-19 patients within 24 h of the first contact. Amar *et al.* [4] used logistic growth regression models to analyze COVID-19 data of Egypt. Khanday [14] showed that Logistic Regression and Multinomial NB algorithms produce better results than RF, Stochastic GB, DT, and Boosting. Yadav *et al.* [25] proposed a Novel Support Vector Regression method rather than employing a simple regression line to assess five tasks differently.

Darapaneni *et al.* [9] used RF and found accuracy for the training data as 97.17% and that for testing data as 94.80%. Kumari *et al.* [15] used DT training techniques for splitting data and Autoregression model to predict the possible number in the future. Gupta *et al.* [12] used RF, Linear Model, SVM, DT, and NN for forecasting and found that RF outperformed the others. Mary *et al.* [16] employed Feature Selection Techniques, SVM, kNN, and NB, and found that SVM outperformed other algorithms. Tiwari *et al.* [22] conducted a comprehensive evaluation and comparison of 37 ML-related studies that covered many ML algorithms. However, all these papers have used limited dataset due to early conduct of their research works. In few papers, vague assumptions have been made. Apart from this, forecast is done for only a week and results is some papers are overestimated.

3   Dataset and Methodology

3.1   Dataset

Our dataset was compiled using multiple sources to ensure its accuracy and completeness. To validate the data and ensure that the most appropriate time-series information was collected, we referred to several sources including Ministry of Health and Family Welfare (MoHFW) [1] and Indian Council of Medical Research (ICMR) [2] of India. The data was collected from as early as January 2020, i.e., since initial days of COVID-19 in the country. We utilized R scripts to fetch and validate the data from the GitHub repository of CSSE, Johns Hopkins University [3] as well. The CSSE database consisted of the data related to not only India but for the entire world. Additionally, we cleaned the data by removing outliers and filling in null values for several states where no cases were reported during the early days.

3.2   Model Configuration

We target training and evaluating the performance of multiple ML algorithms by training on the same data. To meet this goal, our research work primarily utilizes the programming language 'Python', and for comparing the performance of different models based on five evaluation metrics, we have employed PyCaret, Table 4 provided in Appendix contains the configuration details used for training the models.

3.3   Methodology

The collected time-series data contained a total count of active, recovered, and death cases in 36 states and union territories of India. The data also consisted of the cases from some unclassified locations those were handled during the data pre-cleaning phase. The date-range of collected data starts from 25th March 2020 till the day when the lockdown was withdrawn, i.e., 28th July 2020. This data was collected every single day and stored in two formats. One formatting was done as a cumulative data for India. The other format of data was purely time series with daily case counts. The final time-series data was taken as the cumulative sum of confirmed cases ζ^i on every i-th day for every j-th state and union territory S_j of India as per Eq. 1. Later, the collective sum of counts of every S_j is taken to generate the data for the whole country (Eq. 2).

$$\zeta^i = \zeta^i_{S_j} \tag{1}$$

$$\zeta^i_s = \zeta^i_\delta \ \forall \ \delta = \sum_{36}^{j=1} S_j \tag{2}$$

Working towards our goal, in order to forecast the daily count of unique cases, we created a separate column in the dataset η_i for each day using Eq. 3. The proposed approach performed this computation for each day from 25th March 2020 to 28th July 2020 in India.

$$\eta_i = \zeta^i_S - \zeta^{i-1}_S \tag{3}$$

Firstly, data cleaning was performed that included the following three essential steps:

- **Step-1:** Removing or replacing the redundant values from data
- **Step-2:** Removing the $NULL$ and dealing with blank (no) data points
- **Step-3:** Outliers removal

For further analysis, we have split our entire data in a ratio of 70 : 30 as training and testing data, respectively 85 : 30 rows. In order to make the comparison, the 'PyCaret' library has been used, which was developed by data scientist Moez Ali. This library is capable of building many regression models simultaneously.

Once the dataset was ready, we began with the regression-based supervised learning approach to build the ML model. Here, we have taken three folds cross-validation in order to increase the efficiency of the models. At the same time, we sorted the performance of multiple regression algorithms based on the following five evaluation metrics $MAE, MSE, RMSE, MAPE$ & R^2. We have separately studied cumulative data and data consisting of new cases of all the states.

This allowed us to set up various machine-learning models using the generated data. Here the aim was to predict the total number of new cases of COVID-19 in the country. We tried to explore the maximum possible literature works, pondering and analyzing the work as per the state of the art approaches. Here, we trained many regression algorithms, namely Orthogonal Matching Pursuit, Extra Trees Regressor, Ridge Regression, Light GB Machine, and 15 more algorithms mentioned in Table 2 of Sect. 4.

4   Results and Observations

We initially tried exploring the data of COVID-19 spread in India both statewise as well as collectively for the whole country. These graphs typically depict the cumulative number of cases and new cases in all the states of India. The line graphs plotted for this observation had varying trends. After analyzing all graphs, we can observe that results in a few states are dynamically varying, whereas, in some states, the trend is constant. We found that big states like Rajasthan, and Tamil Nadu, as shown in Fig. 1 and Fig. 2, respectively, have fluctuations in new case counts for short duration.

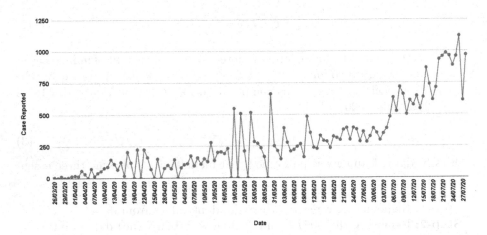

Fig. 1. Fluctuating number of new cases in Rajasthan from 26th March 2020 to 27th July 2020.

However, it can be seen that there is increase in number of new cases reported in a long period of time. One possible reason could be the wide area where it is a challenging task to deploy a complete lockdown in one go. Considering these insights, we found that the large states and those with higher populations decide the overall trend in growth of COVID-19 in the country. On the basis of the hypothesis Fig. 3 shows the reporting of new cases on the daily bases. Similarly in Fig. 1 and Fig. 2, it can observed that the count of newly reported COVID-19 cases has increased over the time. It became more clear when we analyzed the cumulative growth of COVID-19 pandemic throughout the country, as shown in Fig. 4.

We have also noticed that few of the states and UTs show some different trends. Delhi is one among these few. We can see from Fig. 5 that cases in Delhi have actually decreased over the time. We can actually trace here a pattern among growth and control in the COVID-19 cases over the time. As a result of this, we can easily predict the total number of instances in the following days. However, without changing its configurations, it is unlikely that a model can accurately predict the upcoming cases for states like Tamil Nadu as precisely as it can for Delhi. Meanwhile, the decrease in number of new cases emphasizes how the usage of masks and creating awareness, and imposing lockdowns helped in the management of this lethal spread.

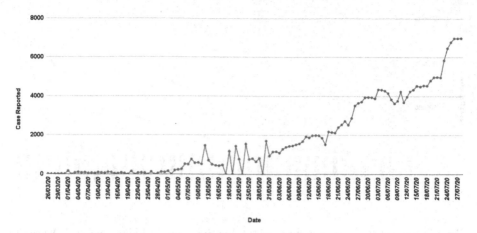

Fig. 2. Sharp increase in number of new cases in Tamil Nadu from 26th March 2020 to 27th July 2020.

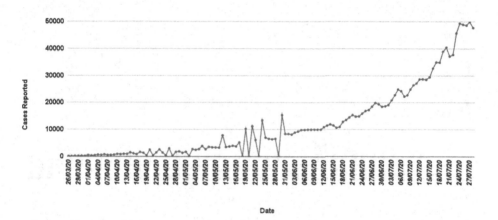

Fig. 3. Fluctuation in number of new cases in India during lockdown from 26th March to 27th July 2020.

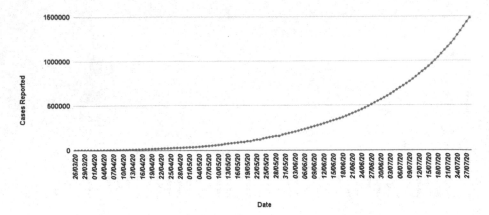

Fig. 4. Constant increase in number of total active cases in India from 26th March to 27th July 2020.

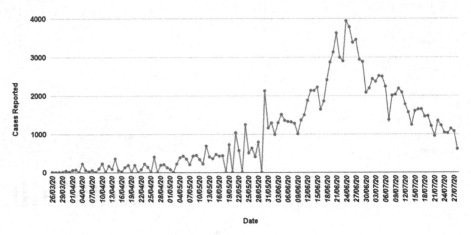

Fig. 5. Trend of COVID-19 daily spread in Delhi between 26th March 2020 and 27th July 2020.

After carefully examining all the graphs, it is clear that the deployment of a number of stringent regulations and public safety measures has been successful because the number of cases has decreased whenever travel restrictions and a lockdown have been put in place. Also, we found that it is very difficult to build one identical model to forecast the results for all different states. Therefore, we tried to collectively analyze the data for the whole country and forecast the results accordingly. It can also be seen in Fig. 4 that there is less fluctuation in data as compared to each individual state or UTs.

After going through the existing literature, we found that the Random forest [5,9,12,14,22], Linear Regression [19,21–23,25], SVM [5,12,16,19,22,23,25], Decision Trees [9,14,22], and a few additional algorithms are frequently used for prediction. Other models, namely Hybrid ARIMA Wavelet [8], SIR [26], SEIRS [13], and SEI_DI_HQHRD [17] have also been used for some other types of predictive analysis for COVID-19. Figure 6 shows the complete distribution of all of the algorithms considered in total $N = 20$ papers.

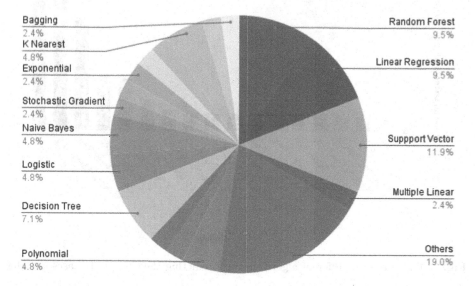

Fig. 6. Distribution of various ML algorithms that have been used in existing papers

In our analysis, we have considered five performance metrics namely MSE, R^2, MAPE, MAE, and RMSE [7] for each model. In addition to that, the time taken in milliseconds (ms) by each of these algorithms, except the LR, is given in Fig. 7. It also shows that OMP is the second-best model in terms of time consumption with just 16.7 ms. On this scale, we found that LR is the worst prediction algorithm which takes as much as 1103.3 ms as shown in Fig. 8, which is much more in comparison to any other algorithms.

After implementing our model, we found that Orthogonal Matching Pursuit [18] outperforms all other algorithms and fits best for the data we prepared in order to forecast the estimated growth of COVID-19 cases in India. To validate our results, we have evaluated and compared $N = 18$ ML algorithms in Table 2. The details related to the model selected as per our research are described in Sect. 3.2.

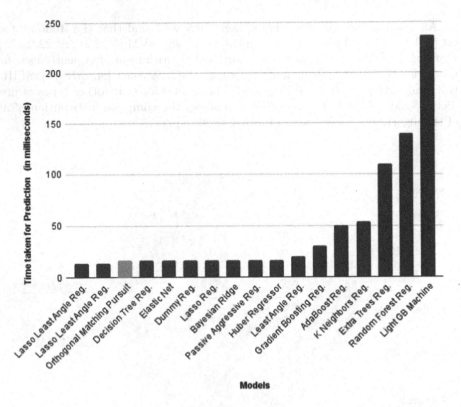

Fig. 7. Time taken by different algorithms for predicting the COVID-19 cases in India

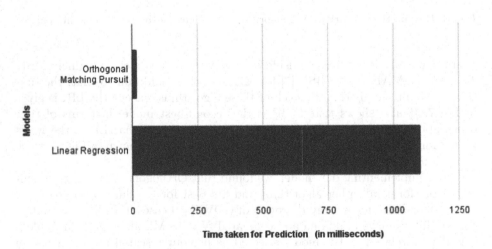

Fig. 8. Comparison of OMP with LR for time consumption

Table 2. Results of all evaluation parameters for 18 different Machine Learning Algorithms.

Model	MAE	MSE	RMSE	R^2	MAPE
Orthogonal Matching Pursuit	1.0776 e3	4.1988 e6	1.5092 e3	−1.1100 e-1	6.7410 e-1
kNN Regression	1.3975 e3	6.2503 e6	1.7938 e3	−3.2040 e-1	4.4610 e-1
Decision Tree Regression	1.4608 e3	6.6125 e6	1.8587 e3	−4.2380 e-1	4.6390 e-1
Extra Trees Regression	1.4677 e3	6.6382 e6	1.8609 e3	−4.2320 e-1	4.6880 e-1
Ridge Regression	1.4689 e3	5.8834 e6	1.9554 e3	−1.6667	9.9870 e-1
Random Forest Regression	1.4710 e3	6.6667 e6	1.8551 e3	−3.9440 e-1	4.6670 e-1
Gradient Boasting Regression	1.4976 e3	6.9410 e6	1.8910 e3	−4.3200 e-1	4.7890 e-1
AdaBoost Regression	1.5014 e3	6.8975 e6	1.8819 e3	−4.1460 e-1	4.7400 e-1
Elastic Net	1.6851 e3	7.0733 e6	2.1302 e3	−1.9207	1.0471
Linear Regression	1.7562 e3	9.6891 e6	2.2832 e3	−1.0814	8.2000 e-1
Light Gradient Boasting Machine	1.8583 e3	9.7754 e6	2.2605 e3	−9.4800 e-1	4.7970 e-1
Lasso Least Angle Regression	1.9754 e3	1.1123 e7	2.3863 e3	−1.0622	5.1250 e-1
Dummy Regression	1.9754 e3	1.1123 e7	2.3863 e3	−1.0622	5.1250 e-1
LASSO Regression	1.9970 e3	1.1112 e7	2.5932 e3	−2.3992	1.1260
Bayesian Ridge	2.4297 e3	1.9990 e7	3.3005 e3	−3.0900	1.1927
Passive Aggressive Regression	3.4993 e3	5.4454 e7	4.8409 e3	−5.0192	7.2030 e-1
Huber Regression	5.2911 e3	1.0623 e8	7.1759 e3	−1.3597 e1	1.9984
Least Angle Regression	3.0938 e9	6.9931 e19	4.8282 e9	−694.03 e11	2.3890 e6

The comparison among similar approaches has been done against the proposed approach in Table 3. Values of evaluation parameters that are not found as per the existing literature for analysis are denoted by cross mark (×). Kumari *et al.* [15] worked on the spread of COVID-19 in various geographic areas of India and proposed a model for forecasting the quantity of confirmed, recovered, and fatal cases. They forecasted the potential number of instances of new COVID cases using multiple linear regression and autoregression. Rustam *et al.* [19] used LR, LASSO, SVM, ES for the prediction of number of newly infected, deaths, and recoveries in the next ten days and found ES outperforms others with values given in Table 3. Chakraborty *et al.* [8] proposed ARIMA, wavelet-based, and hybrid ARIMA wavelet-based model to forecast the number of daily confirmed cases for five countries. However, we have considered the performance of their proposed model for India only. Vashisht *et al.* [23] have estimated the possible rate of active cases in China for the upcoming week and compared SVM, kNN, LR, Polynomial Regression models. They found that PR performs better compared to other algorithm. The evaluation parameters obtained by them for the best model are mentioned in Table 3. It can be clearly seen in the last row of Table 3 that the results obtained by our study are better considering the amount of dataset which has been used for analysis.

Table 3. Evaluation parameters obtained by other Authors

Authors	Model	MAE	MSE	RMSE	R^2
Kumari R. *et al.* [15]	Multiple LR	×	×	3085.43	0.999
Rustam F. *et al.* [19]	Exponential Smoothing	406.08	66.22e4	813.77	0.98
Chakraborty T. *et al.* [8]	ARIMA	16.07	50.83	×	×
Vashisht G. *et al.* [23]	Polynomial Regression	×	×	0.582	0.999
Proposed Model	OMP [18]	1.08e3	4.20e6	1.51e3	−1.11e-1

5   Conclusion and Future Scope

Through this research, we have collected the COVID-19 data for India from various data repositories and processed the same to obtain the most suitable dataset. Later, we have tested as many as eighteen different machine learning algorithms and compared their performance to come up with a best possible solution in order to forecast COVID-19 instances in India. The findings of this paper indicate that the likelihood of India reporting new daily COVID-19 cases is heavily influenced by the trends in its densely populated and larger states. However, it also emphasizes the importance of considering states with dissimilar trends to avoid any potential biases in the final results.

This study opens a wide area of research opportunities and societal benefits like deciding effective management strategies for fatal diseases like COVID-19. The results can be applied to various predictive classifications also, allowing for timely warnings and implementation of appropriate safety measures. Particularly in the light of the current economic downturn, this information is invaluable for the country's financial planning and highlights areas that require immediate attention. Additionally, this work provides more processed data that may be used for determining the allocation of resources towards constructing new hospitals/isolation centers, acquiring COVID-19 test kits, medical equipment, and improving care and treatment.

Our research work can be further enhanced by adding more attributes and comparison based on the neural network driven models that have not yet been tested.

Acknowledgment. We would like to express our gratitude to Dr. Jagriti Saini, Siddheshwari Dutt Mishra, and Mohammad Ahsan Siddiqui from the Department of Computer Science & Engineering, NITTTR, Chandigarh as well as Deepak Jaglan from Central University of Haryana, for their technical support at various stages of this research work.

Appendix

Table 4. Configuration for PyCaret to evaluate the performance of multiple models.

Parameter/Description	Value
Original Data	(85, 5)
Missing Values	FALSE
Numeric Features	3
Transformed Train Set	(85, 24)
Transformed Test Set	(40, 24)
Shuffle Train-Test	TRUE
Fold Generator	TimeSeriesSplit
Fold Number	3
Use GPU	FALSE
Experiment Name	reg-default-name
USI	4a9b
Imputation Type	simple
Numeric Imputer	mean
Iterative Imputation Numeric Model	None
Categorical Imputer	constant
Unknown Categoricals Handling	least_frequent
Transformation Method	None
Transform Target	TRUE
Transform Target Method	box-cox

References

1. Ministry of Health and Family Welfare website (1947)
2. Indian Council of Medical Research website (1949)
3. CSSE - Johns Hopkins University website (2019)
4. Amar, L.A., Taha, A.A., Mohamed, M.Y.: Prediction of the final size for COVID-19 epidemic using machine learning: a case study of Egypt. Infect. Dis. Model. **5**, 622–634 (2020)
5. Benitez-Pena, S., Carrizosa, E., Guerrero, V., Dolores, M.: Short-term predictions of the evolution of COVID-19 in Andalusia. An ensemble method. Preprint (2020)

6. Burdick, H., et al.: Prediction of respiratory decompensation in COVID-19 patients using machine learning: the ready trial. Comput. Biol. Med. **124**, 103949 (2020)
7. Chai, T., Draxler, R.R.: Root mean square error (RMSE) or mean absolute error (MAE). Geosci. Model Dev. Discuss. **7**(1), 1525–1534 (2014)
8. Chakraborty, T., Ghosh, I.: Real-time forecasts and risk assessment of novel coronavirus (COVID-19) cases: a data-driven analysis. Chaos Solitons Fractals **135**, 109850 (2020)
9. Darapaneni, N., et al.: A machine learning approach to predicting COVID-19 cases amongst suspected cases and their category of admission. In: 2020 IEEE 15th International Conference on Industrial and Information Systems (ICIIS), pp. 375–380. IEEE (2020)
10. Devarajan, J.P., Manimuthu, A., Sreedharan, V.R.: Healthcare operations and black swan event for COVID-19 pandemic: a predictive analytics. IEEE Trans. Eng. Manag. 1–15 (2021)
11. Goswami, K., Bharali, S., Hazarika, J.: Projections for COVID-19 pandemic in India and effect of temperature and humidity. Diabetes Metab. Syndr. Clin. Res. Rev. **14**(5), 801–805 (2020)
12. Gupta, V.K., Gupta, A., Kumar, D., Sardana, A.: Prediction of COVID-19 confirmed, death, and cured cases in India using random forest model. Big Data Mining Anal. **4**(2), 116–123 (2021)
13. Kanagarathinam, K., Sekar, K.: Estimation of the reproduction number and early prediction of the COVID-19 outbreak in India using a statistical computing approach. Epidemiol. Health **42**, 1–5 (2020)
14. Khanday, A.M.U.D., Rabani, S.T., Khan, Q.R., Rouf, N., Mohi Ud Din, M.: Machine learning based approaches for detecting COVID-19 using clinical text data. Int. J. Inf. Technol. **12**, 731–739 (2020)
15. Kumari, R., et al.: Analysis and predictions of spread, recovery, and death caused by COVID-19 in India. Big Data Mining Anal. **4**(2), 65–75 (2021)
16. Mary, L.W., Raj, S.A.A.: Machine learning algorithms for predicting SARS-CoV-2 (COVID-19) - a comparative analysis. In: 2021 2nd International Conference on Smart Electronics and Communication (ICOSEC), pp. 1607–1611 (2021)
17. Nabi, K.N.: Forecasting COVID-19 pandemic: a data-driven analysis. Chaos Solitons Fractals **139**, 110046 (2020)
18. Pati, Y.C., Rezaiifar, R., Krishnaprasad, P.S.: Orthogonal matching pursuit: recursive function approximation with applications to wavelet decomposition. In: Proceedings of 27th Asilomar Conference on Signals, Systems and Computers, pp. 40–44. IEEE (1993)
19. Rustam, F., et al.: COVID-19 future forecasting using supervised machine learning models. IEEE Access **8**, 101489–101499 (2020)
20. Schneider, P., Xhafa, F.: Anomaly Detection and Complex Event Processing Over IoT Data Streams: With Application to EHealth and Patient Data Monitoring. Academic Press (2022)
21. Sujath, R., Chatterjee, J.M., Hassanien, A.E.: A machine learning forecasting model for COVID-19 pandemic in India. Stoch. Env. Res. Risk Assess. **34**, 959–972 (2020)
22. Tiwari, S., Chanak, P., Singh, S.K.: A review of the machine learning algorithms for COVID-19 case analysis. IEEE Trans. Artif. Intell. **4**(1), 44–59 (2023)
23. Vashisht, G., Prakash, R.: Predicting the rate of growth of the novel corona virus 2020. Int. J. Emerg. Technol. **11**(3), 19–25 (2020)

24. Wang, P., Zheng, X., Li, J., Zhu, B.: Prediction of epidemic trends in COVID-19 with logistic model and machine learning technics. Chaos Solitons Fractals **139**, 110058 (2020)
25. Yadav, M., Perumal, M., Srinivas, M.: Analysis on novel coronavirus (COVID-19) using machine learning methods. Chaos Solitons Fractals **139**, 110050 (2020)
26. Zhong, L., Mu, L., Li, J., Wang, J., Yin, Z., Liu, D.: Early prediction of the 2019 novel coronavirus outbreak in the mainland china based on simple mathematical model. IEEE Access **8**, 51761–51769 (2020)

Performance Analysis of Different Machine Learning Classifiers for Prediction of Lung Cancer

Taruna Saini[(⊠)] [iD] and Amit Chhabra [iD]

Chandigarh College of Engineering and Technology, Sector-26, Chandigarh 160019, India
Sainitaruna0@gmail.com

Abstract. Cancer is, beyond doubt, among the most significant causes of death today. Cancer continues to be a major mortality factor despite several decades of clinical research and experiments of new treatments. It can occur in any part of the body, including the lungs. Primary lung cancer symptoms frequently lack specificity and could be linked to smoking. In clinical and medical data analysis, the prediction of lung cancer is a difficult task. A subdivision of artificial intelligence, also called "machine learning," employs distinguished analytical, stochastic, and optimization techniques for helping machines to be trained from past understandings and analyze extensive and diverse data sets. As a result, machine learning is widely utilized in the treatment and prediction of cancer. Machine learning (ML) classifiers are useful in contributing to the making of decisions and forecasting the severity of cancer by using cosmic amounts of data. Through the mediums of this study, we have proposed some classification algorithms to deter the existence of lung cancer in a person's body influenced by the symptoms one experiences. Different machine language classifiers are implemented over the Lung cancer dataset. With 93% precision, the accuracy of the SVM classifier has been the highest. A new ensembled model has been introduced with the help of ensemble learning which combines three different models – Logistic Regression (LR), KNN and Random Forest (RF). The accuracy achieved using applied ensemble model is 93.5%.

Keywords: lung cancer · machine learning · classification

1 Introduction

As reported by the WHO, with roughly .82 million new cases, lung cancer was among the most extensive forms of cancer throughout the year 2020 and the second most widespread form of cancer in 2021, with 2.21 million new cases [1]. Lung cancer patients, regardless of being freshly diagnosed or established, have much more serious indications than those with other forms and types of cancers at the same stage of the illness. Several research regarding lung cancer patients has looked at traits over time to see how these develop as the cancer advances. A majority of lung cancer diagnosis records are maintained for future use. A large proportion of pulmonary cancer diagnosis documentation is retained

R. K. Challa et al. (Eds.): ICAIoT 2023, CCIS 1929, pp. 258–276, 2024.
https://doi.org/10.1007/978-3-031-48774-3_18

for future reference. In lung cancer diagnosis, this data is helpful in predicting whether a person has lung cancer or not based on the symptoms one experiences. At this phase, machine learning algorithms can be used to determine whether or not an individual is a lung cancer patient. The total efficiency of the result prediction would be improved by using these categorization approaches [2]. The following two bar graphs, Figs. 1 and 2, depict worldwide cancer rates for men and women, respectively, in 2020.

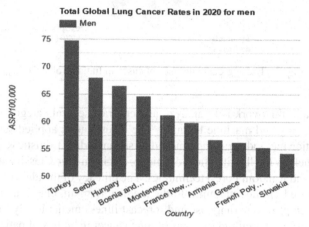

Fig. 1. Total Global Lung Cancer rates in 2020 (men) [3]

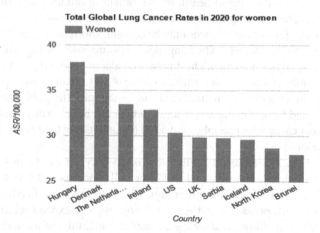

Fig. 2. Total Global Lung Cancer rates in 2020 (women) [3]

Age-standardized rates (ASR) are rates that have been adjusted for age. These are a summarized measure of the rate of mortality which might be present in a population having a standard distribution of age. While making comparisons in populations that vary in age, standardization is necessary since age has a major impact on the risk of mortality from cancer [4]. See Fig. 3 for a comparison of the various types of cancers in India for 2020.

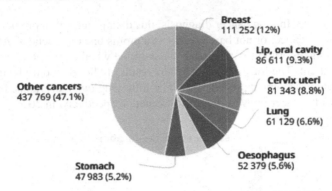

Fig. 3. The aggregate number of cases in India as of 2020 [3].

The main aim this work is to analyze and contrast several categorization methods used in data mining and machine learning. The investigators applied seven commonly used classification methods to a lung cancer data set, including Logistic Regression, Nave Bayesian, Kernel-SVM Classification technique, Decision Tree Classification technique, SVMs, K-NN and Random Forest techniques. The efficacy of the algorithms is analyzed for the topic based on the outcomes. In the medical domain, classification approaches are repeatedly employed to diagnose and forecast illness meticulously [5, 6, 7].

There are different preliminary signs of lung cancer in bodies of patients prior to the disease taking hold [8]. It is also important to consider the cancer type, its stage (the extent of spread), and the overall health of the patient in order to determine treatment options. Surgery, chemotherapy, and radiation therapy are all possible treatments [9].

The following signs of lung cancer may be present, shortness of breath, age, allergy, smoking, fatigue, yellow fingers, wheezing, peer pressure, swallowing difficulty, chronic disease, anxiety, coughing, gender, chest pain, and alcohol consumption [10].

The use of tobacco is a major cause of illness and death. Lung cancer tends to spread in membranes within bronchial trees and epithelium.The pulmonary system can be directly affected by any malignant disease, which can develop anywhere in the lung. Most lung cancer cases affect people aged 55 to 65, and it may take several years for the disease to develop. [11].

Lung cancer could manifest itself in two primary ways. It can be described as non-small cell and small cell. Hybrid small cell/large cell cancer is a phrase used to portray a cancer that holds traits from both groupings. Contrasting to small cell, non-small cell lung cancer is highly prevalent and tends to progress and expand relatively steadily. Nearly identical to smoking, small cell generates sizeable tumors more rapidly and has the prospect to proliferate predominantly across the body. They frequently begin in the bronchi surrounding the chest's center. It affects the lungs of a person and directly causes the lung cancer fatality rate [12].

The three foremost techniques of thwarting are quitting smoking, changing your diet, as well as using chemotherapy. Another prevention involves screening. A methodical analysis of indications and risk factors is carried out to identify potential lung cancer patients. Human cancer is significantly influenced by variables pertaining to surroundings. The environment around us contains a variety of carcinogens. Cancers with a long latency period, which are linked to contact to common environmental carcinogens, present a particularly difficult case due to the complexity of human cancer etiology [13].

The paper is organized as follows: Sect. 2 consists of some of the similar literature in the domain of application of ML Techniques over forecast of lung cancer. Further Sect. 3 describes the methodologies and methods employed in this work, and Sect. 4 depicts the experimental setup used in the lung cancer dataset analysis., Sect. 5 highlights the results obtained and analyzes the outcomes and Sect. 6 provides the conclusion inferred.

2 Related Works

Gargano et al. [15] defined different algorithms to create a Computer-Aided Diagram (CAD) model of lungs based on images. First, a 3D region growth technique is used to calculate a region in the lungs that corresponds to the ribcage zone, and afterward, an approach is used to make a particular region for the artificial life model used in the algorithm, which refers to the incoming "ants". It is assessed whether the existing structural branches include nodules, and abnormal growths in the lung tissue, or if the nodules are related to the pleura, which is the membrane that lines the outside of the lungs, using active shape models. Finally, they employed a "Snakes" and a "dot" improvement algorithm to refine the algorithm and better define the nodules. "Snakes" refer to an image processing technique that uses a contour to locate an object in an image, while the "dot improvement algorithm" likely refers to a technique for improving the accuracy of nodule detection.

Alam et al. [16] suggested a multi-class classifier that provides a successful algorithm that uses SVM to identify, diagnose, and predict the likelihood of lung cancer. It employs a co-occurrence gray level technique used to construct an algorithm that employs image processing algorithms such as image enhancement, identification and classification, recognition, and extraction of several features. For the classification tasks, SVM is used. The binarization approach is used to make predictions. The UCI ML dataset is obtained, and 500 affected and non-infected CT scans are included. Out of a total of 130 photos, the proposed technique classified 126 as infectious and displayed 87 as malignant out of a total of 100 images which were indicated previously. The identifying accuracy is 97 percent, while the forecasting accuracy is 87 percent, according to the experimental investigation.

Moriya et al. [17] proposed a model to find pulmonary chest radiographs nodules. Multiple Gray Level Thresholding approaches are employed for the identification of nodule possibilities. To distinguish between true and false-positive nodules, input parameters and differential images are employed for feature extraction. For the minimization of nodules that are false-positive, previously extracted features are used. The model which was thus generated was tested using a dataset of a total of around 200 radiographs with approximately 300 total lung nodules for testing. Of all the false-positive photos, 235 (75%) had been identified as a typical automated architecture, while 155 (49%) were identified as pulmonary vessels.

Another study by Gurcan et al. [18] suggested an approach that consists of five stages. K-means clustering is used as the first approach to partition areas. Having segregated the pulmonary curvatures, questionable areas are segregated from lung areas using pre-processing techniques, resulting in a binary picture with voids owing to the segregation stage, a technique is employed to cover these gaps as nodule candidates are regarded as solid objects. The suggested structure may include generalized areas and blood vessel-rich lung nodules. Classifications based on rules using 2 and 3-dimensional characteristics are utilized to distinguish this nodule. The suggested approach was tested on a database that included 1454 CT images from an aggregate of 34 patients who had 63 lung nodules that were diagnosed.

Ozawa et al. [19] suggested a simple framework for measuring the impact of CAD on the capacity of a radiologist to detect lung nodules. The suggested method used various image processing methods and methodologies for lung and intrapulmonary architecture classification. On a given image, the Top-hat transformation approach is used to identify the smooth picture for the classification of intrapulmonary features. Adapting this type of approach improved the recognition of CT scans by pulmonary nodule inhabitants.

Gomathi et al. [20] described a CAD configuration for analyzing cancer in Computed Tomography (CT) Scan is constructed. The recognition of the selected area in incoming scans is one of the most standard stages of CAD. The selection of lung areas is accompanied by region segmentation for lungs, and cancer lesions are detected and analyzed using the Fuzzy Possibility CMean (FPCM) clustering techniques. The diagnosing guidelines are formulated using the highest Drawable Circle intensity value. Next, with the help of Extreme Learning Machines (ELM), those rules are put into action to comprehend.

Shankar et al. [21] designed a model to forecast lung cancer. Low-Dose Radiography (LDR) is used to classify lung nodules, and an algorithm is used to optimize them. For the experimental program, a basic CT database containing 50 lung cancer Computed Tomography (CT) Scan pictures was used. These pictures were low-dosed. This approach is consistent with other representations, and the investigational research shows that the formulated representation has the best possible results, with 94.56 percent accuracy, 96.2 percent sensitivities, and 94.2 percent precision, correspondingly.

Thirach et al. [22] devised a methodology in which they employed a CNN approach consisting of 121 layers in combination with transfer learning for the identification of pictures from the chest. To recognize these nodules, the suggested model is trained on two databases. The system has an efficiency of around 74.436.01%, a sensitization of about 74 15 percent, as well as a precision of around 74 9 percent.

Hamburger et al. [23] suggested a tomography lung cancer image-based forecasting model. CNN is employed for the extraction of features. For hazard prediction, the Cox model trains CNN. The dataset, is being used for scientific investigation. Contains 422 pictures for 318 out of a total of 422 patients. Table 1 represents different previous works related to lung cancer.

Table 1. An analysis of works pertaining to lung cancer.

S.No	Authors	Year	Approaches used	Accuracy
1	Gurcan et al. [18]	2002	k-means clustering	84%
2	Moriya et al. [17]	2004	CAD	73%
3	Ozawa et al. [19]	2004	CAD	TP-94%
4	M. Gomathi et al. [20]	2010	FPCM clustering	TP:10, FP:122
5	Alam et al. [16]	2018	SVM	97%
6	Thirach et al. [22]	2018	LDCT	$74.43 \pm 6.01\%$
7	Shankar et al. [21]	2019	ODNN, LDA	94.56%
8	Haarburger et al. [23]	2019	CNN	c-index:0.623
9	Walker et al. [12]	2019	DeepScreener algorithm	78.2%
10	Sallow et al. [28]	2021	SVM, KNN, CNN	SVM - 95.56%, CNN-92.11% KNN-88.40%
11	Xie et al. [29]	2021	AdaBoost, Random Forest, KNN, Neural Network, Naïve Bayes, SVM	Neural Network-94%, SVM-94%, KNN-85%, Naïve Bayes-100%, AdaBoost-63%, Random Forest-89%

3 Proposed Approach

3.1 Dataset

The data has a total of 16 attributes with 309 instances. This dataset comprises of tests that have been classified as yes (lung cancer positive) or no (lung cancer negative). Table 2 reperesents 16 attributes of the dataset used.

3.2 Modules

Figure 4. Represents the implementation flow of the complete process for the prediction of lung cancer, and each process is explained as follows-

Data Collection. The data required for our collection has been taken from a survey of 309 people, of which 39 people did not have lung cancer and 270 people were diagnosed with lung cancer.

Data Preprocessing. Data pre-processing is a key procedure in the data mining approach, and the extent to which the information is processed before implementing the data mining techniques determines its effectiveness. Pre-processing can be defined as the process of preparing a dataset in anticipation of using it with mining methods.

Data Mining. The data from the pre-processing step is evaluated by employing data mining methods in this step.

Table 2. Attributes of the dataset along with their description and data type

S. No	Description	Data Type
1	The gender of the subjects, M or F	String
2	The age of the subjects	Integer
3	2 if they smoke and 1 if they don't	Integer
4	2 if they have yellow fingers and 1 if they don't	Integer
5	2 if they have been suffering from anxiety and 1 if they're not	Integer
6	2 if they are influenced by peer pressure and 1 if they're not	Integer
7	2 if they have a chronic disease and 1 if they don't	Integer
8	2 if they're easily fatigued and 1 if they're not	Integer
9	2 if they have an allergy and 1 if they don't	Integer
10	2 if they wheeze and 1 if they don't	Integer
11	2 if they consume alcohol and 1 if they don't	Integer
12	2 if they cough regularly and 1 if they don't	Integer
13	2 if they experience shortness of breath and 1 if they don't	Integer
14	2 if they have a problem swallowing and 1 if they don't	Integer
15	2 if they have chest pain and 1 if they don't	Integer
16	YES, if they are positive for lung cancer and NO if they're not. This is the column we are trying to predict. It is called the target	String

Table 3. The analogy of Accuracy, Error, Recall Score, Precision Score and F1 Score

Classifier Technique	Accuracy (%)	Recall Score (%)	Precision Score (%)	F1 Score (%)
KNN	88.46	88.46	88.46	88.46
Kernel-SVM	91.03	91.03	91.03	91.03
SVM	93.59	93.59	93.59	93.59
Decision Tree	89.74	89.74	89.74	89.74
Naïve Bayes	88.46	88.46	88.46	88.46
Logistic Regression	92.31	92.31	92.31	92.31
Random Forest	91.03	91.03	91.03	91.03

Data Analysis and Evaluation. The necessary values are computed, analyzed and are evaluated.

Report Generation. Estimated values are calculated after analysis and assessment. This can be visualized graphically for a better understanding.

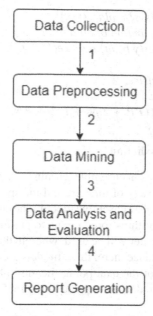

Fig. 4. Implementation flow

3.3 Algorithm

The algorithm provided outlines the steps for analyzing a lung cancer dataset using various machine-learning algorithms. The first step involves defining a dataset, then separating data for testing and training purposes. The third step involves checking for error or missing information. In the fourth step, the target column is converted into a numeric value using Sklearn Label Encoder. In steps five and six, the distribution of data in the feature columns and target column is evaluated using count plots. In step seven, seven machine learning technologies are used to build models In step eight, the confusion matrices are plotted. Finally, in step nine, accuracy of each algorithm is compared. By following these steps, the algorithm provides a framework for analyzing and comparing the performance of various machine-learning technologies on a lung cancer dataset.

Input: Lung cancer Dataset.

Output: Label ε{Yes, No}.

Method:

Step1: Load the data.

Step2: Break the data into training and testing models.

X train, X test, y train, y test = train test split ().

Step3: Check for missing values.

X train.isnull().sum().

Step4: Convert the target column into a numeric value using Sklearn Label Encoder.

y train = le. Transform (y train).

y test = le.transform(y test).

Step5: Evaluating the distribution of data in the feature columns.

sns.countplot(x = X train[sys].replace(key)).
Step6: Evaluating the distribution of data in the target column.
sns.countplot(x = pd.Series(y train).replace([0, 1], [No, Yes])).
Step7: Build models.
Step8:P lot Confusion Matrices.
Step9: Compare Accuracy.
Accuracy = (T.P. + T.N.)/ (T.P. + T.N. + F.P. + F.N.)

3.4 Ensemble Learning Prediction

Various ML techniques are used in ensemble learning to improve estimates on a dataset. A dataset is used to train a variety of models, and the specific assumptions made from each algorithm form the basis of an ensemble model. The ensemble model then combines the outcomes of different algorithms' predictions to produce a complete outcome [32].

In this work, a voting classifier will be used, wherein the ensemble model predicts by the majority of votes. For instance, here, three models are used, and if, in one instance, they each estimate [1, 0, 1] for the data point, the ensemble model's ultimate forecast will be 1, as two of the three models anticipated 1. Figure 5 represents the block diagram of Ensemble Learning.

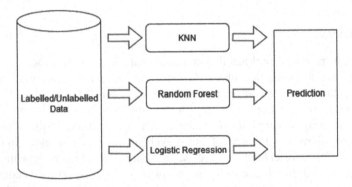

Fig. 5. Ensemble Learning block diagram

Three ML algorithms such as k-Nearest Neighbours, Random Forest and Logistic Regression algorithms are incorporated into the voting classifier utilizing the lung cancer dataset as well as Python Scikit-learn module.

Section 4 portrays the experimental setup used in the analysis of the lung cancer dataset being used in this work.

4 Experimental Learning

Experimental learning is a crucial aspect of any work, as it allows one to test the hypothesis and validate the results. In this study, we have conducted experimental learning by performing data preprocessing and mining on the lung cancer dataset. The dataset was cleaned by removing blank spaces, restoring null values, closing data gaps, and eliminating unneeded values. To analyze the data, we applied seven different classification approaches and evaluated their performance using the confusion matrix.

4.1 Data Preprocessing

The dataset has been cleaned in this stage by eliminating blank space, restoring null values, closing data gaps, and eliminating unneeded values. To screen the data, all of these procedures are applied to the data collection. All these tasks are carried out in the Jupyter Notebook.

4.2 Data Mining

Using the classification approaches listed in the next section, seven distinct classifiers are created.

The evaluation is done with the help of a confusion matrix. The values obtained from a confusion matrix help in calculating various parameters such as –

False Negative (F.N). The total fragment of the positive values which were discovered incorrectly.

True Positive (T.P). The total fragment of the positive values which were discovered correctly.

True Negative (T.N). The total fragment of the negative values which were discovered correctly.

False Positive (F.P). The total fragment of the negative values which were discovered incorrectly.

Accuracy. The probability of the values which were predicted correctly, is given as in Eq. (1)

$$Accuracy = \frac{T.N + T.P}{T.N + T.P + F.P + F.N} \tag{1}$$

Error Rate. The probability of the values which were predicted incorrectly, is given as in Eq. (2)

$$Error\ Rate = \frac{F.P + F.N}{T.P + T.N + F.P + F.N} \tag{2}$$

The next section highlights the results obtained in the study and analyses the outcomes.

5 Results and Discussion

The criteria for measuring a model's performance are usually F1 scores, recall, accuracy, and precision. A confusion matrix has been employed to characterize the performance metrics. The confusion matrix is a table that comprises information about actual and expected classifications received from the algorithm's performance.

5.1 Logistic Regression

An approach for supervised learning called logistic regression is applied to interpret problems entailing binary categorization. Binary classification is modeled mathematically using the logistic function and logistic regression, which has a number of more intricate expansions [30].

$$loglog \frac{p}{1-p} = \beta_\circ + \beta \, (Age) \tag{3}$$

where, $\frac{p}{1-p}$ is the odd ratio, and p is the probability of success.

Using the Logistic Regression approach, the following confusion matrix as displayed in Fig. 6, is obtained.

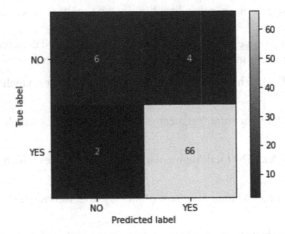

Fig. 6. An image of Confusion Matrix plotting for Logistic Regression

5.2 KNN

By using Euclidean distance amongst data points, KNN finds neighbors among the data [30]. The distance can be calculated using the following formula-

$$d = \sqrt{a_2 - a_1^2 + b_2 - b_1^2} \tag{4}$$

where a_1, a_2, b_1, and b_2 refer to the coordinates of two points in a 2-dimensional space. Plugging in these values in the formula will give you the distance between the two points.

Figure 7 shows the confusion matrix plot after applying KNN.

5.3 SVM

For a variety of classification problems, SVMs (Support Vector Machines) have proven effective. By counting the points on the perimeter of the class characteristics, it tries to determine which class has the best hyperplane.

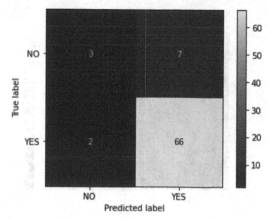

Fig. 7. An image of Confusion Matrix plotting for KNN

The optimal margin classifier can be obtained by

$$min_{\gamma,\omega,\beta} \frac{1}{2}||\omega||^2 \tag{5}$$

In this case, the regularisation parameter that regulates the trade-off between obtaining a low training error and a low testing error.

ω: the weight vector that represents the hyperplane that best separates the classes.

β: the bias term (also known as intercept) that shifts the hyperplane to the appropriate position in the feature space.

$||\omega||$: the magnitude of the weight vector.

$$y^{(i)}\left(\omega^T x^{(i)} + b\right) \geq 1, i = 1, 2, \ldots, m \tag{6}$$

where, $y^{(i)}$: the binary class label of the ith training example.

ω: the weight vector.

$X^{(i)}$: the feature vector of the ith training example.

b: the bias term (also known as intercept).

$(\omega^{(T)} x^{(i)} + b)$: the dot product of the weight vector and the feature vector plus the bias term.

The confusion matrix by employing the SVM approach is picturized in Fig. 8 [30].

5.4 Kernel-SVM

A cumulation of mathematical operations called as the kernel are utilized by SVM techniques. Data is loaded into the kernel, which then converts it all into the appropriate format. Various kernel operations are employed by different SVM algorithms.

$$K(x, y) = e^{-\left(\frac{||x-y||^2}{2\alpha^2}\right)} \tag{7}$$

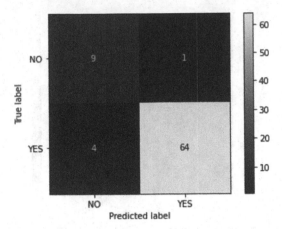

Fig. 8. An image of Confusion Matrix plotting for SVM

where, x and y are input data points, ‖x-y‖ is the Euclidean distance between x and y, and α is a parameter that controls the width of the kernel

This formula is employed whenever there is no previous knowledge about the data.The confusion matrix for the Kernel-SVM approach is displayed using Fig. 9.

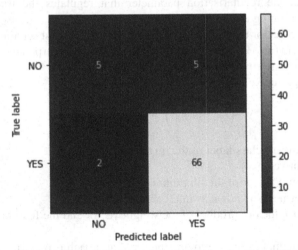

Fig. 9. An image of Confusion Matrix plotting for Kernel-SVM

5.5 Naïve Bayes

This approach makes it possible for all characteristics to have an identical effect on the conclusion [30].

$$Probability\left(\frac{M}{N}\right) = \frac{Probability\left(\frac{N}{M}\right)Probability(M)}{Probability(N)} \tag{8}$$

The confusion matrix plotted using Naïve Bayesian is as given in Fig. 10.

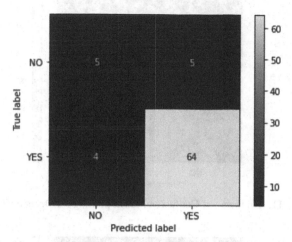

Fig. 10. An image of Confusion Matrix plotting for Naïve Bayes

5.6 Decision Tree

Using straightforward choice principles from the training dataset, the decision tree is intended to construct a model for predicting the desired variable [32][33].

$$Entropy(E) = \sum -p_i loglog p_i \qquad (9)$$

$$Information\, Gain(P, Q) = Entropy(P) - Entropy(P, Q) \qquad (10)$$

$$Ginni\, Index = 1 - \sum p_i^2 \qquad (11)$$

The confusion matrix for decision tree approach is picturized in Fig. 11.

5.7 Random Forest

The primary concept behind ensemble techniques is that a collection of models will result in a potent model. A decision tree using the conventional ML technique is included with the random forest [34].

$$Mean\, Squared\, Error = \frac{1}{N} \sum (f_i - y_i)^2 \qquad (12)$$

where N is the Total number of data points, f_i is the the value returned by the model, y_i is the the actual value for data point i

Using the Random Forest approach, the decision matrix is obtained as in Fig. 12:

Table 4 compares different values of various models and Fig. 13 compares the accuracy of these models. Figure 14 compares their error values. Using different formulas, the error values for each classifier can be calculated and compared.

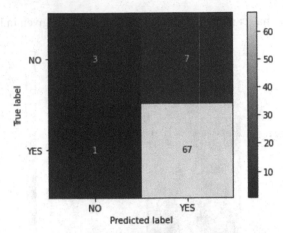

Fig. 11. An image of Confusion Matrix plotting for Decision Tree

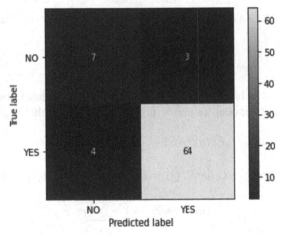

Fig. 12. An image of Confusion Matrix plotting for Random Forest

5.8 Ensemble Learning Model

It is built from three models(LR, KNN and RF). The reason for choosing these models for ensembling could be that each of these models is based on a different approach and has its own strengths and weaknesses. LR is a simple and interpretable linear model. KNN is a non-parametric model that is based on distance metrics and is useful when the decision boundary is complex and non-linear. Random Forest is a decision tree-based model that is useful for handling complex interactions between features and is less prone to overfitting than a single decision tree. By combining these models, the ensemble learning model can leverage the strengths of each model and produce a more accurate and robust prediction.

Table 4. Comparison of False Negative values, True Positive values, True Negative values and False Positive of Classifiers

Classifier Technique	True Positive Values	False Positive Value	False Negative Values	True Negative Values
SVM	64	1	9	4
KNN	66	7	3	2
Logistic Regression	66	4	6	2
Kernel-SVM	66	5	5	2
Decision Tree	67	7	3	1
Naïve Bayes	64	5	5	4
Random Forest	64	3	7	4

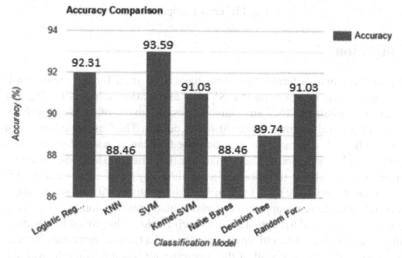

Fig. 13. Accuracy Comparison

The accuracy predicted by the ensemble learning model is 93.59% which is equal to the accuracy predicted by the SVM model [35, 36]. SVM model gives high scores of accuracy, recall, precision and F1 score.

Section 6 provides the conclusion inferred from the study.

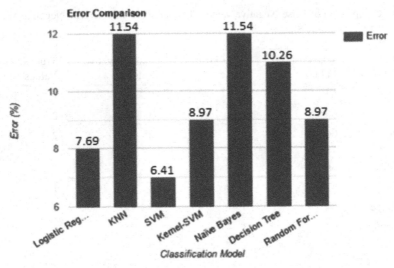

Fig. 14: Error Comparison

6 Conclusion

In the research work, different algorithms have been used for the prediction of lung cancer. Experiments demonstrate that SVM Classification achieves the highest accuracy of all the classifiers. KNN and Naive Bayesian, on the other hand, performed the worst classification since they had the lowest accuracy. The Ensemble learning method yielded 93.5 percent accuracy. We can conclude that cancer researchers can accurately forecast lung cancer symptoms. They can utilize the Ensemble learning results for accurate prediction, which fits the data set better than any other classification technique. This technique would assist them in identifying lung cancer patients so that appropriate action could be taken at the appropriate time. It also prevents non-lung cancer patients from being misclassified as such. This approach could be improved by creating a new algorithm that classifies data with great accuracy and a reduced error rate. Neither of the classifiers were able to satisfy all of the requirements. As a result, a hybrid algorithm with high accuracy and a low margin of error can be built.

References

1. Ferlay, J., et al.: Cancer statistics for the year 2020: an overview. Int. J. Cancer **149**, 778–789 (2021)
2. Mahesh, B.: Machine learning algorithms - a review - International Journal of Science and Research (IJSR). **9**, 381–386 (2020)
3. World Health Organization. International agency for research on cancer (2019)
4. La Vecchia, C., Negri, E., Decarli, A., Fasoli, M., Cislaghi, C.: Cancer mortality in Italy: an overview of age-specific and age-standardised trends from 1955 to 1984. Tumori Journal. **76**, 87–166 (1990)
5. Jacob, J., Mathew, J., Mathew, J., Issac, E.: Diagnosis of liver disease using machine learning techniques. Int Res J Eng Technol **5**, 4 (2018)

6. V. Ramalingam, V., Dandapath, A., Karthik Raja, M.: Heart disease prediction using Machine Learning Techniques : a survey. Int. J. Eng. Technol. **7**, 684 (2018)

7. Zebari, D.A., Zeebaree, D.Q., Abdulazeez, A.M., Haron, H., Hamed, H.N.: Improved threshold based and trainable fully automated segmentation for breast cancer boundary and pectoral muscle in mammogram images. IEEE Access. **8**, 203097–203116 (2020)

8. Min Park, S., et al.: Prediagnosis smoking, obesity, insulin resistance, and second primary cancer risk in male cancer survivors: national health insurance corporation study. J. Clin. Oncol.Clin. Oncol. **25**, 4835–4843 (2007)

9. Melamed, M.R., Flehinger, B.J., Zaman, M.B., Heelan, R.T., Perchick, W.A., Martini, N.: Screening for early lung cancer. Chest **86**, 44–53 (1984)

10. Spiro, S.G., Gould, M.K., Colice, G.L.: Initial evaluation of the patient with lung cancer: Symptoms, signs, laboratory tests, and paraneoplastic syndromes. Chest. **132**, (2007)

11. Cooley, M.E.: Symptoms in adults with lung cancer. J. Pain Symptom Manage. **19**, 137–153 (2000)

12. Qiang, Y., Guo, Y., Li, X., Wang, Q., Chen, H., Cuic, D.: The diagnostic rules of peripheral lung cancer preliminary study based on data mining technique. J. Nanjing Med. Univ. **21**, 190–195 (2007)

13. Karabatak, M., Ince, M.C.: An expert system for detection of breast cancer based on association rules and neural network. Expert Syst. Appl. **36**, 3465–3469 (2009)

14. Causey, J., et al.: [PDF] lung cancer screening with low-dose CT scans using a deep learning approach: Semantic scholar, 2019. arXiv preprint arXiv:1906.00240

15. Cheran, S.C., Gargano, G.: Computer aided diagnosis for lung CT using Artificial Life Models. Seventh International Symposium on Symbolic and Numeric Algorithms for Scientific Computing (SYNASC'05) (2005)

16. Alam, J., Alam, S., Hossan, A.: Multi-stage lung cancer detection and prediction using multi-class SVM classifie. In: 2018 International Conference on Computer, Communication, Chemical, Material and Electronic Engineering (IC4ME2) (2018)

17. Kakeda, S., et al.: Improved detection of lung nodules on chest radiographs using a commercial computer-aided diagnosis system. Am. J. Roentgenol.Roentgenol. **182**, 505–510 (2004)

18. Gurcan, M.N., et al.: Lung nodule detection on thoracic computed tomography images: preliminary evaluation of a computer-aided diagnosis system. Med. Phys. **29**, 2552–2558 (2002)

19. Awai, K., et al.: Pulmonary nodules at chest CT: effect of computer-aided diagnosis on radiologists' detection performance. Radiology **230**, 347–352 (2004)

20. Gomathi, M., Thangaraj, P.P.: Automated CAD for lung nodule detection using CT scans. In: 2010 International Conference on Data Storage and Data Engineering. (2010)

21. S.K., L., Mohanty, S.N., K., S., N., A., Ramirez, G.: Optimal Deep Learning Model for classification of lung cancer on CT images. Future Generation Computer Systems. **92**, 374–382 (2019)

22. Ausawalaithong, W., Thirach, A., Marukatat, S., Wilaiprasitporn, T.: Automatic lung cancer prediction from chest X-ray images using the Deep Learning Approach. In: 2018 11th Biomedical Engineering International Conference (BMEiCON) (2018)

23. Haarburger, C., Weitz, P., Rippel, O., Merhof, D.: Image-based survival prediction for lung cancer patients using CNNS. In: 2019 IEEE 16th International Symposium on Biomedical Imaging (ISBI 2019) (2019)

24. Gift, A.G., Stommel, M., Jablonski, A., Given, W.: A cluster of symptoms over time in patients with lung cancer. Nurs. Res. Res. **52**, 393–400 (2003)

25. Krech, R.L., Davis, J., Walsh, D., Curtis, E.B.: Symptoms of lung cancer. Palliat. Med.. Med. **6**, 309–315 (1992)

26. Birring, S.S.: Symptoms and the early diagnosis of lung cancer. Thorax **60**, 268–269 (2005)

27. Hopwood, P., Stephens, R.J.: Symptoms at presentation for treatment in patients with lung cancer: implications for the evaluation of palliative treatment. Br. J. Cancer **71**, 633–636 (1995)

28. Mustafa Abdullah, D., Mohsin Abdulazeez, A., Bibo Sallow, A.: Lung cancer prediction and classification based on correlation selection method using machine learning techniques. Qubahan Academic Journal. **1**, 141–149 (2021)

29. Xie, Y., et al.: Early lung cancer diagnostic biomarker discovery by machine learning methods. Trans. Oncol. **14**, 100907 (2021)

30. Ibrahim, I., Abdulazeez, A.: The role of machine learning algorithms for diagnosing diseases. J. Appl. Sci. Technol. Trends. **2**, 10–19 (2021)

31. Ali, M.M., Paul, B.K., Ahmed, K., Bui, F.M., Quinn, J.M.W., Moni, M.A.: Heart disease prediction using supervised machine learning algorithms: performance analysis and comparison. Comput. Biol. Med.. Biol. Med. **136**, 104672 (2021)

32. Lappalainen, H., Miskin, J.W.: Ensemble learning. In: Advances in Independent Component Analysis, pp. 75–92 (2000)

33. Verma, R., Chhabra, A., Gupta, A.: A statistical analysis of tweets on covid-19 vaccine hesitancy utilizing opinion mining: an Indian perspective. Social Netw. Anal. Mining **13**(1), (2022). https://doi.org/10.1007/s13278-022-01015-2

34. Gupta, S., Chhabra, A., Agrawal, S., Singh, S.K.: A comprehensive comparative study of machine learning classifiers for Spam Filtering. In: Nedjah, N., Pérez, G.M., Gupta, B.B. (eds.) International Conference on Cyber Security, Privacy and Networking (ICSPN 2022), pp. 257–268. Springer International Publishing, Cham (2023). https://doi.org/10.1007/978-3-031-22018-0_24

35. Bharany, S., Sharma, S., Alsharabi, N., Tag Eldin, E., Ghamry, N.A.: Energy-efficient clustering protocol for underwater wireless sensor networks using optimized glowworm swarm optimization. Front. Marine Sci. **10**, 1117787 (2023)

36. Kaushik, K., et al.: A machine learning-based framework for the prediction of Cervical Cancer Risk in women. Sustainability. **14**, 11947 (2022)

Localization Improvements in Faster Residual Convolutional Neural Network Model for Temporomandibular Joint – Osteoarthritis Detection

K. Vijaya Kumar[1]([⊠]) [ID] and Santhi Baskaran[2]

[1] Department of Computer Science and Engineering, Puducherry Technological University, Puducherry, India
mkvijayakumaramphd@gmail.com
[2] Department of Information Technology, Puducherry Technological University, Puducherry, India
santhibaskaran@ptuniv.edu.in

Abstract. To recognize the Osteoarthritis of the Temporomandibular Joint (TMJ-OA) from panoramic dental X-ray images, deep learning algorithms are widely used these days. Among others, an Optimized Generative Adversarial Network with Faster Residual Convolutional Neural Network (OGAN-FRCNN) was recently achieved better FRCNN training by providing more synthetic images for TMJ-OA recognition. However, the localization of a small condyle OA region was ineffective because of the complex background, occlusion and low-resolution images. Hence, this article proposes the OGAN with Progressive Localization-improved FRCNN (PLFRCNN) model for TMJ-OA recognition. First, the OGAN can augment the number of panoramic X-ray scans. Then, ResNet101 can extract the Feature maps (F-maps) at multiple levels, followed by the Feature Pyramid Network (FPN) with a Region-of-Interest (RoI)-grid attention for multiscale F-map extraction. Those F-maps are given to the Modified Region Proposal Network (MRPN), which applies a multiscale convolution feature fusion and an Improved Non-Maximum Suppression (INMS) scheme for creating the Region Proposals (RPs) with more information. To resolve the localizing variance and obtain the proposal F-maps, the improved RoI pooling based on bilinear interpolation merges both F-maps and RPs. Moreover, the fully connected layer is used to classify those F-maps into corresponding classes and localize the target Bounding Box (BB). Additionally, the BB regression in the TMJ-OA localization stage is enhanced by the new Intersection-Over-Union (IOU) loss function. Finally, the test outcomes reveal that the OGAN-PLFRCNN model attains an accuracy of 98.18% on the panoramic dental X-ray corpus, in contrast to the classical CNN models.

Keywords: Temporomandibular joint · Osteoarthritis · Panoramic imaging · OGAN-FRCNN · RoI pooling · Region proposal network · IOU · Bounding box · Non-maximum suppression

R. K. Challa et al. (Eds.): ICAIoT 2023, CCIS 1929, pp. 277–288, 2024.
https://doi.org/10.1007/978-3-031-48774-3_19

1   Introduction

Osteoarthritis (OA) is the foremost prevalent kind of osteoporosis that affects the TMJ. The primary reason for OA is severe stress on joints. Owing to ache, crepitus and localized paraspinal soreness in the joint, TMJ-OA presents as reduced lower jaw movement [1]. It is identified when a radiographic scan demonstrates functional bone displacement. In persons with juvenile idiopathic arthritis, orthopantomography can be utilized to measure the condylar and Ramal irregularities of the maxilla [2]. There is a broad variety of TMJ conditions and diagnoses, some of which may or may not be painful [3].

Panoramic imaging is frequently utilized in premature levels of treatment once bone abnormalities in the TMJ are found [4]. Nevertheless, the TMJ contains tiny bone contours at the joint, which are hidden by the cranium. Morphological abnormalities in the TMJ are usually overlooked in basic tests. While there aren't many doctors who can diagnose OA purely from radiographs, it takes time to send images to diagnostic experts. An AI-based approach for autonomously treating TMJ-OA has been developed to solve some of the challenges faced by doctors in diagnosing and treating the condition [5–7]. Several X-ray image analysis techniques have been made available thanks to several developments in the AI paradigm [8]. The extraction of certain X-ray scan areas and the identification of abnormalities are only two examples of the many processes that CNN technology has accomplished. Numerous research studies on the study of dental X-ray images have been conducted, including ones on osteoporosis, sinusitis and teeth using panoramic image analysis [9]. However, panoramic X-ray analysis-based investigations are ineffective.

Consequently, a model [10] was created to recognize the mandibular condyle by categorizing panoramic dental X-ray pictures using image recognition algorithms. First, the panoramic scans were gathered and two distinct frameworks have been applied: (a) FRCNN for TMJ-OA and near structures (joint fossa and condyle) recognition and (b) CNN for predicting whether the recognized structure region has any abnormality according to the TMJ shape. As well, the VGG16, ResNet and Inception frameworks were modified for predicting the occurrence of TMJ-OA. In contrast, the number of scans was not sufficient, resulting in less accuracy in identifying objects. As a result, an OGAN was created [11], which produces fake panoramic dental X-ray images. In this GAN model, the generator was used to make synthetic scans, while the discriminator was trained to distinguish between false and real scans. Additionally, utilizing the Elephant Herding Optimization (EHO) method based on clan and separation factors, the GAN's most effective hyperparameters were selected. The freshly created synthetic panoramic images were then added to the real database to create the training and test sets. The condylar area was extracted from the learning sets using the FRCNN, which was then used to spot abnormalities in the images. With the use of test sets for TMJ-OA recognition, the trained FRCNN model was further verified. Conversely, the accuracy of FRCNN was lower while recognizing targets in panoramic scans, which encompass analogous objects. The noise and alike background textures in the scans affected the recognition performance.

To tackle these issues, the FRCNN is improved by considering the FPN and RoI-grid attention mechanisms [12] to produce multiscale F-maps with rich contextual information. According to this model, the training pictures are passed to ResNet101 to extract multiscale traits from the dental panoramic scans. The RoI-grid attention technique is used to transfer such traits to the FPN, which encodes deeper properties from sparse points using a unified formulation of both the attention-based and graph-based point operators. Then, those traits are combined at several levels to provide a multiscale F-map that is more informative and dramatically improved network performance. Additionally, the F-map is provided to the RPN for RP creation after predicting the class and BB. Such RPs and the F-maps are concatenated in the RoI pooling layer to obtain proposal F-maps, which are classified by the fully connected layers for TMJ-OA classification and localization. On the other hand, the complex background, occlusion and low-resolution images degrade the efficiency of localizing a small condyle OA region in panoramic X-ray scans.

Therefore in this manuscript, the OGAN-PLFRCNN model is proposed for classifying TMJ-OA and localizing the condyle OA region efficiently. This model employs the MPRN rather than the standard RPN to alleviate the variation in localizing the condyle OA region and reduce the BB loss. In this MRPN, a multiscale convolutional feature fusion is used instead of single-layer convolutional F-maps for creating the RPs with more information after predicting the class and BB. Also, an INMS scheme is applied to prevent the loss of overlapping objects. After that, the F-maps from the FPN and the RPs from the MRPN are concatenated by the RoI pooling process, which is enhanced by the bilinear interpolation to obtain the proposal F-maps. Such F-maps are passed to the fully connected layer to get the target class and the accurate location of the target BB. Moreover, the improved IOU loss function is adopted in the TMJ-OA localization process for enhancing BB regression. Thus, this model supports the effective localization of small condyle OA and prevents the loss of overlapping objects.

The remaining sections are outlined as follows: The earlier studies on the segmentation of panoramic scans are discussed in Sect. 2. Section 3 explains the OGAN-PLFRCNN model, while Sect. 4 illustrates its efficacy. Section 5 summarizes the work and suggests further development.

2   Literature Survey

Yang et al. [13] presented the YOLOv2 algorithm to identify and categorize cysts and tumors of the jaw in panoramic scans. First, panoramic scans were collected and annotated into various labels, like dentigerous cysts, odontogenic keratocysts, ameloblastoma and no tumor. Then, various image augmentation methods were applied to increase the number of training scans. After that, YOLOv2 was used to get the BB and corresponding labels. But, its localization error was high, resulting in less accuracy. Takahashi et al. [14] designed a YOLOv3 classifier to identify dental prostheses and replace teeth. On the other hand, identification was hard for tooth-colored prostheses and replacements when solely tooth photographs were utilized.

Cha et al. [15] applied a Deep Neural Network (DNN) to segment the maxillary sinus and mandibular canal from the dental panoramic scans. But, the overfitting problem was not avoided and the number of scans considered was limited. Leo and Reddy [16] presented a Hybrid Neural Network (HNN) for identifying dental caries and categorizing the caries areas. Initially, the dental scans were pre-processed and segregated into their constituent pixels. After that, distinct traits were mined and learned by the HNN to categorize the caries labels. Conversely, the learning quality was impacted by the limited scans. Aljabri et al. [17] presented DenseNet121, VGG16, InceptionV3 and ResNet50 to categorize the canine impaction classes from dental panoramic scans. But, it has a limited number of annotated scans. As well, the accuracy was influenced by the low-resolution and poor-quality scans.

Rohrer et al. [18] developed a multi-label segmentation of dental restorations on panoramic scans by a tiling scheme. First, the actual scans were cropped into several segments, which were then fed to the U-Net model to obtain the related segmentation mask. But, its generalizability to different databases was not analyzed. Also, it was tested on a limited database and used more hyperparameters, which impact the accuracy. Bayrakdar et al. [19] presented a DCNN framework relying on the U-Net to segregate apical diseases in dental panoramic scans. Conversely, solely single imagery equipment and typical variables were considered for scanning, resulting in a limited number of scans.

Zhu et al. [20] designed a new deep-learning model named CariesNet to delineate various caries degrees from panoramic scans. Initially, high-quality panoramic scans with well-delineated caries tumors were collected. After that, CariesNet was built as a U-shaped network with an extra full-scale axial attention unit for segmenting various caries categories. But, the efficiency of the moderate caries delineation was comparatively lower because the edges between deep and moderate caries were comparatively blurred, resulting in the misclassification of moderate caries.

3   Proposed Methodology

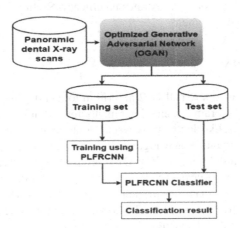

Fig. 1. Overall flow of the proposed study.

This section briefly describes the OGAN-PLFRCNN model for TMJ-OA classification and localization. Figure 1 portrays the pipeline of an entire study. At first, a panoramic dental X-ray scan corpus is gathered from the open source and augmented by the OGAN.

Next, the corpus is partitioned into learning and test collections. The learning collection is applied to train the novel PLFRCNN classifier, while the test collection is utilized to validate the trained PLFRCNN for mandibular condylar localization and TMJ-OA classification.

3.1 Progressive Localized-Improved FRCNN Model

To boost the efficiency of BB regression and classification, a novel localization-improved FRCNN model, namely PLFRCNN, is developed, as illustrated in Fig. 2.

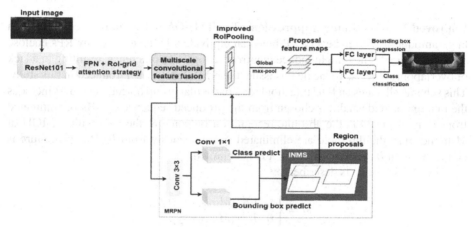

Fig. 2. Architecture of PLFRCNN model.

Modified RPN Unit. In this mrpn unit, multiscale convolutional feature fusion and inms are applied for creating RPS. There are frequently significant image characteristics missing from single-layer convolutional f-maps. It is challenging to adequately describe the characteristics of tiny objects due to the convolution and pooling procedures of frcnn, which result in smaller f-maps. The frcnn utilizes only the outcome of the conv3 unit as the f-map for the following setup.

The dimension of the F-maps created in all convolutional layers is varied, whereas the dimension of the F-map of Conv4 is unaltered and the dimension of the F-map of Conv3 and Conv5 is altered to the dimension of the Conv4 F-map. The max-pooling through subsampling is implemented for the F-map of Conv3 and the upsampling is utilized to enhance the quality of the F-map of Conv5 to make them stable with the Conv4 F-map. Eventually, the concatenated F-map is acquired via accumulating the F-maps involved in the sub-sampling of the outcome of Conv3 and upsampling via the outcome of Conv5 and the F-maps of Conv4. The local response regularization is utilized before fusing 3-layer convolutional F-maps to process all F-maps; therefore, the stimulated ranges of the F-map are identical. The fused F-map comprises rich and abstract semantic data. As a result, this PLFRCNN model combines the characteristics captured via the Conv3, Conv4 and Conv5 units. The process of multiscale convolutional feature fusion is illustrated in Fig. 3.

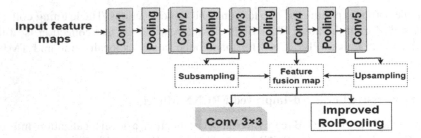

Fig. 3. Process of multiscale convolutional feature fusion.

Improved Non-Maximum Suppression. The TMJ-OA recognition process creates a huge amount of RPs and every RP holds an equivalent grade and nearby RPs enclose appropriate grades that might induce incorrect recognition outcomes, resulting in a lack of overlapping objects. To tackle this issue, the NMS scheme is introduced in this study. This scheme allocates an IOU threshold for a particular group object, the BB M includes the maximum grade and it is chosen from the produced sequence of BBs B, eliminated from B and situated in the absolute recognition outcome R, the BBs with the IOU of M higher than the threshold are eliminated simultaneously from B. This procedure is continued until B is empty and the set D is obtained.

The NMS scheme is described by

$$s_i = \begin{cases} s_i, \ IoU(M, B_i) < p \\ 0, \ IoU(M, B_i) \ge p \end{cases} \tag{1}$$

In Eq. (1), p denotes the threshold of IOU. It is observed that the NMS directly modifies the BB grade of the nearby class to 0, resulting in a lack of overlapping objects. So, the INMS scheme is adopted, which regrades the BB. When a BB overlays with M maximum, it can receive a minimum grade. When the degree of overlay is low, the grade is unaltered. So, the INMS is described by

$$s_i = \begin{cases} s_i, & IoU(M, B_i) < p \\ s_i \times (1 - IoU(M, B_i)), & IoU(M, B_i) \ge p \end{cases} \tag{2}$$

So, the poorer grades of boxes are eliminated. Simultaneously, because the BB with the maximum grade is M, they are reunited when there are BBs with maximum grades or IOUs with M are higher than 0.9. The locations of the reunited BBs are united by the weighted-mean scheme.

Improved RoI Pooling. After obtaining the RPs from the MRPN, these are fed to the RoI pooling layer to fuse with the F-maps and get the proposal F-maps. Because the dimensions of retrieved RPs depend on the actual scan, they have to be translated to the F-map and the F-map related to every RP is split into $k \times k$ bins and the max-pooling is executed for all bins. So, the outcome of the RoI pooling is often $k \times k$. But, in the task of translating RPs from the actual scan to the F-map, the coordinates of the translated RPs on the F-map are normally decimal. As well, in the task of splitting RPs, the rounding process is implemented for all $k \times k$ bins. It causes the variation of the

F-map to be translated to the actual scan to be higher, thus the localization of the BB degrades precision.

To combat this issue, bilinear interpolation is applied, which comprises linear interpolation in two directions. The values of the points Q_{11}, Q_{12}, Q_{21} and Q_{22} are known, whereas the aim is to find the value at the point P. In the x direction, a linear interpolation of Q_{11} and Q_{21} has the value of $R_1(x, y_1)$. Likewise, a linear interpolation of Q_{12} and Q_{22} is executed to find the value of $R_2(x, y_2)$:

$$f(R_1) \approx \frac{x_2 - x}{x_2 - x_1} f(Q_{11}) + \frac{x - x_1}{x_2 - x_1} f(Q_{21}) \tag{3}$$

$$f(R_2) \approx \frac{x_2 - x}{x_2 - x_1} f(Q_{12}) + \frac{x - x_1}{x_2 - x_1} f(Q_{22}) \tag{4}$$

Also, linearly interpolating with R_1 and R_2 in the y direction to get:

$$f(P) \approx \frac{y_2 - y}{y_2 - y_1} f(R_1) + \frac{y - y_1}{y_2 - y_1} f(R_2) \tag{5}$$

To avoid the localizing variation, the RoI pooling rounding process is avoided. Once the RoI pooling process is enhanced, the backpropagation is tuned simultaneously. It is noticed that the standard pooling task BB contains a relevant compensation from the actual object and the rounding process induces the estimated BB of the RP to not equal the ground truth, providing in a condyle OA being lost. However, using the improved RoI pooling procedure, tiny condyle OAs are precisely localized. Further, the RoI pooling outcome is passed to the fully connected layer to get the target class and localize the target BB.

Improved IOU Error Factor. In the BB regression process, the most basic criteria for determining the variance between the estimated BB and the ground truth box is IoU, which is defined by

$$IOU = \frac{S_{recognition\ out} \cap S_{Ground\ truth}}{S_{recognition\ out} \cup S_{Ground\ truth}} \tag{6}$$

To achieve precise localization, TMJ-OA recognition depends on BB regression. But, the utilization of IoU-based L_1-norm or L_2-norm error values in the task of regression is not acceptable. To overcome this issue, the Modified IOU (MIOU) is defined by

$$MIOU = IOU - \frac{C - (A \cup B)}{C} \tag{7}$$

The selected region is S, where the minimum region is $C(C \subseteq S)$ that covers A and B. The value of IOU is $[0, 1]$ and the value of MIOU is $[-1, 1]$. The highest range is 1 if two areas overlap and the minimum value is -1 if there is no overlapping between two areas. Thus, MIOU is an appropriate distance metric and since MIOU presents the minimum region C covering A and B, it not only focuses on the overlapping regions, yet also on other non-overlapping regions, though A and B do not overlap, they are modified.

The determination of overlapping regions is similar to IOU. While determining the lowest closure region C, solely the highest and lowest coordinate ranges of the two BBs are required. The rectangle surrounded through such two coordinates is C. The coordinates of the top left and bottom right edges are utilized to define all BBs. The estimated BB is denoted by $B = (x_1, y_1, x_2, y_2)$ and the BB of ground truth is denoted by $B^* = (x_1^*, x_2^*, y_1^*, y_2^*)$. The regions S^* and S of B^* and B are determined by

$$S^* = \left| (x_2^* - x_1^*) * (y_2^* - y_1^*) \right| \tag{8}$$

$$S = \left| (x_2 - x_1) * (y_2 - y_1) \right| \tag{8a}$$

The overlap region S^I of B^* and B is calculated as:

$$S^I = \begin{cases} \left| (x_2^I - x_1^I) * (y_2^I - y_1^I) \right|, & x_2^I > x_1^I, y_2^I > y_1^I \\ 0, & Or\ else \end{cases} \tag{9}$$

In Eq. (9), x_1^I, x_2^I, y_1^I and y_2^I are described by

$$x_1^I = \max(x_1, x_1^*), x_2^I = \min(x_2, x_2^*) \tag{10a}$$

$$y_1^I = \max(y_1, y_1^*), y_2^I = \min(y_2, y_2^*) \tag{10b}$$

The minimum rectangle $C = (x_1^C, x_2^C, y_1^C, y_2^C)$ covers B^* and B is discovered by

$$x_1^C = \min(x_1, x_1^*), x_2^C = \max(x_2, x_2^*) \tag{11a}$$

$$y_1^C = \min(y_1, y_1^*), y_2^C = \max(y_2, y_2^*) \tag{11b}$$

The region S^C of C is determined as:

$$S^C = \left| (x_2^C - x_1^C) * (y_2^C - y_1^C) \right| \tag{12}$$

The IOU is obtained as:

$$IOU = \frac{S^I}{S + S^* - S^I} \tag{13}$$

The MIOU is calculated based on Eq. (7) as:

$$MIOU = IOU - \frac{S^C - (S + S^* - S^I)}{S^C} \tag{14}$$

So, the regression error value of the BB is defined as follows:

$$L_{MIOU} = 1 - MIOU \tag{15}$$

In Eq. (15), L_{MIOU} indicates non-negative and $L_{MIOU} \subseteq [0, 2]$. It is evident that if the L_n-norm is used as an error value, the local best is not essentially the best IOU.

Also, contrasted with IOU, the L_n-norm is prone to the dimension of the object. Even if the mean error between the estimated BB and ground truth box is equal, the overlay of the two boxes varies in the optimization since the sizes of the estimated BBs vary. However, IOU is an idea of a percentage and is not prone to dimension. It demonstrates there is a challenge between adjusting the L_n-norm error and the actual criteria of the IOU. MIOU exemplifies the overlay between the estimated BB and ground truth box in this condition, focusing on overlying and non-overlying areas concurrently. According to the loss function L_{MIOU}, it can adjust the location of the regression box, which solves the issue of varying error values and the real criteria of MIOU and circumvents the shortcoming of directly utilizing IOU as an error value.

Hence, this PLFRCNN is trained and utilized for categorizing unknown scans into healthy or TMJ-OA and appropriately localizing the small mandibular condyle areas.

4 Experimental Results

This section assesses the efficacy of the OGAN-PLFRCNN model by implementing it in MATLAB 2019b. In this experiment, 116 panoramic dental X-ray scans are acquired from https://data.mendeley.com/datasets/hxt48yk462/. Those scans are augmented by 6,000 scans by the OGAN. Of these, 65% of the scans are taken for training and 35% of the scans are taken for testing. Additionally, a comparative analysis is conducted between the proposed and earlier models implemented on the considered corpus: FRCNN [10], OGAN-FRCNN [11], YOLOv3 [14], DNN [15], HNN [16] and CariesNet [20] regarding the following metrics:

- Accuracy: It defines the percentage of appropriate classification over the total scans assessed.

$$Accuracy = \frac{True\ Positive\ (TP) + True\ Negative\ (TN)}{TP + TN + False\ Positive\ (FP) + False\ Negative\ (FN)} \quad (16)$$

In Eq. (16), TP is the number of healthy scans appropriately classified as healthy, TN is the number of TMJ-OA scans appropriately classified as TMJ-OA, FP is the number of TMJ-OA scans inappropriately categorized as healthy and FN is the number of healthy scans inappropriately categorized as TMJ-OA.

- Precision: It calculates the properly classified scans at TP and FP rates.

$$Precision = \frac{TP}{TP + FP} \quad (17)$$

- Recall: It measures the number of scans that are properly classified at TP and FN rates.

$$Recall = \frac{TP}{TP + FN} \quad (18)$$

- F-score (F): It is computed by

$$F = \frac{2 \times Precision \times Recall}{Precision + Recall} \quad (19)$$

Table 1. Comparison of proposed and existing TMJ-OA recognition and localization models.

Metrics	DNN	HNN	YOLOv3	CariesNet	FRCNN	OGAN-FRCNN	OGAN-PLFRCNN
Precision	0.831	0.858	0.867	0.88	0.906	0.922	0.9839
Recall	0.84	0.865	0.88	0.894	0.92	0.945	0.9776
F-score	0.836	0.862	0.874	0.887	0.913	0.934	0.9808
Accuracy (%)	83.64	85.93	87.18	89.17	92.21	94.59	98.18

Table 1 presents the results achieved by various models for TMJ-OA recognition and localization.

Figure 4 displays the efficacy of various models on the panoramic dental X-ray corpus for TMJ-OA recognition and localization. It declares that the success rate of the OGAN-PLFRCNN model regarding precision, recall and f-score is greater than that of all other earlier models because it enhances the BB regression and prevents variance in mandibular condyle localization. Accordingly, it is understood that the precision of the OGAN-PLFRCNN is 18.4%, 14.7%, 13.5%, 11.8%, 8.6% and 6.7% superior to the DNN, HNN, YOLOv3, CariesNet, FRCNN and OGAN-FRCNN models, respectively. The recall of the OGAN-PLFRCNN is 16.4%, 13%, 11.1%, 9.4%, 6.3% and 3.4% improved than the DNN, HNN, YOLOv3, CariesNet, FRCNN and OGAN-FRCNN models, respectively. Also, the f-score of the OGAN-PLFRCNN is 17.4%, 13.8%, 12.3%, 10.6%, 7.4% and 5.1% better than the DNN, HNN, YOLOv3, CariesNet, FRCNN and OGAN-FRCNN models, respectively.

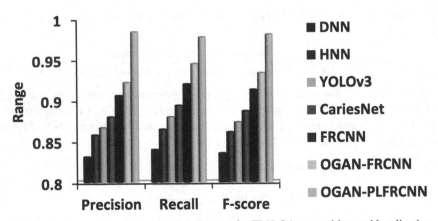

Fig. 4. Analysis of Precision, recall & F-score for TMJ-OA recognition and localization.

Figure 5 depicts the accuracy of various models tested by the panoramic dental X-ray corpus for TMJ-OA recognition and localization. It realizes that the accuracy of the OGAN-PLFRCNN is 17.4% superior to the DNN, 14.3% superior to the HNN, 12.6% superior to the YOLOv3, 10.1% superior to the CariesNet, 6.5% superior to the FRCNN and 3.8% superior to the OGAN-FRCNN models. Based on these analyses, it is established that the OGAN-PLFRCNN model maximizes the accuracy of recognizing

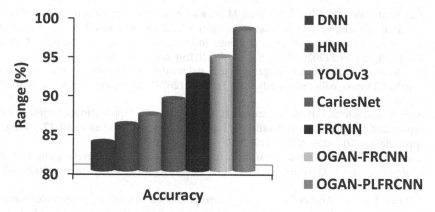

Fig. 5. Analysis of Accuracy of TMJ-OA recognition and localization.

the TMJ-OA and localizing the small mandibular condyle effectively compared to the other earlier models.

5 Conclusion

In this study, the OGAN-PLFRCNN model was presented for TMJ-OA classification and localization by resolving the variation in small condyle OA localization. At first, the panoramic X-ray database was augmented by the OGAN and given to the FPN with ROI-grid attention for extracting F-maps at various scales. Then, such F-maps were passed to the MRPN, wherein multiscale convolution feature fusion and INMS generated the RPs with more rich information. Afterward, both F-maps and RPs were concatenated by the improved RoI pooling process to get the proposal F-maps. Moreover, such F-maps were learned by the fully connected layer for TMJ-OA recognition and localization. Also, the MIOU was used to enhance the BB regression during localization. At last, the extensive experiments revealed that the OGAN-PLFRCNN model on the panoramic dental X-ray corpus has 98.18% accuracy than the other models for TMJ-OA recognition and localization.

References

1. Covert, L., Mater, H.V., Hechler, B.L.: Comprehensive management of rheumatic diseases affecting the temporomandibular joint. Diagnostics. **11**(3), 409 (2021). https://doi.org/10.3390/diagnostics11030409
2. Piancino, M.G., et al.: Cranial structure and condylar asymmetry of patients with juvenile idiopathic arthritis: a risky growth pattern. Clin. Rheumatol. **37**(10), 2667–2673 (2018). https://doi.org/10.1007/s10067-018-4180-5
3. Han, M.D., Lieblich, S.E.: Anatomy and pathophysiology of the temporomandibular joint. In: Michael Miloro, G.E., Ghali, P.E., Larsen, P.W. (eds.) Peterson's Principles of Oral and Maxillofacial Surgery, pp. 1535–1550. Springer International Publishing, Cham (2022). https://doi.org/10.1007/978-3-030-91920-7_51

4. Tsai, C.M., Wu, F.Y., Chai, J.W., Chen, M.H., Kao, C.T.: The advantage of cone-beam computerized tomography over panoramic radiography and temporomandibular joint quadruple radiography in assessing temporomandibular joint osseous degenerative changes. J. Dental Sci. **15**(2), 153–162 (2020). https://doi.org/10.1016/j.jds.2020.03.004
5. Heo, M.S., et al.: Artificial intelligence in oral and maxillofacial radiology: what is currently possible? Dentomaxillofacial Radiol. **50**(3), 1–13 (2021). https://doi.org/10.1259/dmfr.202 00375
6. Asiri, S.N., Tadlock, L.P., Schneiderman, E., Buschang, P.H.: Applications of artificial intelligence and machine learning in orthodontics. APOS Trends Orthodontics **10**, 17–24 (2020). https://doi.org/10.25259/APOS_117_2019
7. Chen, Y.W., Stanley, K., Att, W.: Artificial intelligence in dentistry: current applications and future perspectives. Quintessence Int. **51**(3), 248–257 (2020). https://doi.org/10.3290/j.qi. a43952
8. Vollmer, A., et al.: Artificial intelligence-based prediction of oroantral communication after tooth extraction utilizing preoperative panoramic radiography. Diagnostics. **12**(6), 1–13 (2022). https://doi.org/10.3390/diagnostics12061406
9. AbuSalim, S., Nordin Zakaria, Md., Islam, R., Kumar, G., Mokhtar, N., Abdulkadir, S.J.: Analysis of deep learning techniques for dental informatics: a systematic literature review. Healthcare **10**(10), 1892 (2022). https://doi.org/10.3390/healthcare10101892
10. Kim, D., Choi, E., Jeong, H.G., Chang, J., Youm, S.: Expert system for mandibular condyle detection and osteoarthritis classification in panoramic imaging using R-CNN and CNN. Appl. Sci. **10**(21), 7464 (2020). https://doi.org/10.3390/app10217464
11. Vijayakumar, K., Santhi, B.: Optimized adversarial network with faster residual deep learning for osteoarthritis classification in panoramic radiography. Int. J. Intell. Eng. Syst. **15**(6), 191–200 (2022). https://doi.org/10.22266/ijies2022.1231.19
12. Yan, D., et al.: An improved faster R-CNN method to detect tailings ponds from high-resolution remote sensing images. Remote Sens.. **13**(11), 1–18 (2021). https://doi.org/10. 3390/rs13112052
13. Yang, H., et al.: Deep learning for automated detection of cyst and tumors of the jaw in panoramic radiographs. J. Clin. Med. **9**(6), 1–14 (2020). https://doi.org/10.3390/jcm9061839
14. Takahashi, T., Nozaki, K., Gonda, T., Mameno, T., Ikebe, K.: Deep learning-based detection of dental prostheses and restorations. Sci. Rep. **11**(1), 1–7 (2021). https://doi.org/10.1038/ s41598-021-81202-x
15. Cha, J.Y., Yoon, H.I., Yeo, I.S., Huh, K.H., Han, J.S.: Panoptic segmentation on panoramic radiographs: Deep learning-based segmentation of various structures including maxillary sinus and mandibular canal. J. Clin. Med. **10**(12), 1–14 (2021). https://doi.org/10.3390/jcm 10122577
16. Leo, L.M., Reddy, T.K.: Learning compact and discriminative hybrid neural network for dental caries classification. Microprocess. Microsyst. **82**, 1–5 (2021). https://doi.org/10.1016/j.mic pro.2021.103836
17. Aljabri, M., et al.: Canine impaction classification from panoramic dental radiographic images using deep learning models. Inform. Med. Unlocked. **30**, 1–10 (2022). https://doi.org/10.1016/ j.imu.2022.100918
18. Rohrer, C., Krois, J., Patel, J., Meyer-Lueckel, H., Rodrigues, J.A., Schwendicke, F.: Segmentation of dental restorations on panoramic radiographs using deep learning. Diagnostics. **12**(6), 1–8 (2022). https://doi.org/10.3390/diagnostics12061316
19. Bayrakdar, I.S., et al.: A U-Net approach to apical lesion segmentation on panoramic radiographs. Biomed. Res. Int. **2022**, 1–7 (2022). https://doi.org/10.1155/2022/7035367
20. Zhu, H., Cao, Z., Lian, L., Ye, G., Gao, H., Wu, J.: CariesNet: a deep learning approach for segmentation of multi-stage caries lesion from oral panoramic X-ray image. Neural Comput. Appl. **35**, 16051–16059 (2021). https://doi.org/10.1007/s00521-021-06684-2

Semi-Automated Diabetes Prediction Using AutoGluon and TabPFN Models

Vikram Puri[1]([✉]) [ID], Aman Kataria[2] [ID], and Bhuvan Puri[3] [ID]

[1] Center of Visualization and Simulation, Duy Tan University, Da Nang, Vietnam
purivikram@duytan.edu.vn
[2] CSIR-CSIO, Chandigarh 160030, India
[3] Lovely Professional University, Jalandhar, India

Abstract. Since the previous decade, the number of diabetic patients has increased significantly, which raises the risk of additional complications such as heart attacks, renal failure, decreased vision, and nerve damage. If this illness is detected and treated in its early stage, patients can be saved from a life-threatening disease. The discipline of artificial intelligence (AI), which is rapidly expanding, has possibilities that could revolutionize how this serious ailment is diagnosed and managed. At present, AI can only forecast diabetes using manually entered data. This paper proposes a semi-automated AI model that allows us to measure in automated and manual ways. Additionally, the two distinct AI models, AutoGluon and TabPFN, are employed to train AI models and are assessed using statistical metrics. Additionally, it also compares with the four traditional models w.r.t evaluation performance. Feature importance is then utilized to determine which feature is more advantageous.

Keywords: Camera Vision · IoT · Machine Learning · Diabetes · OpenCV

1 Introduction

Diabetes is a long-term condition brought on by either insufficient insulin production by the pancreas or improper insulin utilization by the body. A hormone called insulin controls how much sugar is in the blood. A common complication of untreated diabetes is hyperglycemia. Hyperglycemia is a severe condition that can harm several body systems, including the nervous and circulatory systems. As stated in [1], diabetes is a major issue in developed and developing countries. The pancreas secretes the hormone insulin, which stimulates the absorption of glucose from food into the bloodstream. An inability of the pancreas to produce enough of this hormone leads to diabetes. Some of the potential complications of diabetes include coma, renal and retinal failure, pathological destruction of pancreatic beta cells, cardiovascular dysfunction, cerebral vascular dysfunction, peripheral vascular disease, sexual dysfunction, joint failure, weight loss, ulcer, and pathogenic effects on immunity [2].

As a result, there have been a lot of studies in disease forecasting, and today we employ decision-support algorithms and smart methods to make predictions.

R. K. Challa et al. (Eds.): ICAIoT 2023, CCIS 1929, pp. 289–295, 2024.
https://doi.org/10.1007/978-3-031-48774-3_20

In the medical field, for example, decision-support models have been used to diagnose illnesses like diabetes. Complications with macrovascular and capillary health, as well as ocular problems and renal failure, are more likely when diabetes is diagnosed and predicted later than it should be due to inadequate blood glucose control [3, 4]. In [5], authors suggested a framework for classifying diabetes prediction supported by machine learning (ML) models. The performance is improved by using grid search and other hyperparameter optimization methods. Some research [6, 7] suggested diabetes classifiers using various ML models. ML model and fuzzy logic were combined to improve prediction performance [8]. Support vector classifier (SVC) and artificial neural network (ANN) are the ML models considered for this investigation. Most studies concentrate on the established ML-enabled framework that categorizes diabetes prediction [9–12].

Moreover, inputs considered for these models are traditional. To overcome these issues, a semi-automated prediction system is proposed. The contribution of the study is as follows:

a. Trained diabetes prediction dataset through two different models such as AutoGluon and TabPFN.
b. Compared the performance of AutoGluon and TabPFN with four traditional models such as Logistic Regression (LR), Support Vector Classifier (SVC), Random Forest (RF) and Decision Tree (DT).
c. The input for the AI model is split into automated (Camera Vision) and manual (User Interface).
d. Also, the more advantageous features are highlighted for further research.

The rest of the paper is discussed as follows Sect. 2 discusses the proposed approach methodology, Sect. 3 evaluates the proposed approach through different statistical parameters, and Sect. 4 concludes the overall study.

2 Methodology

Researchers have recently developed various AI-based prediction techniques [13]. However, the only foundation for these techniques is the training and assessing various machine learning (ML) and deep learning (DL) models. In this study, a proposed architecture is made from collecting real-time data to forecast the user's diabetes. The architecture and workflow of the suggested solution is shown in Fig. 1.

2.1 Data Collection

Blood glucose, insulin, blood pressure, and age are a few factors that can be used as features to train an AI model and parameters to identify diabetes in a patient. The initial model must be trained on a single dataset before user data can be collected, following which it can repeatedly be trained using the user data. In this study, there are two methods to collect patient data.

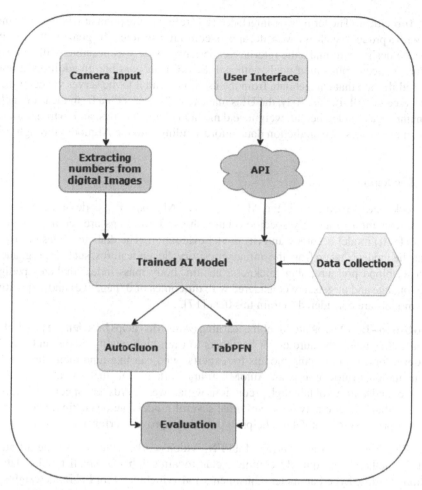

Fig. 1. Workflow of the proposed solution

Capturing: A procedure called capture is used to recognize the text area in an image and extract the text from it. In the first stage, the contour function [14] (OpenCV library) is used to locate the rectangular object needed to recognize the text area. The precision of the detection process can be increased by passing the function to the precise location parameters. In the next stage, the captured area through the contour function consists of alphanumeric characters, which need to be extracted through the optical character recognition (OCR) technique. Image resizing and grayscale conversion form the method for identifying and extracting the number from the image. The bilateral filter should then remove any extraneous details from the grayscale image. The undesired details are first removed to extract the numeric image from the main image. Then the edge detection approach is used to locate the boundaries of the necessary item. The numeric from the image is then extracted using EasyOCR [15] for further processing. This study applies this technique to extract the user's real-time blood glucose and blood pressure.

User Interface: The term "user interface" (UI) refers to the point at which a user inter-acts with a product, such as a website, app, or computer system. The goal of effective user interface design is to make the interface experience as easy as possible while allowing the user to accomplish their goals with as little effort as possible. In addition, it works as a middleware that can get data from the user and send it to the server via centralized, distributed or API. In this study, the UI is made to collect user data from the user, includ-ing name, age, gender, height, weight, and medical characteristics, and to transform that data into the necessary prediction data before sending it to the AI model through API.

2.2 Prediction

The backbone of prediction is the AI models. An AI model is a code or algorithm that uses information to identify specific relationships. Two things are required with any AI model (i) model accuracy and (ii) model evaluation. The diabetes dataset [16] that trains the model is taken from the online repository. Seven features, such as pregnancies, glucose, blood pressure, skin thickness, insulin, body mass index, diabetes pedigree function, age and targets, are considered with outcomes of 0 (negative) and 1 (positive). Two models are considered to train this data [17].

AutoGluon–Tabular. Auto Machine learning (AutoML) helps broaden the possibilities because they lower the hurdles for beginners to train high-quality model and speed up the development of working models for experts, who can fine-tune them through data augmentation. AutoGluon is an AutoML framework for DL that streamlines all ML activities and helps us obtain high predictions with fewer affords. For specific workloads, such as tabular neural networks, ensemble several models, auto stacking, and rigorous data pre-processing, AutoGluon helps to provide improved performance [18].

TabPFN. The trained transformer TabPFN, which works with state-of-the-art classi-fication methods, can quickly complete classification jobs for small tabular datasets without needing hyperparameter adjustment. It also includes TabPFN in its weights and receives training and test samples as set-valued inputs before producing predictions for the entire test set in a single forward pass [19].

3 Testbed and Analysis

In this section, the experimental testbed and analysis of the results from the experiment are discussed as follows:

3.1 Experimental Testbed

Two experiment testbeds are used (i) to extract the image data and (ii) to train the dataset with two different models. Table 1 represents the experimental configuration of the proposed study.

Table 1. Experimental Configuration of the Proposed Study.

Testbed	Configuration	Libraries
Testbed-1 (extract images)	Ubuntu-21.04, 16 GB RAM, 512 SSD	OpenCV, easyOCR
Testbed-2 (Model training)	Ubuntu-21.04, 16 GB RAM, 512 SSD	sklearn, pandas, AutoGluon, TabPFN

3.2 Analysis

Four statistical parameters are used to evaluate the trained model: accuracy score, precision, recall and F1 score. The accuracy score is not enough to decide the model performance. With the support of another parameter, it can evaluate the performance more clearly. In the case of the AutoGluon, the accuracy, precision, recall, and F1 scores are 0.779, 0.781, 0.543 and 0.641, respectively, in Table 2. However, TabPFN model accuracy and precision are less than the AutoGluon but recall and F1-score are higher. The recall and F1 scores of TabPFN and AutoGluon-Tabular are 0.72:0.543 and 0.73:0.641, respectively, as represented in Table 2.

Table 2. Evaluation and comparison of AutoGluon-Tabular and TabPFN with other models on the Diabetes Dataset

Model	Accuracy Score	Precision	Recall	F1- Score
LR	0.694	0.577	0.481	0.525
SVC	0.733	0.644	0.537	0.585
RF	0.720	0.607	0.574	0.590
DT	0.629	0.457	0.444	0.457
AutoGluon-Tabular	**0.779**	**0.781**	**0.543**	**0.641**
TabPFN	**0.759**	**0.73**	**0.72**	**0.73**

From Table 2, it is evident that the accuracy score, precision, recall and F1-score of LR, SVC, RF and DT are lower than those values of AutoGluon and TabPFN respectively. Figure 2 describes the graphical representation of the above-mentioned results.

The performance, whenever the algorithm produces prediction on a disturbed version of the information where the numerical values of this characteristic have been arbitrarily jumbled across rows, is represented by the significance rating of a feature. According to the significance score for glucose, which is 0.177821012, the predictive performance of the models decreases by this amount when there is a shift. Similarly, in the case of other features. 'p99_high' and 'p99_low' represent the importance of genuine features lies between the p99_high and p99_low (confidence level of 99%). Table 3 represents the feature importance of AutoGluon-Tabular on the dataset. The highest importance feature of AutoGluon is the glucose in terms of importance, standard deviation, p_value, p99_high and p_low and the rest of the features such as BMI, age, blood pressure, pregnancies, skin thickness, insulin, and diabetes pedigree function.

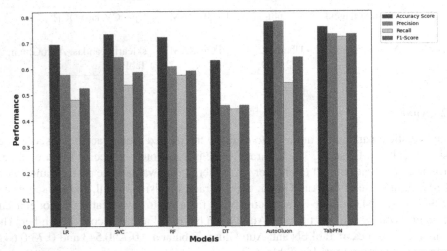

Fig. 2. Graphical representation of evaluation and comparison of AutoGluon-Tabular and TabPFN with other models on the Diabetes Dataset

Table 3. AutoGluon-Tabular Features Importance

Model	Importance	stddev	p_value	n	p99_high	p99_low
Glucose	0.177821012	0.01631	8.40E-06	5	0.211408	0.144234
BMI	0.091828794	0.00484	9.26E-07	5	0.101803	0.081854
Age	0.076264591	0.00503	2.27E-06	5	0.086634	0.065896
Blood Pressure	0.0692607	0.00525	3.95E-06	5	0.080084	0.058437
Pregnancies	0.059533074	0.00447	3.82E-06	5	0.068755	0.050311
SkinThickness	0.057976654	0.00288	7.34E-07	5	0.063918	0.052035
Insulin	0.055252918	0.00709	3.19E-05	5	0.069862	0.040644
DiabetesPedigreeFunction	0.049416342	0.00876	0.000114	5	0.067465	0.031368

4   Conclusion

The most important aspect of the healthcare system is the prompt diagnosis and treatment of diseases. AI can potentially improve patient assistance and serve as a solid support for the healthcare system. A semi-automated technique to predict diabetes in patients at an early stage is suggested in this study. The system can automatically and manually detect the patient's parameters. It can also directly identify the patient's diabetes from the machine using a camera. The dataset is trained and validated through two different models, AutoGluon and TabPFN.

More human input parameters can be replaced with automatic ones in the upcoming work, and a decentralized system can be built to keep patient data more private.

References

1. Misra, A., et al.: L: Diabetes in developing countries. J. Diabetes **11**(7), 522–539 (2019)
2. Vaishali, R., Sasikala, R., Ramasubbareddy, S., Remya, S., Nalluri: Genetic algorithm based feature selection and MOE Fuzzy classification algorithm on Pima Indians Diabetes dataset. In: 2017 international conference on computing networking and informatics (ICCNI), pp. 1–5. IEEE (2017)
3. World Health Organization. Definition, diagnosis and classification of diabetes mellitus and its complications: report of a WHO consultation. Part 1, diagnosis and classification of diabetes mellitus (No. WHO/NCD/NCS/99.2). World health organization (1999)
4. Temurtas, H., Yumusak, N., Temurtas, F.: A comparative study on diabetes disease diagnosis using neural networks. Expert. Syst. **36**, 8610–8615 (2009)
5. Hasan, M.K., Alam, M.A., Das, D., Hossain, E., Hasan, M.: Diabetes prediction using ensembling of different machine learning classifiers. IEEE Access **8**, 76516–76531 (2020)
6. Krishnamoorthi, R., et al.: A novel diabetes healthcare disease prediction framework using machine learning techniques. J. Healthc. Eng. 1–10 (2022)
7. Verma, A.K., Biswas, S.K., Chakraborty, M., Boruah, A.N.: A transparent machine learning algorithm to manage diabetes: TDMSML. Adv. Comput. Intell. **3**, 5 (2023)
8. Ahmed, U., et al.: Prediction of diabetes empowered with fused machine learning. IEEE Access **10**, 8529–8538 (2022)
9. Kataria, A., Ghosh, S., Karar, V.: Data prediction of optical head tracking using self healing neural model for head mounted display. J. Sci. Ind. Res. **77**, 288–292 (2018)
10. Kataria, A., Ghosh, S., Karar, V.: Data prediction of electromagnetic head tracking using self healing neural model for head-mounted display. Rom. J. Sci. Inf. Technol. **23**(5), 354–367 (2020)
11. Banerjee, K., et al.: A machine-learning approach for prediction of water contamination using latitude, longitude, and elevation. Water **14**(5), 1–20 (2022)
12. Kataria, A., Puri, V.: AI and IoT-based hybrid model for air quality prediction in a smart city with network assistance. IET Networks **11**(6), 221–233 (2022)
13. Olisah, C.C., Smith, L., Smith, M.: Diabetes mellitus prediction and diagnosis from a data pre-processing and machine learning perspective. Comput. Methods Programs Biomed. **220**, 1–12 (2022)
14. Gollapudi, S.: OpenCV with Python. Learn Computer Vision Using OpenCV: With Deep Learning CNNs and RNNs 31–50 (2019)
15. EasyOCR: https://www.jaided.ai/easyocr/. Accessed 15 Jan 2023
16. Diabetes Dataset: https://www.kaggle.com/datasets/mathchi/diabetes-data-set. Accessed 15 Jan 2023
17. Zhu, T., Li, K., Herrero, P., Georgiou, P.: Personalized blood glucose prediction for type 1 diabetes using evidential deep learning and meta-learning. IEEE Trans. Biomed. Eng. **70**(1), 193–204 (2022)
18. Erickson, N., et al.: Autogluon-tabular: Robust and accurate automl for structured data. arXiv preprint arXiv:2003.06505 (2020)
19. Hollmann, N., Müller, S., Eggensperger, K., Hutter, F.: TabPFN: A transformer that solves small tabular classification problems in a second. arXiv preprint arXiv:2207.01848 (2022)

Malnutrition Detection Analysis and Nutritional Treatment Using Ensemble Learning

Premanand Ghadekar⬤, Tejas Adsare⬤, Neeraj Agrawal⬤, Tejas Dharmik⁽✉⁾⬤, Aishwarya Patil⬤, and Sakshi Zod⬤

Vishwakarma Institute of Technology, 666, Upper Indiranagar, Bibwewadi, Pune, Maharashtra 411 037, India
{tejas.adsare20,tejas.adsare20}@vit.edu

Abstract. Malnutrition is a condition that arises when an individual's diet contains excessive amounts of certain nutrients or insufficient amounts of one or more of the essential nutrients. The proposed system uses an ensemble learning model of the CNN, the transfer learning algorithms such as Inception-v3, VGG16 and VGG19 were combined together with the help of ensemble learning to enhance classification, prediction, function approximation, etc. The model takes input images and classifies them as normal, wasting, stunting, and obesity. The goal of the proposed system is to identify malnutrition and its types and provide treatments for each category and ways to prevent malnutrition which will assist in lowering the danger of mortality, health and physical problems by using the appropriate treatments or precautions. In conclusion, the proposed system is a significant step towards identifying and treating malnutrition effectively. By using ensemble learning algorithms, it can accurately classify different types of malnutrition and provide appropriate treatments to those affected.

Keywords: Analysis · CNN · Deep learning · Ensemble learning · InceptionV3 · Nutritional · Malnutrition · VGG16 · Vgg19

1 Introduction

Malnutrition is a serious health issue that affects millions of people worldwide. It results from the inadequate intake of essential nutrients from food, causing various health issues such as weakened immune systems, stunted growth, and an increased risk of infections. Early detection and treatment of malnutrition are essential to prevent the long-term consequences of this condition. The conventional methods for detecting malnutrition include physical examination, body weight measurement, and blood tests. Nevertheless, these methods are often costly, time-consuming, and may not always produce precise outcomes. As a result, there is a necessity for a more dependable and efficient approach to identifying malnutrition.

Deficiency of consumption of different nutrients can lead to malnutrition which is one of the main health issues that persist all over the world. Underlying child mortality and morbidity can be considered one of the major factors for malnutrition in infants.

© The Author(s), under exclusive license to Springer Nature Switzerland AG 2024
R. K. Challa et al. (Eds.): ICAIoT 2023, CCIS 1929, pp. 296–310, 2024.
https://doi.org/10.1007/978-3-031-48774-3_21

Undernutrition is thought to be the cause of about 45% of fatalities in children who belong to the age group of 0–5 years. According to a survey 42.5% of children who are under the age of four to five in India were underweight, 19.8% were wasted, 6.4% were seriously wasted, and 48.0% were stunted. According to the National Family Health Survey (NFHS–4), 35.8% of children were underweight, 21.0% were severely wasted, and 38.4% were stunted.

Although stunted and underweight children have decreased slightly over the past ten years, the worrisome rate of wasted youngsters has not decreased. Children suffer more from malnutrition, and detecting it can assist lower the risk of mortality and physical or developmental problems by taking the required actions. In India, 20% of children suffer from wasting,48% of children suffer from stunting, and being underweight affects 43% which depicts the severity of this health issue that prevails all over the country. Using the machine learning technique to detect malnutrition assists in the healthcare sector. Categorizing malnutrition into their categories helps provide the treatment for that particular type of malnutrition that has been detected. When malnutrition in children is identified, it might be easier for individuals and healthcare professionals to take preventative action and lessen the negative effects on children. The research paper includes an innovative approach for detecting malnutrition using ensemble learning. We will evaluate the effectiveness of various transfer learning algorithms such as Inception-v3, VGG16 and VGG19, in detecting malnutrition in a patient dataset. Once malnutrition is detected, appropriate nutritional treatment can be provided to the patient. In this paper, various nutritional treatment recommendations are also explored. It takes input in the form of images and assigns weightage to various objects in images that help in differentiating kinds of malnutrition.

The main goal of this paper is to advance the development of more effective and trustworthy techniques for detecting and treating malnutrition through machine learning methods. The suggested approach has the potential to increase the precision and rapidity of malnutrition detection while also delivering individualized nutritional recommendations to patients, resulting in improved health outcomes.

2 Literature Review

The proposed model by Lakshminarayanan et al. used CNN to detect malnutrition in children. Alex Net architecture was used. Learning rate is 0.001 and accuracy is 96%. ReLU activation function was used to improve the rate of performance. Dataset included 500 images where 90% were used for training purposes and 10% for testing [1].

The proposed model by Kadam et al. summarizes a system that takes the image as an input and detects malnutrition using the TensorFlow algorithm in underage children. Here first of all the processing of the images is done, the features are extracted from the images which are used for disease analysis. The extracted features from the input images were then compared to those in the training dataset. With the help of this, one can conclude that the image color feature reasonably matched the training data [2].

Shetty et al. contributed to a system that used Agent Technology and Data mining technology. The techniques used were Multilayer-perceptron and Bayesian Networks but it resulted in inefficiency of the mall and the redundancy of the data because of the small size of the dataset [3].

Dhanamjayulu et al. aimed to detect individuals affected by malnutrition by analyzing facial images. The model utilized a combination of factors, including weight, age, gender, and BMI, to determine malnutrition status. To detect faces within the images, the researchers used a Multi-task Cascaded Convolutional Neural network. Additionally, a residual neural network was built and trained to estimate BMI values based on the facial images. The model seeks to establish a relationship between facial weight and appearance to estimate BMI values, ultimately enabling the identification of individuals affected by malnutrition. [4].

In their 2022 publication, Islam et al. proposed a model for detecting malnutrition among women in Bangladesh. The model employs five machine learning-based algorithms, namely decision tree, artificial neural network, Naïve Bayes, support vector machine, and random forest, to identify individuals who may be suffering from malnutrition. The model is designed to analyze various factors and indicators of malnutrition and provide accurate predictions based on the collected data [5]. From this paper it can be concluded that the Random Forest based classifier provides the accuracy of 81.4% for underweight women and for obese women it shows the accuracy of 82.4%. Combining MLR-RF based methods can more accurately detect malnourished women. The proposed approach aims to reduce the burden on healthcare services while also helping to identify women who are at a higher risk of malnutrition. By using machine learning algorithms to analyze various indicators of malnutrition, the model can accurately predict which individuals may require additional attention or support [5].

Talukder and Ahammed used five different machine learning algorithms to analyze data on children's health and nutrition. The algorithms used are Linear Discriminant Analysis, K-Nearest Neighbors, Support Vector Machine, Random Forest, and Logistic Regression. The model is designed to detect malnutrition in children aged 0–5 in Bangladesh. The goal is to identify malnourished children and provide them with the necessary treatment and support [6]. Out of which, it was concluded that Random Forest provided better accuracy as compared to other algorithms.

Najaflou and Rabieiindetected the malnutrition in the children of the age group of 6–12 years Decision Tree algorithm and AdaBoost Algorithm model was used for the recommendation system of the nutritional diet. The accuracy of the model was 90.27%. The dataset consisted of 1001 images out of which 806 were facing the problem of underweight. BMI was the main factor that was used in order to detect malnutrition [7].

Theilla et al. summarizes that Patients which are in intensive care units (ICU) are at greatest risk of malnutrition However, GLIM, a recognized diagnostic criterion for malnutrition, has not been authenticated for the patients in intensive care unit. SGA (Subjective Global Assessment) is an authenticated tool. Physician access to the nutritional status of inpatients is considered the gold standard [8].

Yin et al. developed the system of detecting malnutrition based on classification of tree-based machine learning models for cancer patients. The system included a dataset with 16 types of cancer. The GLIM criteria were used for the diagnosis purpose. A k test was performed to check the results obtained from the system and compare with the actual result. It helped in the pretreatment identification of malnutrition [9].

Browne et al. developed a system for the poverty and malnutrition prevalence based on the multivariate Random Forest. To improve the accuracy of the detection, a five-fold cross-validation approach was employed in conjunction with the use of two key parameters in the Random Forest algorithm: the maximum number of trees and the down-sampling rate for features. A sequence prediction framework was used for the better classification of the dataset [10].

Ahirwar et al. designed a system that utilizes several machine learning algorithms to predict the occurrence of malnutrition disease. In this paper, they use the WEKA tool to evaluate classifiers in a comparative manner in order to increase classification accuracy. The findings of this study on the malnutrition dataset demonstrate that linear regression processing effectiveness and prediction accuracy are superior to k-nearest neighbor, multilayer perceptron regression decision tree, and methods [11].

Kavya et al. created a novel machine learning-based application that can forecast the occurrence of malnutrition and anemia. This system is unique and can provide critical insights into the likelihood of these conditions. They employ technologies like "Visual Studio" for the front end and "SQL server" for the back end to construct the real-time application. They are both effective tools for using the real-time application. A team of researchers has created a system that describes various classification techniques that are applicable for identifying malnutrition and anemia in children who are younger than five years old [12].

Nyarko et al. developed a system using machine learning algorithms to predict undernutrition among children under five in Ethiopia. They collected data from the 2016 Ethiopian Demographic and Health Survey and used various machine learning models, with the xgbTree algorithm achieving the highest accuracy. The proposed system can facilitate earlier interventions and treatments to prevent undernutrition and its associated complications [13].

Wajgi and Wajgi have developed a system that uses machine learning to detect malnutrition in infants. They collected data on 550 infants and trained various machine learning models, with the random forest model achieving the highest accuracy of 92.6% in detecting malnutrition. The proposed system can facilitate earlier interventions and treatments to prevent malnutrition and associated complications [14].

3   Existing System

On the basis of numeric data, various algorithms were used to identify malnutrition in binary format (whether the person will have a deficiency or not). Malnutrition was identified based on age, height, and weight criteria. BMI was also the main factor that was used to detect malnutrition. To estimate BMI values, a residual neural network was created and trained. The association between body weight and facial features was predicted, as was the estimate of BMI using images of human faces [4]. With the help of images, the conventional neural network was used to detect and predict whether a child is affected by malnutrition or not. For classification purposes, Alexnet was used to identify patterns in images, as well as for recognizing faces and objects. The classification here is binary: whether or not a person will suffer from malnutrition or a nutritional deficiency [2]. The existing system is unable to classify all types of malnutrition into a single model.

Table 1. Comparative table between existing and proposed system.

Approach	Advantages	Limitations
Physical Examination and Body Weight Measurement	Non-invasive and simple to execute	This approach has limited capabilities to identify mild or moderate malnutrition and may produce inaccurate results in the presence of edema and dehydration. Additionally, it requires skilled professionals to perform the examination accurately
Blood Tests	This technique offers detailed information on nutrient deficiencies in the body	This method is invasive, costly, time-consuming, and necessitates specialized equipment and professionals
Machine Learning Algorithms (Random Forests, Decision tree and Neural Networks)	Machine learning algorithms provide non-invasive, quick, and precise results. They can process large datasets and identify multiple factors contributing to malnutrition	To provide accurate results, machine learning algorithms need a lot of high-quality data. The model may overfit the data, and its effectiveness depends on expertise in machine learning techniques
Ensemble Learning (Combining Multiple Machine Learning Models)	Several machine learning models can be combined through ensemble learning to improve the reliability and accuracy of predictions. It can handle missing data and identify multiple factors contributing to malnutrition	However, ensemble learning can be computationally expensive
Proposed Approach (Ensemble Learning for Malnutrition Detection and Nutritional Treatment)	The proposed approach of using ensemble learning for malnutrition detection and nutritional treatment aims to improve the accuracy and speed of malnutrition detection. An ensemble learning system is developed that takes input images and classifies them into four categories: normal, wasting, stunting, and obesity. After detecting malnutrition, nutritional treatment is provided	The proposed approach also requires a large amount of high-quality data and may be computationally expensive. It also requires expertise in machine learning techniques to be implemented effectively

No additional research is done to determine how the disease will affect the individual. Malnutrition analysis is not provided because the treatment section is missing. Table [1] states the comparison between the existing and proposed system.

4 Methodology/Proposed System

The methodology of this proposed system for having deficiencies (Malnutrition) in people consists of three stages, first is the designing and training of the malnutrition detection model and second is the Recognition of Malnutrition type and further using the data of the person having that disease of Malnutrition will be useful for recommending the treatment and further suggest the future impact.

A. Malnutrition Type Classification

In this system, the design and test of a model for images have been done. The dataset includes various types of malnutrition deficiencies like Obesity, Stunting, Wasting, and normal where users can form the required words or sentences.

In this system, the implementation of ensemble learning with the use of a CNN algorithm and the use of transfer learning such as InceptionV3 and VGG16 has taken place. To train models, technologies used are TensorFlow and Keras. For implementation of the proposed system, it is further divided into five sub-groups that include the collection of datasets of 4 different classes, the implementation of the model, the extraction and training of the datasets, the interfacing model with an application for recommending treatment and the impact of the type of malnutrition.

The first stage decides the base of the entire model and how it is to be implemented. The system starts with passing the image, then processing the input, and then passing it through the deep learning architecture. It is further explained in detail.

4.1 Collection and Classes of Datasets

For training the model, the dataset has been created. The dataset contains images that belong to 4 classes. These are Obesity, Stunting, Wasting, and normal. In Table 2 it shows the four classes with its labels. Firstly, the dataset contains images that have been collected of different sizes.

Table 2. Types with its labels

Labels	Malnutrition Type
0	Obesity
1	Normal
2	Stunting
3	Wasting

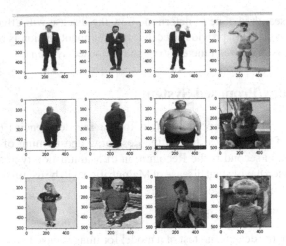

Fig. 1. Images from Dataset.

In Fig. 1 it depicts the overview of the images that are present in the dataset.

The neural networks get inputs of the images which are of the same size, they all need to be resized before inputting them to the CNN. Next, the dataset was split into separate training and testing sets. For training of models, the dataset is created with the help of data augmentation techniques using the augmenter library from the python module for increasing the images up to 4000. The images were too collected from different healthcare websites.

4.2 Implementation of the Model

In the implementation of the model as given in Fig. 2 firstly it takes the input in the form of images. After preprocessing of images and resizing it to 224x224 size. It is passed through the deep learning architecture; it detects it in 4 different classes along with recommending a deficient person about treatment and its future impact by inputting the values of height and weight.

The project flow diagram is given.

In this diagram, the images in Fig. 3. Are collected of 4 types and image processing is done i.e., remove the images which are of no use. So, after loading the dataset, generation of images is done using the python module known as augmenter, which produces about 4000 images. Prediction is done by applying different algorithms and one can input height and weight to know the recommendation, treatment and its impact. Convolutional neural network (CNN) is used in image processing that is designed to process pixel data. After the images are passed through the system, the image is processed, and the datasets are loaded. A total of 3 convolutional, 3 Max Pooling, 1 flatten and 2 dense layers comprise the CNN [2].

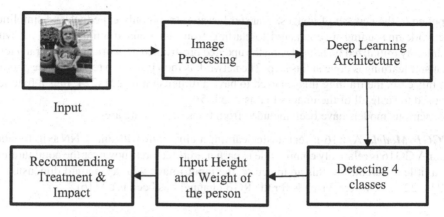

Input

Fig. 2. Methodology of Malnutrition Detection System.

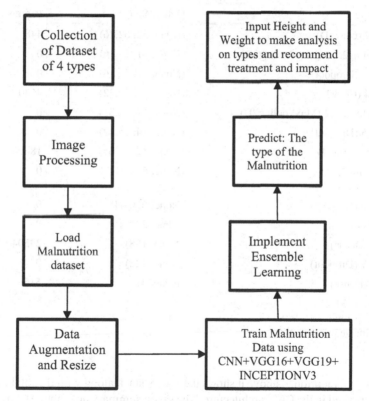

Fig. 3. Flowchart for Detection of Types of Malnutrition

4.3 Transfer Learning

The task is first learned using the pretrained network, after which the final layers are substituted with the new, smaller collection of pictures, allowing the classification to be applied to the new set of images. Transfer learning commonly expedites and simplifies network fine-tuning as compared to starting from zero and training a network with randomly initialized weights. Using the updated pre-trained network and the parameters, transfer learning is done at this step. The network is then trained to recognize the photos. In this case, the training images need to have a dimension of 224x224, thus a function is used to scale all of the images to this size [15].

Various models have been included from transfer learning are-

VGG16 Model. As a 16-layer transfer learning architecture with only CNN as its foundation, VGG16 is relatively comparable to earlier architectures, however the configuration is a little different. For this architecture, the input image with a standard dimension of 224 x 224 x 3, where 3 stands for the RGB channel has been used [16].

Table 3. Summary of CNN classifier

Layer (type)	Output Shape	Params
conv2d_3 (conv2D)	(None, 62, 62,16)	448
max_pooling2d_3 (MaxPooling2D)	(None, 31, 31,16)	0
dropout_4 (Dropout)	(None, 31, 31,16)	0
conv2d_4 (Conv2D)	(None, 29, 29,32)	4640
max_pooling2d_4 (MaxPooling2D)	(None, 14, 14,32)	0
dropout_5 (Dropout)	(None, 14, 14,32)	0
conv2d_5 (Conv2D)	(None, 12, 12,64)	18496
max_pooling2d_5 (MaxPooling2D)	(None, 6, 6,64)	0
dropout_6 (Dropout)	(None, 6, 6,64)	0
flatten_1 (Flatten)	(None, 2304)	0
dense_2 (Dense)	(None, 128)	295040
dropout_7 (Dropout)	(None, 128)	0
dense_3 (Dense)	(None, 4)	516
Total param: 319,140 Trainable params: 319,140 Non-Trainable params: 0		

In Table 3 mentioned above it shows the layers stack along with the total number of parameters used in the CNN architecture, The given summary describes a Convolutional Neural Network (CNN) classifier with multiple layers. It consists of several convolutional and pooling layers, followed by dropout layers to reduce overfitting. The network then flattens the output and passes it through fully connected dense layers. Finally, the output is transformed through dropout layers and a final dense layer, resulting in a classification with 4 classes. The network architecture has a total of 516 parameters.

InceptionV3. The Inception-v3 is a convolutional neural network that consists of 48 layers. The ImageNet database consists of pretrained network versions trained on over thousands of images. Several animals, a keyboard, a mouse, and a pencil are among the 1000 different item categories that the pre-trained network can classify photographs into. Therefore, the network contains extensive feature representations for various photographs. The input image for the network is 299 by 299 pixels in size [17].

VGG19 Model. A cutting-edge object-recognition model called VGG can accommodate up to 19 layers. VGG, which was designed as a deep CNN, performs better than baselines on several tasks and datasets outside of ImageNet. In comparison to VGG16, VGG19 has a few more convolutional relu units in the network's center [18].

Ensemble Learning. Ensemble learning is a method for solving specific computational intelligence problems by strategically generating and combining a number of models, such as classifiers or experts. The main goal of ensemble learning is to enhance classification, prediction, function approximation, etc. In this system 4 models were ensemble to make the accuracy of the model better [19].

4.4 Training and Testing of Model

The main step after using different models is the ensemble learning to train the model. For training the CNN, transfer learning algorithms such as CNN, Inception-v3, VGG16 and VGG19 were combined together with the help of ensemble learning. The model is run on 50 epochs and saves the model. After training all the models the training and validation accuracy and their loss are given in Table 4.

5   Results and Discussions

The proposed system used images of different types of malnutrition including in a folder such as obesity, wasting, stunting and normal. For training of models 80% of data is used whereas 20% is for testing. Dataset images are resized to 224 x 224 x 3 for training and testing purposes the taken images as input for classification. Following are the comparison between different CNN architecture with ensemble learning models.

Table 4. Comparison of accuracy of different algorithms

Sr. no	Algorithms	Training accuracy	Validation Accuracy
1	CNN	94.3%	57.14%
2	InceptionV3	92.3%	81%
3	VGG16	55.26%	35.29%
4	VGG19	88.5%	80.25%
5	Ensemble model	93%	84%

Table 4 compares the performance of different algorithms based on their training and validation accuracy. The results show that VGG16 had the lowest accuracy with a training accuracy of 55.26% and a validation accuracy of 35.29%. The CNN model had the highest training accuracy of 94.3%, but its validation accuracy was only 57.14%. InceptionV3 and VGG19 performed better than VGG16 with training accuracies of 92.3% and 88.5%, respectively. However, the best performing model was the Ensemble model with a training accuracy of 93% and a validation accuracy of 84%.

Precision Recall and F1 score description-

Table 5. Comparison of Precision, Recall and F1 score of CNN algorithms

Classes	Precision	Recall	F1 score
Normal	0.43	0.87	0.57
Obesity	0.55	0.37	0.44
Stunting	0.67	0.63	0.65
Wasting	0.71	0.15	0.25

Table 5 displays the evaluation metrics for the CNN model, which were generated for four different types of classes. These evaluation metrics are used to assess the performance of the CNN model in classifying the different types of classes.

Table 6. Comparison of Precision, Recall and F1 score of ensemble model

Classes	Precision	Recall	F1 score
Normal	0.56	0.88	0.60
Obesity	0.70	0.55	0.54
Stunting	0.68	0.69	0.60
Wasting	0.80	0.59	0.39

Table 6 displays the precision, recall, and F1 score for the Ensemble model across different categories. These evaluation metrics are used to evaluate the performance of the Ensemble model in classifying instances from each category.

Prediction of images on test data-

In Fig. 4., The model is inputted with an image and it predicts the image as a stunt. In this Fig. 5, the model is inputted with an image and it predicts the image as obesity. Ensembled learning model gives good accuracy as compared with other architectures.

B. **Analysis and Impact on Prediction of Malnutrition Deficiency**

The analysis section included the categorization of the three types according to the height and weight provided by the user.

```
1/1 [==============================] - 0s 67ms/step
stunting
The predicted image corresponds to "2"
```

Fig. 4. Stunting Image Prediction

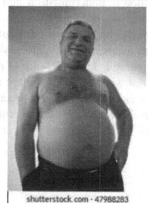

shutterstock.com · 47988283

```
1/1 [==============================] - 0s 206ms/step
obesity
The predicted image corresponds to "1"
```

Fig. 5. Obesity Image Prediction

The dataset includes obesity wasting and normal categories based on the height and weight it classifies between different categories. It includes about 300 obesity data, 75 normal, and 35 wasting. The dataset was trained using a random forest algorithm. It includes 255 females and 245 males.

In Fig. 6. Pie chart it shows the distribution of females and males according to data and how much of them are affected with the malnutrition type.

In Fig. 7. it shows number of trees vs accuracy. The number of trees depends on the number of rows present in the dataset, which are further combined to form the final decision of the model. It provides about 74% accuracy on the basis of values of height and weight using a random forest algorithm.

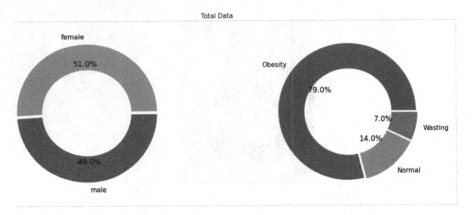

Fig. 6. Pie chart of males, females and types of malnutrition types.

The impact section included the possibility of deficiency diseases of protein, vitamins, and minerals if the prediction is undernutrition and similarly for the obese it includes the possibility of hemosiderosis or hypervitaminosis as well as hypertension. Moreover, breathing problems can also be faced with difficulty in physical functioning. The treatment consists of replacing meals with low-calorie shakes or meal bars and eating more plant-based foods and getting at least 150 min of physical activity to prevent further weight gain. The impact of wasting includes feeling tired as well as weaker and depressed. The treatment includes adding ready-to-use therapeutic food to the diet The impact of the stunning is a frequent illness as well as recurrent undernutrition. It prevents the children from reaching their physical and cognitive strengths.

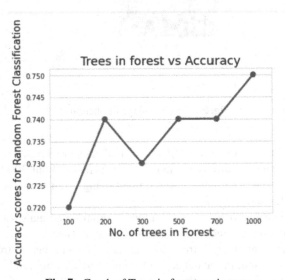

Fig. 7. Graph of Trees in forest vs Accuracy

The treatment includes giving vitamin A with zinc and plant-sourced foods in the diet. Furthermore, an exact nutritional diet chart can be obtained from the doctor to seek medical advice.

6 Conclusion

Malnutrition is incredibly prevalent and has impacted numerous nations worldwide in one or more ways. Malnutrition detection or prediction will assist the government or health services in implementing preventative measures. Types of malnutrition are found using the convolutional neural network (CNN or ConvNet) technique. Normal and deficient people's pictures are the input. InceptionV3, VGG16, and VGG19 is a CNN architecture that does classification tasks and looks for patterns in photos to identify faces and objects. The system predicts whether or not a person would suffer from which type of malnutrition utilizing parametric settings. Ensemble learning is used to take the average of the 3 CNN architecture algorithms used. Additionally, it does an analysis of predicted output and provides proper treatment and impact to the person. The system can provide the doctor assistance for the proper guidance of the medical treatment. Furthermore, it can also add the diet tracker option which will track their daily consumption and notify them. The lab test facility can be made available for them via which they can test and come to know about more precision value of the different vitamins.

References

1. Lakshminarayanan, A.R., Rajeswari V.P.B., Parthasarathy, S.: Malnutrition detection using CNN. J. Med. Syst. **45**(1), 1–12 (2021). https://doi.org/10.1007/s10916-020-01688-1
2. Kadam, N., Dabhade, V., Baravkar, R., Saravade, V., Chaitanya, P.: Detect malnutrition in underage children by using TensorFlow algorithm of artificial intelligence. Int. J. Eng. Dev. Res. **6**(12), 390–393 (2019)
3. Shetty, S., Lobo, L.G., Monisha, K.N., Shetty, S.K.N.: A survey on detection and diagnosis of malnutrition. Int. J. Emerg. Technol. Eng. Res. **10**(1), 172–177 (2022)
4. Chittathuru, D., et al.: Identification of malnutrition and prediction of BMI from facial images using real-time image processing and machine learning. IET Image Process. **16**(3), 647–658 (2022). https://doi.org/10.1049/ipr2.12222
5. Islam, M.M., et al.: Application of machine learning based algorithm for prediction of malnutrition among women in Bangladesh. Int. J. Cogn. Comput. Eng. **3**, 46–57 (2022). https://doi.org/10.1007/s42452-021-0579-9
6. Talukder, A., Ahammed, B.: Machine learning algorithms for predicting malnutrition among under-five children in Bangladesh. Nutrition **78**, 110861 (2020). https://doi.org/10.1016/j.nut.2020.110861
7. Najaflou, S., Rabiei, M.: Recommended system for controlling malnutrition in Iranian children 6 to 12 years old using machine learning algorithms. Health Inf. Manag. **4**(1), 27–33 (2021)
8. Theilla, M., Rattanachaiwong, S., Kagan, I., Rigler, M., Bendavid, I., Singer, P.: Validation of GLIM malnutrition criteria for diagnosis of malnutrition in ICU patients: an observational study. Clin. Nutr. **40**(5), 3578–3584 (2021). https://doi.org/10.1016/j.clnu.2020.12.021
9. Yin, L., et al.: Classification tree-based machine learning to visualize and validate a decision tool for identifying malnutrition in cancer patients. JPEN J. Parenter. Enteral Nutr. **45**(8), 1736–1748 (2021). https://doi.org/10.1002/jpen.2070

10. Browne, C., Matteson, D.S., McBride, L., Hu, L., Liu, Y., Sun, Y.: Multivariate random forest prediction of poverty and malnutrition prevalence. PLoS ONE **16**(9), e0255519 (2021). https://doi.org/10.1371/journal.pone.0255519

11. Khan, R., Ahirwar, M., Shukla, P.K.: Predicting malnutrition disease using various machine learning algorithms. Int. J. Sci. Technol. Res. **8**(11), 2420–2423 (2019)

12. Priya, K., Chaithra, M.L., Ganavi, M., Shreyaswini, P., Prathibha, B.: An innovative application to predict malnutrition and anemia using ML. Int. J. Adv. Res. Comput. Commun. Eng. **10**(7), 5589–5595 (2021)

13. Bitew, F.H., Sparks, C.S., Nyarko, S.H.: Machine learning algorithms for predicting undernutrition among under-five children in Ethiopia. Public Health Nutr. **25**(2), 269–280 (2022). https://doi.org/10.1017/S1368980021004262

14. Wajgi, R., Wajgi, D.: Contributed to malnutrition detection in infants using machine learning approach. AIP Conf. Proc. **2424**, 040006. (2022) https://doi.org/10.1063/5.0076876

15. https://www.analyticsvidhya.com/blog/2021/10/understanding-transfer-learning-for-deep-learning/

16. https://www.mathworks.com/help/deeplearning/ref/vgg16.html

17. https://www.mathworks.com/help/deeplearning/ref/inceptionv3.html

18. https://www.mathworks.com/help/deeplearning/ref/vgg19.html

19. https://www.analyticsvidhya.com/blog/2018/06/comprehensive-guide-for-ensemble-models/

Multitudinous Disease Forecasting Using Extreme Learning Machine

S. Anslam Sibi[1](✉) , S. Nikkath Bushra[3] , M. Revathi[1] , A. Beena Godbin[2] ,
and S. Akshaya[1]

[1] St.Joseph's Institute of Technology, Chennai, Tamil Nadu, India
anslam.sibi@gmail.com
[2] Vellore Institute of Technology, Chennai, Tamil Nadu, India
[3] SRM Institute of Science and Technology, Kattankulathur, Chennai, India

Abstract. In today's world, Machine learning and Artificial intelligence are applied to a wide range of healthcare services. Predictive modeling is implemented in this system with the basis of symptoms entered by the user, the system will perform the disease prediction. For implementing the disease prediction, it implements a Support vector, Extreme learning, Random Forest classifier, KNN and Logistic Regression for analysis. The developed system predicts Malaria, Heart, and Parkinson using Extreme Learning. The proposed system uses a KNN, Random Forest, XG Boost, Extreme Learning for predicting disease with better accuracy. There will be several parameters asked as input from the user related to the disease selected by the user. With this input, the proposed system will determine whether the user has the disease. After applying our expertise, the algorithm with the highest accuracy rate is chosen for each ailment. The extreme learning algorithm has provided an accuracy of 93% compared with other machine learning algorithms in disease prediction. This proposed system will help a lot of people for predicting multiple diseases at the same point.

Keywords: Support Vector machine · Extreme learning · Random Forest classifier · KNN and Logistic Regression · Diabetes · Heart · Parkinson

1 Introduction

Records are considered an asset and in this virtual world, plenty and fast data are generated and stored as records from various filed. In the healthcare industry, all patient-related information is considered data. The proposed system is developed for forecasting people's illnesses in the healthcare sector. There are many prediction systems available online they all perform only one disease prediction. Like one system diagnosis of diabetes, one for cancer diagnosis, and one for skin or any other kind of diagnosis. There might not be a single prediction system for the diagnosis of more than one disease at once. Because of this, the proposed system is developed to diagnose people with rapid and accurate disorder predictions based on the symptoms they enter as input. The proposed system is utilized to anticipate a few illnesses. We'll look into the diagnosis of diabetes, heart, and

R. K. Challa et al. (Eds.): ICAIoT 2023, CCIS 1929, pp. 311–323, 2024.
https://doi.org/10.1007/978-3-031-48774-3_22

Parkinson's disease analyses in this tool. As future updates, many serious diseases may be covered. Machine learning algorithms like CNN, Random Forest, Logistic Regression, Support vector machine, and Logistic Regression for intense studying machines for predicting multiple diseases. To preserve the version's behaviour Python pickling is used. This system gets various parameters to investigate the disease and make it possible to discover the condition more effectively and correctly. The finished model is saved as a pickle document in python.

There has been a lot of analysis of current practices in the healthcare industry, but just one disease is predicted at a time. One system might diagnose diabetes, another one for diabetic retinopathy and yet another might predict heart disease. Most systems focus on a single diagnosis. The hospital organization should employ numerous systems for looking at patient health records. The methods used in the proposed system help analyse even the simplest, most uncommon diseases. From a single website, a user can predict many diseases. For performing the prediction, whether or not the user is now suffering from a specific illness, the user does not need to travel to several areas.

2 Related Work

Depending on the kind of symptoms a person is displaying, this system [1] forecasts their sickness. In this case, the system evaluates the user's symptoms using the Random Forest algorithm, Decision Tree algorithm, and Naive Bayes algorithm, three different types of supervised algorithms. The user's symptoms can range from 1 to 5 in number. An Excel spreadsheet is used to capture the raw data (a CSV file). A list of various symptoms and the illnesses they may cause can be found in the CSV file. The CSV file's massive data is then used for additional testing and various analytical applications. The Python module's Numpy module will now analyse the CSV file that the Pandas module had previously read.

Predicting liver illness is the primary objective of this paper [2]. In this study, ML models are created by balancing the uneven data using a variety of pre-processing approaches, then they are forecasted using the RF algorithm. The models are run in Anaconda using Jupyter Notebook with in-built libraries. In the northeast of Andhra Pradesh, India, there are 416 liver patients and 167 non-liver patients who make up the data set. Numerous over- and under sampling strategies are employed to balance the data set since it is unbalanced. Samples from minority classes are replicated in over-sampling, whereas samples from majority classes are removed in under sampling. A supervised machine learning (ML) model called Random Forest is applied to classification and regression issues. It consists of numerous decision trees that are applied to various aspects of the provided data set, and prediction is accomplished by averaging the results from each decision tree. Performance increases as the number of trees increases.

The process of diagnosing a disease or categorising a sample according to the many disease classes is known as disease prediction. Any medical sample would include a number of features, and the procedure would need to assess how well the sample features matched those of various illness groups. This method [3] would predict diseases based on the similarity value obtained. Simple illness prediction involves fitting or matching each attribute to the presence of each symptom associated with a particular disease. The difficulty with this system is that the majority of diseases have similar symptoms.

Typhoid fever, for instance, has the unique symptoms of stomach pain and vomiting while general fever has similar symptoms including warmth, body ache, and fatigue. Therefore, the maximum parameter must be taken into account while calculating the similarity between characteristics. Additionally, how well a disease prediction performs depends on how many samples are available in the training class.

This model's [4] objective is to estimate a person's risk of developing heart disease, diabetes, or a coronavirus based on their responses to a series of questions. The same thing is done with machine learning. The first step in converting raw data into a useable dataset is data purification. Then a data analysis to determine the importance of each attribute. The features are located and transformed into ML-acceptable form throughout this phase. Every model is put through the aforementioned steps in order to forecast heart, diabetes, and cancer-related illnesses. The dataset cannot be used directly since it has missing values and has not been cleaned. The symptoms, nation, age, region, etc., are examples of features. The dataset with missing and null values is not used. Height, age, and gender are some of the characteristics considered in the model. After the datasets were cleaned and analysed, various machine learning models were used. Using the logistic regression model, all datasets are processed. Then, three logistic regression models are used to make the prediction. This aids in the model's better training.

Several machine learning algorithms, including logistic regression, decision trees, random forests, support vector machines, and adaptive boosting, are used in this paper [5]. The datasets for breast cancer, diabetes, and heart disease are all included in this system. A data dictionary of the relevant attributes is also examined together with the dataset in the Python environment. The suggested strategy has a prediction accuracy of 87.1% for Heart Disease using Logistic Regression, 85.71% for Diabetes using a Support Vector Machine (linear kernel), and 98.57% for Breast Cancer using the AdaBoost classifier. Automation of processes like data munging, feature selection, and model fitting for best prediction accuracy is part of the project's future scope and improvement. A further benefit could come from the usage of pipeline structures for data pre-processing.

This study [6] is founded on the analysis of heart disease prognosis. Future potentials can be predicted using the current data set in the prediction analysis approach. An earlier SVM classifier is used in this study's prediction analysis. Among all machine learning algorithms, the SVM algorithm is regarded as one of the simplest. As no assumptions are made regarding the underlying information distribution, a decision tree is recognised as a non-parametric supervised learning approach. The samples in this study are categorised based on the closest patterns found in the feature space. Feature vectors are kept along with the labels of the training photographs during the training process. During the categorization process, the unlabelled question point is distributed according to the labels of its k-nearest neighbours. The majority share cote is used to describe the object in accordance with the labels of its neighbours. When k = 1, the object is essentially categorised by the class of the object that is closest to it. When just two classes are still extant in such a case, k is recognised as an odd integer. When performing multiclass classification, there may be a tie if k is an odd whole number.

There are two participant groups in this study [7]. The first two groups are SVM and Dynamic KNN. Heart Disease UCI | Kaggle uses this dataset to forecast diseases. This dataset includes 14 important features and more than 303 rows of data about the heart.

Using Clinical.com, the sample size is calculated with the following parameters: alpha = 0.05, power = 0.8, and beta = 0.2. For two different groups, this system collects a total of 20 samples from each group. The training dataset and the testing dataset are the two halves of the dataset. This system collects 20 samples for training data and 20 samples for testing data from both groupsA machine with a Windows operating system and a 50 GB hard drive is employed. Python is the language used, and Jupyter Notebook has 8 GB of RAM (Anaconda). An Intel i5 is the one in use. Treetops, chol, ancient peak, exang, slope, and that are independent variables for predicting heart disease. Improved accuracy values are dependent variables. The study was conducted using IBM SPSS version 26. For accuracy and loss, the Independent Sample test and Independent Sample Effect sizes are computed. The examination of the Dynamic KNN and Support Vector Machine shows that the Dynamic KNN appears to perform and have greater accuracy than the Support Vector Machine. Due to various issues with the feature extraction of text-based data in physical examination data, there are some issues with hypertension and hyperlipidaemia prediction studies.

This study's [8] suggestion is to extract features using a convolutional neural network (CNN), and then build the prediction model using the gradient lifting tree approach. Between the predicted and actual results, there is a 0.0277 variance. The bias in collective prediction is 0.0394. Consequently, using a CNN for feature extraction in the prediction of hypertension and hyperlipidaemia can lessen the bias in the prediction and enhance the predictive power. Text mining frequently uses the feature vectoring technique Term Frequency-Inverse Document Frequency (TF-IDF) to reflect the significance of terms in the corpus. The fundamental tenet of TF-IDF is that a word or phrase is regarded to have good class distinguishing power and is suitable for use in classifying if it frequently (TF) appears in one article and infrequently (IDF) appears in other articles. A CNN-based feature extraction technique is suggested. The physical examination canter provides the examination data for feature processing, and the prediction step is performed by the gradient lifting tree algorithm. CNNbased feature extraction has a positive impact on disease prediction, according to experiments. However, the CNN-based strategy for predicting hypertension and hyperlipidaemia index is still not ideal in illness.

This heart disease prediction system [9] uses machine learning to determine how a condition will develop. The initial step is to gather data on heart illness; in this paper, Kaggle's data collection is employed. It is necessary to get ready for machine learning after gathering the dataset and visualizing the data. This process of downscaling, standardizing, and normalizing data is known as data pre-processing. Scikit-Leam libraries are used in order to accomplish this. Next, the features are chosen. Performance can be adversely affected by irrelevant features. The following characteristics were taken from a big collection of heart disease characteristics. In this study, the authors assessed and contrasted seven different kinds of machine learning algorithms. Logistic regression, linear discriminate analysis, KNN and CART decision trees, Gaussian Naive Bayes, support vector machines, and random forest classifiers are some examples of machine learning techniques. Calculating the mean, mean absolute error, FAR, FRR, accuracy, precision, and other metrics will show how well an algorithm performs. Regression model performance is assessed using Mean Absolute Error (MEA).

Gaussian Naive Bayes has the lowest mean error in comparison to Decision Tree Classifier, which has a large (poor) mean error (ignoring the negative sign owing to absolute value). Among all of them, it also has the lowest standard deviation from the mean error. The implementation of an ontology-based disease prediction system for illnesses and symptoms is recommended in this study [10]. To make the material simpler to interpret, a domain ontology that is completely dedicated to human diseases must be developed. In this study, Protege is utilized to create ontologies, and SPARQL queries are used to access the ontology through the Apache Jena fuseki server. The dataset may be readily uploaded into the server under this setup, and Apache Jena Fuseki Server is used to conduct queries. To suit our purposes, the ontology file can be readily transformed into a different format. The benefits of the Jena Fuseki Server, how to install Apache Jena Fuseki, how to upload datasets to a server, and how to create and use SPARQL queries are all covered in this paper, which will be helpful to ontology developers. Ontology and SPARQL integration can increase the precision of disease prediction.

This article [11] reviews various machine learning techniques for diagnosing liver disease in humans. Using Naive Bayes, KNN, and Logistic Regression are the three ML methods. All of the models were used to implement the system, and their performance was assessed. This investigation includes a reduction in fatalities and improvement in diagnosis increased comfort when preparing for a liver illness. More precise diagnoses of liver illness by experts. In response to the classification of liver disorders and the need to reduce the burden on experts, an improvised algorithm or medical tool was developed for quick diagnosis of liver infections in the early stages of machine learning tactics. These various algorithms' accuracy has been evaluated against a variety of performance indicators. Comparing the accuracy of the logistic regression model to those of the other methods, it achieved a maximum accuracy of 75%.

This system [12] is primarily designed to investigate how many people have histori- cally utilised artificial intelligence to forecast or identify cardiac disease. It is unfortunate that the data from multiple studies do not provide a detailed source in the part on ANNs of this paper, where the developer gathered five articles employing ANNs to forecast heart disease. The BRBES results show a sensitivity of 79% and a specificity of 63%. This paper only proposes the flow of its system architecture; it does not propose the parameters used in the neural network part of its system and algorithm. However, it mentions how to adjust the weights they use, and the constrained optimization function is the fmincon function in Matlab. The study employs the three methods of logistic regression, classification, regression tree, and multi-layer perceptron to predict coronary artery disease (MLP). MLP is the most pertinent to ANNs. A supervised network, MLP is a common backpropagation technique for neural networks taught using feed-forward neural networks. A sensitivity of 93% and a specificity of 46% are the outcomes of MLP.

3 Proposed Model

In Fig. 1, three diseases that are unrelated to one another—heart, diabetes, and Parkin- son's—are considered. Any machine learning difficulty must be prepared for as its initial step. Data for heart disease, diabetes, and Parkinson's disease were therefore extracted from the UCI dataset, PIMA dataset, and Kaggle, respectively. Each of these datasets is regarded as an input dataset for the suggested system. Each input data should be displayed following the import of the dataset as input. After the input data has been visualised, it must be pre-processed in order to look for missing values, outliers, and to separate the data into training and testing data. On the training dataset, the system applies a Support vector, Extreme learning, Random Forest classifier, CNN, and Logistic Regression for analysis.

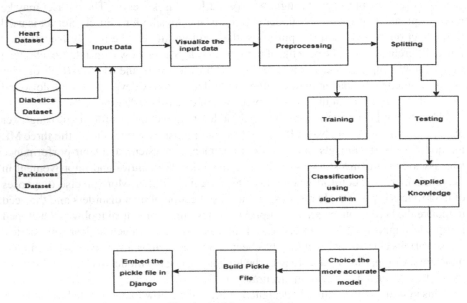

Fig. 1. Architecture Design

The Extreme Learning machine is chosen for highest accuracy rate for each ailment. Python item structures are serialised and deserialized using Pickle, often known as marshalling or knocking down or flattening. Serialization is the process of converting a memory-based item into a byte transfer that can be saved on disc or sent over a network. For each illness, a pickle document is created, which we then incorporated with the Django framework to produce the version for the website.

4   Methodology

Different machine learning algorithms are used in this disease prediction system. The methodology used is the KNN algorithm, Random Forest classifier, XGBoost algorithm, and Extreme learning algorithm. The best accuracy-yielding algorithm is used for reciting the selected disease.

4.1 KNN Algorithm

The K-NN algorithm performs as follows:

Step 1: Begin by choosing the K value, for instance, k = 5.
Step 2: Next, calculate the Euclidean distance among the points. Calculations are made as follows:

$$Euclidean\,Distance = \sqrt{(X2 - X1)^2 + (Y2 - Y1)^2}$$

Step 3: Determine the nearest neighbor's Euclidean distance.
Step 4: Count how many data points are there in each category.
For instance, two values for category B and three values for category A were discovered.
Step 5: Give the new point to the category with the most neighbours. Since Category A, for instance, has the most neighbours, assign the new data point to Category A.
Step 6: KNN model is now complete.

4.2 Random Forest Classifier Algorithm

N decision trees can be combined to create a random forest, which can then be used to provide predictions for each tree that was created in the first phase.

As for how the random forest works

Step 1: From the training set, select K data points randomly.
Step 2: Make decision trees that are connected to the selected k data points after making the choice (Subsets).
Step 3: Choose the N-th node for the decision trees you want to create after that.
Step 4: Repetition of steps 1 and 2
Step 5: Locate the predictions made by each decision tree, then assign the new data points to the category that h as the highest support.

4.3 XGBoost Algorithm

The XGBoost algorithm operates as follows:

Step 1: Create a single leaf tree.
Step 2: Calculate the residuals using the provided loss function after the target variable's average has been computed as the first tree's forecast. The forecast from the first tree is then used to determine the residuals for subsequent trees.
Step 3: Determine the similarity score using the algorithm

$$Similarity\,Score = Gradient\frac{Gradient^2}{Hessian + \lambda}$$

where the squared sum of the residuals (Gradient2) is a regularisation hyperparameter and the number of residuals equals the Hessian.

Step 4: Using the similarity score, choose the correct node. More homogeneity is seen when the similarity score is higher.

Step 5: Apply the similarity score to the information gain. Information gained reveals how much homogeneity is obtained by splitting the node at a specific place and helps to distinguish between old and new similarities. The formula used to compute it is:

$$Information\ Gain = Left\ Similarity + Right\ Similarity - Similarity\ for\ Roots$$

Step 6: Making the desired-length tree using the aforementioned technique Playing with the regularisation hyperparameter allows for trimming and regularisation.
Step 7: Using the Decision Tree created, forecast the residual values.
Step 8: The new set of residuals is calculated as follows

$$New\ Residuals = Old\ Residulas + \rho \sum Predicted\ Residuals$$

where ρ the learning rate represents.

Step 9: Go back to step 1 and carry out step 1 again for each tree.

4.4 Extreme Learning

Extreme Learning Machines (ELMs) are single-hidden layer feedforward neural networks (SLFNs) that can train faster than gradient-based learning techniques. It is similar to a conventional hidden layer neural network but does not include learning. Because it does not engage in repetitive tuning, this sort of neural network is quicker and performs better in terms of generalisation than networks trained using the backpropagation method (Fig. 2).

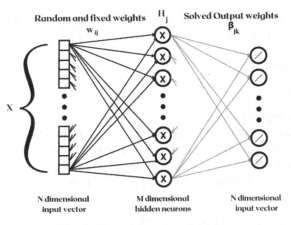

Fig. 2. Extreme learning algorithm

Considering:

- Training set H(w,b,x)
- Hidden node output function H;

- L (xi,,ti)|xi € Rd, ti € Rm, i = 1,...,N = number of hidden nodes

  Implementation of ELM is done by these three steps:

1. Specify the hidden nodes' parameters at random (w, b)
2. Computation of hidden layer to generate output matrix H.
3. Figure out the output weights.

5 Results and Discussion

Figure 3, 4 and 5 shows how, the user can select which type of forecasting is required. There are three different diseases predicted in this system. According to the user's need the user can toggle and select between Diabetics, Heart, and Parkinson's disease prediction. This system gets various parameters to investigate the disease and make it possible to discover the condition more effectively and correctly. For every single disease, medical parameters that are required to conclude that disease is extracted as the input from the users.

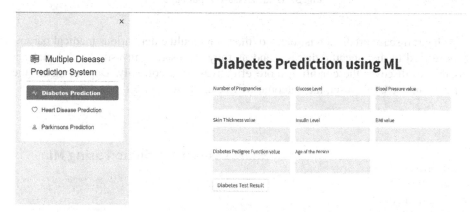

Fig. 3. Diabetes Disease Input page

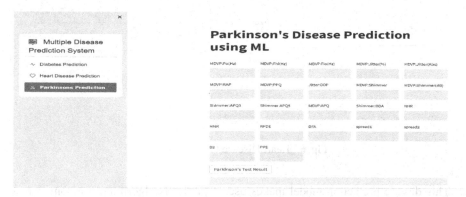

Fig. 4. Parkinson's Disease Input page

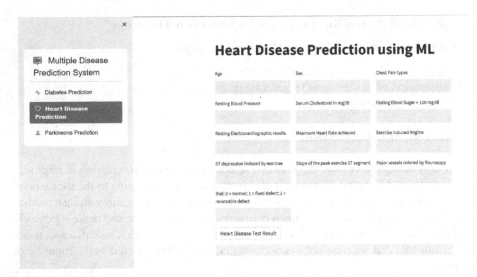

Fig. 5. Heart Disease Input page

If heart disease prediction is selected, the user should enter various medical parameters related to that specific disease, and as they are used to investigate and make it possible to discover the condition more effectively and correctly. Depending on the settings entered by the user, the outcome will be as shown in Figs. 6 and 7.

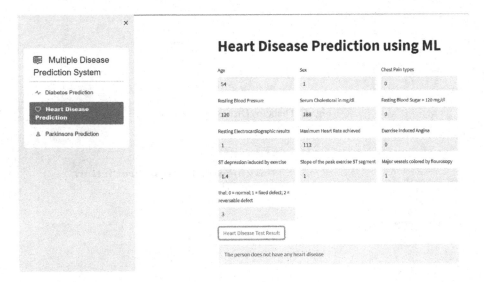

Fig. 6. Heart Disease Output case 1

This disease prediction model makes use of the KNN algorithm, Random Forest, XGBoost algorithm, and Extreme learning to achieve the highest level of accuracy. The

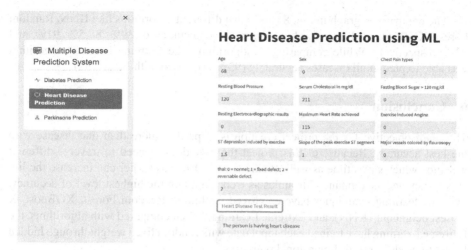

Fig. 7. Heart Disease Output case 2

disease-specific parameter that the patient contributes will reveal whether or not the patient has the condition in issue. If the value is outside of the range, is invalid, or is empty, a warning message will show, urging the user to provide an appropriate value. For instance, age, sex, chest pain types. Resting blood pressure, serum cholesterol in mg/dl, maximum heart rate achieved, exercise induced angina, ST depression by exercises and other three are the medical parameters required to deduct heart disease mention in Fig. 7. The parameters will reflect the required value range. After applying our expertise, we will choose the algorithm with the highest accuracy rate for each ailment. Then, for the output of the model to be shown on a website, we created a pickle file for each ailment and combined it with the Django framework. Each algorithm's level of accuracy with the average of all the three diseases is shown in Fig. 8.

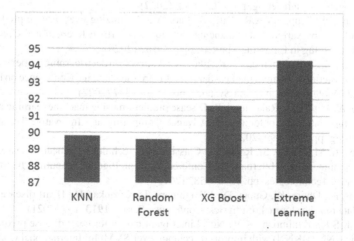

Fig. 8. Comparison graph of KNN, Radom Forest, XGBoost and Extreme Learning algorithms

The comparison graph in Fig. 8 shows that different algorithms like KNN, Random Forest, XGBoost, and Extreme Learning provide accuracy of 89%, 88.5%, 91%, and 93%, respectively. While comparing all algorithms, the Extreme Learning algorithm was able to predict all diseases accurately (93% compared with the others).

6 Conclusion

The main objective for the project creation is to predict more than one disease with the best accuracy. Because of this project the user doesn't need to traverse different websites which saves time as well. Predicting the diseases earlier can increase the life expectancy and save financial troubles as well. To attain the highest level of accuracy, machine learning techniques have employed, including Random Forest, XGBoost, K closest neighbour (KNN), and Extreme Learning. When compared with algorithms, the Extreme Learning have trained each input date with random fixed weight through hidden layers to produce multi-dimensional output.

In several studies, data from private hospitals were used. De-identification of individual patient data may be done to obtain larger datasets. The classifiers produced would be more accurate if more data were provided. This is because having more data also means having more diversity. The generalization error decreases as the model grows more general and is trained on more samples. Finding reliable medical information is challenging. Therefore, if the databases were made available to the general public, researchers would have access to more data. The list of diseases can be expanded in the future by the current API. Make the system as user-friendly as possible and offer a chatbot for common questions.

References

1. Pandey, A.K., Tripathi, A., and Ranjan, R.: Disease prediction using machine learning (Health Buddy). In: 2nd International Conference on Advance Computing and Innovative Technologies in Engineering, pp. 1003–1006 (2022)
2. Ambesange, S., Uppin, V.A.R., Patil, S., Patil, V.: Optimizing liver disease prediction with random forest by various data balancing techniques. In: IEEE International Conference on Cloud Computing in Emerging Markets, pp. 98–102 (2020)
3. Fazlur, S.A.H., Thillaigovindan, S.K., Illa, P.K.: A survey on deep learning model for improved disease prediction with multi medical data sets. In: 3rd International Conference on Electronics and Sustainable Communication Systems, pp. 1530–1537 (2022)
4. Sharma, A., Pathak, J., Rajakumar, P.: Disease prediction using machine learning algorithms. In: 2nd International Conference on Advance Computing and Innovative Technologies in Engineering, pp. 995–999 (2022)
5. Chakarverti, M., Yadav, S., Rajan, R.: Classification technique for heart disease prediction in data mining. In: 2nd International Conference on Intelligent Computing, Instrumentation and Control Technologies, pp. 1578–1582 (2019)
6. Fadnavis, R., Dhore, K., Gupta, D., J Waghmare, J., Kosankar, D.: Heart disease prediction using data mining. Journal of Physics: Conference Series, **1913**, 1–7 (2021)
7. Reddy, K.S.K., Kanimozhi, K.V.: Novel intelligent model for heart disease prediction using dynamic KNN (DKNN) with improved accuracy over SVM. In: International Conference on Business Analytics for Technology and Security, pp. 1–5 (2022)

8. Xie, S., Fan, H.: Research on CNN to feature extraction in diseases prediction. In: International Conference on Computer Network, Electronic and Automation, pp. 197–202 (2019)
9. Kumari, A., Mehta, A.K.: A novel approach for prediction of heart disease using machine learning algorithms. In: Asian Conference on Innovation in Technology, pp. 1–5 (2021)
10. Kumar, R., Thakur P., Chauhan, S.: Special disease prediction system using machine learning. In: International Conference on Machine Learning, Big Data, Cloud and Parallel Computing, pp. 42–45 (2022)
11. Srivastava, A., Kumar, V.V., Vivek, V.: Automated prediction of liver disease using machine learning (ML) algorithms. In: Second International Conference on Advances in Electrical, Computing, Communication and Sustainable Technologies, pp. 1–4 (2022)
12. Peng, C.C., Huang, C.W., Lai, Y.C.: Heart disease prediction using artificial neural networks: a survey. In: IEEE 2nd Eurasia Conference on Biomedical Engineering, Healthcare and Sustainability, pp. 147–150 (2020)

Optimized Ensembled Predictive Model
for Drug Toxicity

Deepak Rawat[1]($\boxtimes$) (iD), Meenakshi[1] (iD), and Rohit Bajaj[2] (iD)

[1] Department of Mathematics, Chandigarh University, Mohali, Punjab, India
rawatdeepak1982@gmail.com
[2] Department of Computer Sciences, Chandigarh University, Mohali, Punjab, India

Abstract. Healthcare is one of the most important concerns for living beings. Prediction of the toxicity of a drug is a great challenge over the years. It is quite an expensive and complex process. Traditional approaches are laborious as well as time-consuming. The era of computational intelligence has started and gives new insights into drug toxicity prediction. The quantitative structure-activity relationship has accomplished significant advancements in the field of toxicity prediction. Nine machine learning algorithms are considered such as Gaussian Process, Linear Regression, Artificial Neural Network, SMO, Kstar, Bagging, Decision Tree, Random Forest, and Random Tree to predict the toxicity of a drug. In the study, we developed an optimized regression model (Optimized KRF) by ensembling Kstar and Random Forest algorithm. For the mentioned machine learning models, evaluation parameters are assessed. The 10-fold cross-validation is applied to validate the model. The optimized model gave a coefficient of correlation, coefficient of determination, mean absolute error, root mean squared error, and accuracy of 0.9, 0.81, 0.23, 0.3, and 77% respectively. Further, the Saw score is calculated in two aspects as W-Saw score and the L-Saw score. The W-Saw score for the optimized ensembled model is 0.83 which is the maximum and L-Saw score is 0.27 which is the lowest in comparison to other classifiers. Saw score provides the strength to an ensemble model. These parameters indicate that the optimized ensembled model is more reliable and made predictions that were more accurate than earlier models. As a result, this model could be efficiently utilized to forecast toxicity.

Keywords: Machine learning · Regression · Accuracy · Saw score · Toxicity prediction · Optimization

1 Introduction and Background

Toxicity means the extent to which a drug compound is toxic to living beings. Prediction of toxicity is a great challenge [1]. Toxicity can cause death, allergies, or adverse effects on a living organism, and it is associated with the number of chemical substances inhaled, applied, or injected [2]. There is a narrow gap between the effective quality of a drug and the toxic quality of the drug. A drug is required to help in illness, diagnosis of a disease, or prevention of disease [3]. The development of a new drug or chemical compound is quite an expensive and complex process.

© The Author(s), under exclusive license to Springer Nature Switzerland AG 2024
R. K. Challa et al. (Eds.): ICAIoT 2023, CCIS 1929, pp. 324–335, 2024.
https://doi.org/10.1007/978-3-031-48774-3_23

A subset of artificial intelligence is machine learning [4]. It is a study of computer algorithm that is automatically improved through experience. Machine learning algorithm creates models based on training data to make predictions without explicit programming [5]. It can learn and enhance the ability to decision-making when introduced to new data. So with the help of these algorithms, models can gain knowledge from experience and enhance their capacity for acting, planning, and thinking [6]. The field of health care has made substantial use of machine learning techniques [7].

Feature selection is a method for choosing pertinent features from a dataset and removing irrelevant features [8]. Feature selection is employed to demonstrate the ranking of each feature with the variances. The input variables used in machine learning models are called features. Essential and non-essential features are part of the input variables [9]. The irrelevant and non-essential features can make the optimal model weaker and slower. Two main feature selection techniques are supervised and unsupervised. Algorithms are essential to anticipate toxicity in the age of artificial intelligence [10]. These techniques make it easier for models to infer intended outcomes from historical data and incidents. Every machine learning technique must ensure an optimal model that will predict the desired outcome best [11].

The ensemble method is a technique that combines multiple base classifiers to generate the best prediction model [12]. The ensembling technique focuses on considering a number of the base model into account and optimizing/averaging these models to provide one final model instead of constructing an individual model and expecting it to predict the paramount outcome [13].

2 Literature Review

In this section, the related work based on various techniques used in machine learning models is deliberated. Ai utilized SVM and the Recursive Feature Elimination (RFE) approach, he created a regression model [14]. Hooda et al. introduced a better feature selection ensemble framework for classifying hazardous compounds, using imbalanced and complex pharmacological data of high dimensions to create an improved model [15]. The Real Coded Genetic Algorithm was used by Pathak et al. to assess the significance of each feature, and k cross-validation was employed to assess the resilience of the best prediction model [16]. Collado et al. worked on a class balancing problem and provided an effective solution for class imbalance datasets to predict toxicity [17]. Cai et al. discussed the challenge in the analysis of high dimensional data in ML and provided effective feature selection methods to improve the learning model [18]. Austin et al. assessed the impact of missing members on the accuracy of the forecast and looked at the impacts missing members had on a voting-based ensemble and a stacking-based ensemble [19]. Invasive ductal carcinoma (IDC) stage identification is very time-consuming and difficult for doctors, as Roy et al. explained, thus they created a computer-assisted breast cancer detection model employing ensembling [20]. Takci et al. discussed the problem of the prediction of heart attack is necessary, especially in low-income countries, and determined the ML model to predict heart attacks [21]. Gambella et al. presented mathematical optimized models for advanced learning. The strengths and weaknesses of the models are discussed and a few open obstacles are highlighted [22]. Tharwat et al. proposed a new version of Grey Wolf optimization to adopt prominent features and to reduce the computational time for the process. These encouraging findings mark a significant step forward in the development of a completely automated toxicity test using photos of zebrafish embryos employing machine learning techniques and the next iteration of GWO [23].

The rest of the paper is organized as Sect. 3 explains the research methodology and the results with discussions are explained in Sect. 4. Finally, the last Sect. 5 concludes the work performed.

3   Proposed Methodology

Computer-aided models are examined in this research. Nine machine learning algorithms are considered such as Gaussian Process, Linear Regression, Artificial Neural Network, SMO, Kstar, Bagging, Decision Tree, Random Forest, and Random Tree to predict toxicity. We developed an optimized regression model (Optimized KRF) by ensembling Kstar and Random Forest algorithm. For the mentioned machine learning models, evaluation parameters are assessed. The results in terms of accuracy are compared and assessed. Ten folds of cross-validation are used to create a robust model. The proposed methodology's workflow procedure is depicted in Fig. 1.

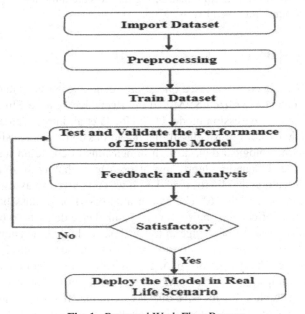

Fig. 1. Proposed Work Flow Process

Figure 2 represents the methodology for an ensembled model. Classifier -1 and classifier- 2 are applying a lazy and eager algorithm for prediction. Further ensembling is performed using different algorithms.

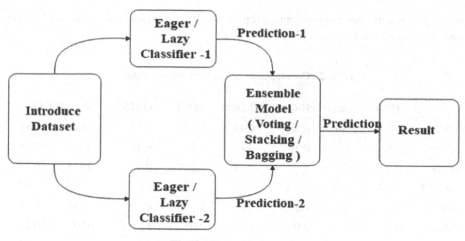

Fig. 2. Ensembled Model

4 Results and Discussion

In this paper, the toxicity dataset is acquired from UCI machine learning datasets "UCI Machine Learning Repository: QSAR aquatic toxicity Data Set" and is used to assess how well learning models perform. The dataset consists of 546 occurrences and 9 attributes (one class attribute and eight predictive attributes). Table 1 lists the specifics of the ranking-related attributes. The ranking of important features is done using the correlation attribute evaluator method.

Table 1. Ranking of the features

Feature Selected	Feature Description	Ranking
TPSA(Tot)	Topological polar surface area	3
SAacc	Surface area acceptors	6
H-050	Number of hydrogen atoms	7
MLOGP	Moriguchi LOGP values	1
RDCHI	Represents topological index	2
GATS1p	Represents molecular polarisability	8
nN	Number of nitrogen atoms	5
C-040	Number of carbon atoms	4

The coefficient of correlation in Table 2 is calculated with the help of Eq. (1) and mentioned as:

$$\rho_{PQ} = \frac{n\Sigma PQ - \Sigma P \Sigma Q}{\sqrt{\left[n\Sigma P^2 - (\Sigma P)^2\right]\left[n\Sigma Q^2 - (\Sigma Q)^2\right]}} \tag{1}$$

where ρ_{PQ} is correlation coefficients, n represents the size, P, and Q are selected features and Σ is the summation symbol.

Table 2. Coefficient of correlation among features

	TPSA	SAacc	H-050	MLOGP	RDCHI	GATS1p	nN	C-040
TPSA	1	0.86	0.66	−0.46	0.52	0.17	0.61	0.41
SAacc	0.86	1	0.77	−0.4	0.57	0.21	0.5	0.45
H-050	0.66	0.77	1	−0.49	0.28	0.06	0.47	0.15
MLOGP	−0.46	−0.4	−0.49	1	0.33	−0.38	−0.29	−0.1
RDCHI	0.52	0.57	0.28	0.33	1	0.05	0.34	0.41
GATS1p	0.17	0.21	0.06	−0.38	0.05	1	0.07	0.14
nN	0.61	0.5	0.47	−0.29	0.34	0.07	1	0.29
C-040	0.41	0.45	0.15	−0.1	0.41	0.14	0.29	1

We have considered 9 machine learning algorithms such as Gaussian Process, Linear Regression, Artificial Neural Network, SMO, Kstar, Bagging, Decision Tree, Random Forest, and Random Tree to predict toxicity. Parameters are evaluated for the mentioned machine learning models. We calculated and compared how accurate each model is to select the best predictive model. The model is validated using the tenfold cross-validation method.

In the study, we developed an optimized regression model (Optimized KRF) by ensembling Kstar and Random Forest algorithm. Further Saw score is calculated in two aspects as W-Saw score and the L-Saw score. Gaussian Process, Linear Regression, Artificial neural Network, SMO, Kstar, Bagging, Decision Tree, Random Forest, and Random Tree achieved 53%, 58%, 50%, 57%, 64%, 60%, 54%, 63%, 60% accuracy respectively. The optimized model gave a coefficient of correlation, coefficient of determination, mean absolute error, root mean square error, and accuracy of 0.9, 0.81, 0.23, 0.3, and 77% respectively.

The W-Saw score for the optimized ensembled model is 0.83 which is the maximum and the L-Saw score is 0.27 which is the lowest in comparison to other classifiers. Saw score provides the strength to an ensemble model. Table 3 shows the state of art parameters evaluated and Fig. 3 represents the Comparison of the coefficient of correlation and determination graphically.

Table 3. Comparison of different models for state of art parameters

Classifier	Coefficient of Correlation (R)	Coefficient of Determination (R^2)	Mean Absolute Error	Root Mean Squared Error
Gaussian Process	0.54	0.29	0.47	0.5
Linear Regression	0.59	0.35	0.42	0.5
Artificial Neural Network	0.44	0.19	0.5	0.6
SMO	0.59	0.35	0.43	0.5
Kstar	0.58	0.34	0.36	0.5
Bagging	0.61	0.37	0.4	0.5
Decision Tree	0.37	0.14	0.46	0.6
Random Forest	0.63	0.4	0.37	0.5
Random Tree	0.46	0.21	0.4	0.6
Optimized Ensembled KRF	0.9	0.81	0.23	0.3

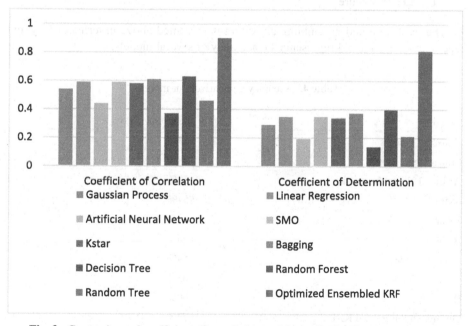

Fig. 3. Comparison of coefficient of correlation and determination for several models

Algorithm: Prediction and Ensembling

---

1. Read Dataset => {D}   # Intrusion data

2. {D} train => {D} [0: y]

3. {D} test => {D} [y+1: n]

4. SET PD_1, PD_2 # Define prediction

    PD_1, PD_2

5. Define Prediction {D}, type = {LAZY} {EAGER} as predictions

6. Return Classifier {type}. predict ({D});

7. Define Voting Classifier (calculations = calculate, type = {hard}, {soft},

    {PD_1}, {PD_2})

8. Voting Classifier. Fit ({PD_1}, {PD_2})

9. Return Voting Classifier. {type}. predict ({D});

10. Ensemble = Voting Classifier (calculate, type, predictions);

11. End Procedure

The prediction and ensembling algorithm is presented above in terms of lazy and eager classifiers. Table 4 represents the accuracy of several models.

Table 4. Accuracy comparison for models

Classifier	Accuracy
Gaussian Process	53%
Linear Regression	58%
Artificial Neural Network	50%
SMO	57%
Kstar	64%
Bagging	60%
Decision Tree	54%
Random Forest	63%
Random Tree	60%
Optimized Ensembled KRF	77%

Figure 4 depicts a comparison of accuracy for several models graphically. The saw score is a multi-attribute score based on the concept of weighted summation. This will seek weighted averages of rating the performance of each alternative. W-Saw score in

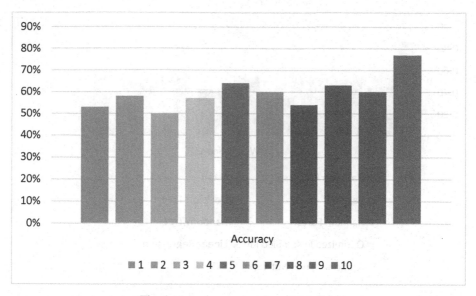

Fig. 4. Accuracy comparison for models

Table 5 will be the highest score among all alternatives and is recommended as shown in Eq. (2). Figure 5 represents W-Saw scores for different models.

Highest Score Recommender

$$W - Saw = \frac{\sum_{i=1}^{n} r_i}{n} \tag{2}$$

Table 5. W-Saw score comparison for models

Classifier	W-Saw Score
Gaussian Process	0.45
Linear Regression	0.51
Artificial Neural Network	0.38
SMO	0.50
Kstar	0.52
Bagging	0.53
Decision Tree	0.35
Random Forest	0.55
Random Tree	0.42
Optimized Ensembled KRF	0.83

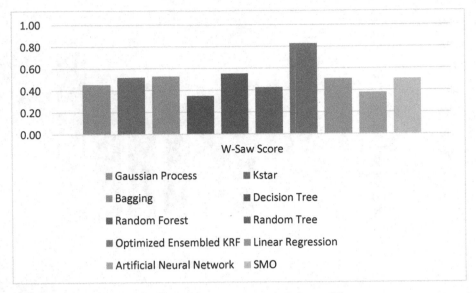

Fig. 5. W-Saw score comparison for models

L-Saw Score in Table 6 is the score evaluated among alternatives and the lowest score is recommended as shown in Eq. (3). Figure 6 represents L-Saw scores for different models.

Lowest Score Recommender.

$$L - Saw = \frac{\sum_{j=1}^{n} r_j}{n} \tag{3}$$

Table 6. L-Saw score comparison for models

Classifier	L-Saw Score
Gaussian Process	0.49
Linear Regression	0.46
Artificial Neural Network	0.55
SMO	0.47
Kstar	0.43
Bagging	0.45
Decision Tree	0.53
Random Forest	0.44
Random Tree	0.50
Optimized Ensembled KRF	0.27

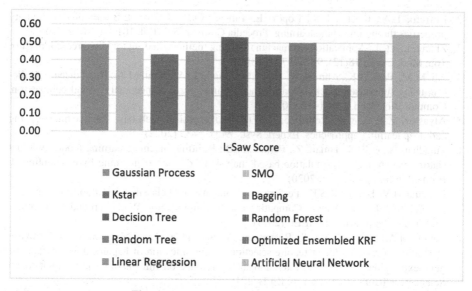

Fig. 6. L-Saw score comparison for models

5 Concluding Remarks and Scope

To reduce the period and complexity of toxicity prediction, we have to develop intelligent systems for living beings so that they can reveal the possibilities of toxicity. Machine learning has significance in toxicity prediction. In this study, nine machine learning algorithms were taken into account such as Gaussian Process, Linear Regression, Artificial neural Network, SMO, Kstar, Bagging, Decision Tree, Random Forest, and Random Tree to predict the toxicity of a drug. In the study, we developed an optimized regression model (Optimized KRF) by ensembling Kstar and Random Forest algorithm. Parameters are evaluated for the mentioned machine learning models. The technique of tenfold cross-validation is used to validate the model. The optimized ensembled model gave a correlation coefficient, coefficient of determination, mean absolute error, root mean square error, and accuracy of 0.9, 0.81, 0.23, 0.3, and 77% respectively. Further Saw score is calculated in two aspects as W-Saw score and the L-Saw score. The W-Saw value in the ensembled model is 0.83 which is the maximum and the L-Saw value for the ensembled model is 0.27 which is the lowest in comparison to other classifiers. Saw score provides the strength to an ensemble model. These parameters indicate that the optimized ensembled model is more reliable and made predictions more accurately than earlier methods. The study can be extended to the ensembling of other classifiers to get higher accuracy and fewer errors. Other techniques can be applied for feature selection, class balancing, and optimization.

References

1. Karim, A., et al.: Quantitative toxicity prediction via meta ensembling of multitask deep learning models. ACS Omega **6**(18), 12306–12317 (2021)

2. Borrero, L.A., Guette, L.S., Lopez, E., Pineda, O.B., Castro, E.B.: Predicting toxicity properties through machine learning. Procedia Comput. Sci. **170**, 1011–1016 (2020)
3. Zhang, L., et al.: Applications of machine learning methods in drug toxicity prediction. Curr. Top. Med. Chem. **18**(12), 987–997 (2018)
4. Ali, M.M., Paul, B.K., Ahmed, K., Bui, F.M., Quinn, J.M., Moni, M.A.: Heart disease prediction using supervised machine learning algorithms: performance analysis and comparison. Comput. Biol. Med. **136**, 1–10 (2021)
5. Alyasseri, Z.A.A., et al.: Review on COVID-19 diagnosis models based on machine learning and deep learning approaches. Expert. Syst. **39**(3), 1–32 (2022)
6. Tuladhar, A., Gill, S., Ismail, Z., Forkert, N.D.: Building machine learning models without sharing patient data: a simulation-based analysis of distributed learning by ensembling. J. Biomed. Inform. **106**, 1–9 (2020)
7. Ramana, B.V., Boddu, R.S.K.: Performance comparison of classification algorithms on medical datasets. In: 9th Annual Computing and Communication Workshop and Conference, CCWC 2019, pp. 140–145. IEEE (2019)
8. Latief, M.A., Siswantining, T., Bustamam, A., Sarwinda, D.: A comparative performance evaluation of random forest feature selection on classification of hepatocellular carcinoma gene expression data. In: 3rd International Conference on Informatics and Computational Sciences, ICICOS 2019, pp. 1–6. IEEE (2019)
9. Babaagba, K.O., Adesanya, S.O.: A study on the effect of feature selection on malware analysis using machine learning. In: 8th International Conference on Educational and Information Technology, ICEIT 2019, pp. 51-55. ACM, Cambridge (2019)
10. Feng, L., Wang, H., Jin, B., Li, H., Xue, M., Wang, L.: Learning a distance metric by balancing kl-divergence for imbalanced datasets. IEEE Trans. Syst. Man Cybern. Syst. **49**(12), 2384–2395 (2018)
11. Rawat, D., Pawar, L., Bathla, G., Kant, R.: Optimized deep learning model for lung cancer prediction using ANN algorithm. In: 3rd International Conference on Electronics and Sustainable Communication Systems, ICESC 2022, pp. 889–894. IEEE (2022)
12. Sarwar, A., Ali, M., Manhas, J., Sharma, V.: Diagnosis of diabetes type-II using hybrid machine learning based ensemble model. Int. J. Inf. Technol. **12**, 419–428 (2020)
13. Uçar, M.K.: Classification performance-based feature selection algorithm for machine learning: P-score. IRBM **41**(4), 229–239 (2020)
14. Ai, H., et al.: QSAR modelling study of the bioconcentration factor and toxicity of organic compounds to aquatic organisms using machine learning and ensemble methods. Ecotoxicol. Environ. Saf. **179**, 71–78 (2019)
15. Hooda, N., Bawa, S., Rana, P.S.: B2FSE framework for high dimensional imbalanced data: a case study for drug toxicity prediction. Neurocomputing **276**, 31–41 (2018)
16. Pathak, Y., Rana, P.S., Singh, P.K., Saraswat, M.: Protein structure prediction (RMSD $\leq$ 5 Å) using machine learning models. Int. J. Data Min. Bioinform. **14**(1), 71–85 (2016)
17. Antelo-Collado, A., Carrasco-Velar, R., García-Pedrajas, N., Cerruela-García, G.: Effective feature selection method for class-imbalance datasets applied to chemical toxicity prediction. J. Chem. Inf. Model. **61**(1), 76–94 (2020)
18. Cai, J., Luo, J., Wang, S., Yang, S.: Feature selection in machine learning: a new perspective. Neurocomputing **300**, 70–79 (2018)
19. Austin, A., Benton, R.: Effects of missing members on classifier ensemble accuracy. In: IEEE International Conference on Big Data, ICBD 2020, pp. 4998–5006. IEEE (2020)
20. Roy, S.D., Das, S., Kar, D., Schwenker, F., Sarkar, R.: Computer aided breast cancer detection using ensembling of texture and statistical image features. Sensors **21**(11), 1–17 (2021)
21. Takci, H.: Improvement of heart attack prediction by the feature selection methods. Turk. J. Electr. Eng. Comput. Sci. **26**(1), 1–10 (2018)

22. Gambella, C., Ghaddar, B., Naoum-Sawaya, J.: Optimization problems for machine learning: a survey. Eur. J. Oper. Res. **290**(3), 807–828 (2021)
23. Tharwat, A., Gaber, T., Hassanien, A.E., Elhoseny, M.: Automated toxicity test model based on a bio-inspired technique and adaBoost classifier. Comput. Electr. Eng. **71**, 346–358 (2018)

Classification for EEG Signals Using Machine Learning Algorithm

Shirish Mohan Dubey[1]([✉]), Budesh Kanwer[2], Geeta Tiwari[1], and Navneet Sharma[3]

[1] Poornima College of Engineering, Jaipur, India
{shirish.dubey,geeta.tiwari}@poornima.org
[2] Poornima Institute of Engineering and Technology, Jaipur, India
[3] IIS (Deemed to Be University), Jaipur, Rajasthan, India
navneet.sharma@iisuniv.ac.in

Abstract. Electroencephalography (EEG) is a non-invasive technique that is used to record the electrical activity of the brain. EEG signals are widely used in the diagnosis of various neurological and psychiatric disorders. EEG signals are complex and noisy, and thus, it is difficult to classify them accurately. In this paper, we have evaluated the performance of two popular machine learning algorithms, namely, Random Forest (RF) and Support Vector Machine (SVM), for classifying EEG signals. The performance of the algorithms was evaluated on a publicly available dataset of EEG signals. The analysis has been done on Bonn University EEG database; the analysis of methodologies signifies that the proposed improved random forest method performs superior to that of conventional random forest as well as support vector machine-based approach.

Keywords: Machine Learning · EEG · Epilepsy · Ictal · Epileptic Seizures · Support Vector Machine · Random forest

1 Introduction

This section discusses two thing first EEG (electroencephalography) which measures the electrical activity in the brain and second Epilepsy diseases.

1.1 EEG and Epilepsy

Earlier detection and diagnosis of diseases on time saves human life. In the modern world, computer aided technologies are used by radiologists or physicians to detect diseases in an earlier manner. From the past decades, Electroencephalography (EEG) signals have been used to detect and analyze the behavior of human beings and to detect diseases which are related to the human brain. Epilepsy is a neurological disorder that affects the brain's electrical activity and can result in seizures, which can have a significant impact on a patient's life. One of the most common methods for detecting epilepsy is through electroencephalography (EEG), which measures the electrical activity in the brain. However, analyzing EEG signals to detect epilepsy can be challenging, as the

R. K. Challa et al. (Eds.): ICAIoT 2023, CCIS 1929, pp. 336–353, 2024.
https://doi.org/10.1007/978-3-031-48774-3_24

signals are often complex and contain a lot of noise. In this paper, we propose a methodology for detecting epilepsy from EEG signals using shearlet transform-based feature extraction and SVM and Random Forest classifier, with and without GA based feature optimizer.

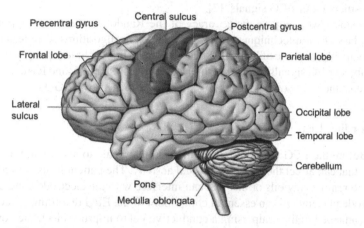

Fig. 1. The brain surface anatomy in the left cerebral hemisphere [1]

The nervous system is made up of the Central Nervous System (CNS) and the Peripheral Nervous System (PNS) [2, 3]. Figure 1 represents the cross-sectional view of the brain. The CNS, which is the main concern of this study, comprises the brain and the nerves which are the control center of the body. Nerve cells respond to information by altering the flow of electrical currents across their membranes, which results in the creation of electric and magnetic fields that may be detected on the scalp.' Small electrodes are applied to the scalp in order to measure the electric fields. Electroencephalograms (EEGs) are used to capture the potentials between electrodes (EEG). Electrical activity is recorded using EEG; tiny magnetic fields generated by the neurons of the brain are measured using Magneto encephalography (MEG). The World Health Organization, the World Bank, and the Harvard School of Public Health collaborated on the Global Burden of Disease research; there are around 600 known kinds of neurological infections starting from headaches to epilepsy [2–4]. It is assessed by the World Health Organization that these issues have their impact on more than 1 billion individuals around the world (World Health Organization, 2006). It is assessed that there are around 5.5 million individuals with Epilepsy in India. As indicated by the epilepsy establishment, the population with epilepsy in the USA is in excess of 2 million and around 65 million individuals are burdened around the world [4].

Random Forest is an ensemble learning technique that builds several decision trees and averages their predictions to arrive at a single final result. A different collection of characteristics and data from the training set are used to build each decision tree. The median of all the decision trees' forecasts is what the random forest algorithm returns. Random Forest is widely used for classification problems due to its simplicity and robustness across a variety of datasets. [5, 6].

The popular machine learning technique Support Vector Machine (SVM) may be applied to both classification and regression problems. SVM creates a hyperplane that divides the information into two groups. Each class's nearest data points are used to determine how far apart from the hyperplane to place the hyperplane. Since SVM is robust against both high-dimensional data and noisy data, it has found extensive usage in the classification of EEG signals. [7].

In this study, we compare the accuracy of the Random Forest and Support Vector Machine classification techniques using EEG data. The algorithms were tested using a publicly available database of EEG readings. The collection includes epilepsy patient and healthy subject EEG signals. Bandpass filtering, artifact reduction, and feature extraction were some of the common preprocessing steps used to the raw EEG data.

1.2 Recording EEG

Making Before the EEG recording, the patient is instructed to avoid caffeine, alcohol, and drugs that can affect the brain's electrical activity. The patient is also asked to wash their hair to remove any oils or dirt that can interfere with the electrode contact.

Electrode placement is an essential component of the EEG recording process. Electrodes are attached to the scalp using a conductive gel to improve electrode contact and reduce electrical impedance. The number and placement of electrodes depend on the clinical indication for the EEG recording [7, 8].

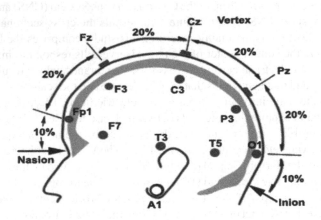

Fig. 2. Electrode's placement in side view [1]

Standard EEG recording uses a 10–20 electrode placement system. This system divides the scalp into specific regions and assigns electrode positions based on their distance from these regions. For example, the Fp1 electrode is placed 10% of the distance between the Fpz and T3 electrodes. EEG signals are recorded using an amplifier connected to the electrodes on the scalp. The amplifier amplifies the small electrical signals generated by the brain and filters out any noise or interference. During the recording, the patient is typically instructed to sit or lie down in a relaxed state with their eyes closed or open. The recording typically lasts 20–40 min and includes a variety of stimuli, including flashing lights or hyperventilation, to trigger specific brain activity. The recorded EEG data is analyzed to identify any abnormalities or patterns of electrical activity that may indicate a neurological condition. EEG data can be analyzed in the time domain or frequency domain. The EEG recording process involves several steps, including preparation, electrode placement, signal acquisition, and data analysis. Electrode placement is an essential component of the EEG recording process, and the number and placement of electrodes depend on the clinical indication for the EEG recording. The recorded EEG data is analyzed to identify any abnormalities or patterns of electrical activity that may indicate a neurological condition. EEG is a safe and non-invasive technique that has revolutionized the diagnosis and monitoring of neurological disorders (Fig. 2).

1.3 Applications of EEG

The Electroencephalography (EEG) is a non-invasive method used to record the electrical activity of the brain. EEG signals are characterized by their frequency content, which can provide valuable information about brain activity. In this paper, we will discuss the frequency distribution and applications of EEG signals (Fig. 3).

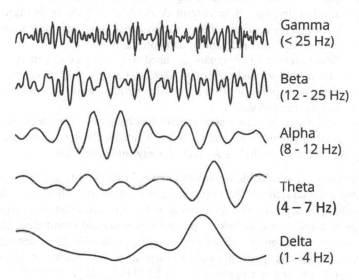

Fig. 3. Single–sided amplitude spectrum of different frequency band of a signal [1]

EEG signals are composed of different frequency components, which are traditionally categorized into five frequency bands: delta, theta, alpha, beta, and gamma. The frequency bands are defined based on the frequency range of the EEG signal. The frequency distribution of EEG signals varies depending on the brain activity being recorded. For example, during deep sleep, the EEG signals are dominated by the delta band, while during wakefulness, the alpha and beta bands are more prominent. The frequency distribution of EEG signals is also affected by age, with children having higher amplitudes in the delta and theta bands, and adults having higher amplitudes in the alpha and beta bands. EEG signals have a wide range of applications in clinical and research settings. EEG signals can provide valuable information about brain activity and are used to diagnose and monitor a variety of neurological conditions, including epilepsy, sleep disorders, and brain injury. EEG signals are also used in neuroscience research to investigate brain function and cognitive processes.

Some of the main applications of EEG signals are discussed below:

1. Epilepsy Diagnosis and Monitoring: EEG signals are widely used to diagnose and monitor epilepsy. During an epileptic seizure, the EEG signals show characteristic patterns of abnormal activity, such as spikes and sharp waves. EEG signals can also be used to determine the type and location of the seizure activity, which can help in the selection of appropriate treatment options.
2. Sleep Studies: EEG signals are used to study sleep patterns and diagnose sleep disorders, such as sleep apnea and insomnia. EEG signals are used to identify the different stages of sleep, including NREM (non-rapid eye movement) and REM (rapid eye movement) sleep. During NREM sleep, the EEG signals are dominated by the delta band, while during REM sleep, the EEG signals are characterized by a desynchronized pattern.
3. Brain-Computer Interface (BCI): EEG signals are used to develop BCIs, which are systems that allow individuals to control devices using their brain signals. BCIs are being developed for a variety of applications, including assistive technology for individuals with disabilities and gaming and entertainment.
4. Cognitive Neuroscience: EEG signals are used to investigate cognitive processes, such as attention, memory, and perception. EEG signals are used to measure changes in brain activity in response to stimuli or tasks, which can provide insight into the underlying neural mechanisms.
5. Neurofeedback: EEG signals are used in neurofeedback, which is a technique used to train individuals to control their brain activity. Neurofeedback has been used to treat a variety of conditions, including ADHD, anxiety, and depression.

EEG signals provide valuable information about brain activity and are widely used in clinical and research settings. EEG signals are characterized by their frequency distribution, which varies depending on the brain activity being recorded. EEG signals have a wide range of applications, including epilepsy diagnosis and monitoring, sleep studies, BCI development, cognitive neuroscience, and neurofeedback. Understanding the frequency distribution and applications of EEG signals is essential for interpreting EEG recordings and developing new applications for EEG technology.

2   Related Works

This section deals with current state of art techniques to study EEG and Epilepsy through EEG.

2.1   EEG and Epilepsy

Recent diagnostic strategies for epilepsy have centered on creating machine/deep learning model (ML/DL)-based electroencephalogram (EEG) techniques since epilepsy is one of the most chronic neurological illnesses. However, conventional ML/DL models struggle to produce a consistent and satisfying diagnostic outcome for EEG due to its low amplitude and nonstationary properties. This work conducts a comprehensive systematic evaluation of the latest breakthroughs in EEG-based ML/DL systems for epileptic seizure identification in an effort to fill this knowledge gap. [7] The latest advances in epileptic seizure diagnosis are discussed, as well as the many statistical feature extraction methods, ML/DL models, their respective performances, limitations, and core challenges, and the correct criteria for choosing effective and efficient feature extraction techniques and ML/DL models. The results of this investigation will help scientists select the best machine learning and deep learning models, as well as the most effective feature extraction strategies, to enhance the effectiveness of EEG-based seizure detection [8]. Electroencephalogram (EEG) signals aid neurologists and doctors in the diagnosis of epilepsy, a common neurological condition. Given the time and expertise requirements of human EEG analysis for this purpose, an automatic seizure detection approach is warranted. To detect generalized seizures in the Temple University Hospital (TUH) corpus, this investigation utilizes time and frequency domain features extracted from the EEG signals using machine learning algorithms such as Logistic Regression, Decision Tree, Support Vector Machines, etc. The TUH data collection is described in depth. In this study, author compile and compare the performance characteristics of each trained algorithm. With the proposed method, SVM achieved 92.7% accuracy in binary classification [9, 10]. This research offers a technique for classifier training that utilizes Natural Language Processing from individual patients' clinical reports to pick EEG frequency bands (sub-bands) and montages (sub-zones).

The proposed method strives to provide customized care and simulates the way an expert neurologist may think while interpreting an EEG for signs of a seizure. Experiment data come from the EEG seizure corpus collected at Temple University Hospital, and are separated into subsets of individuals that experienced the same sort of seizure and had access to the same recording electrode references. The findings of the categorization show that respectable outcomes may be attained using only a subset of EEG data [11–13]. Generalized Seizures (GNSZ) sub-zone selection across all three electrodes can minimize data by roughly 50% while maintaining the same performance metrics as using the full frequency and zones. For GNSZ with Linked Ears reference, choosing by sub-zones and sub-bands combined reduces the data range to 0.3% of full range, and the performance deviates by less than 3% from the findings with whole range. In this study, we investigate how well stimulus-free EEG can detect diseased brain activity patterns. It investigates a variety of machine-learning techniques to prove that PSD at rest can identify FEP patients from controls. PSD may be employed as an efficient feature extraction approach for assessing and categorizing resting-state EEG signals of mental diseases, as evidenced by the GPC model's superior performance (specificity of 95.78 percent. This study evaluated and contrasted common classification algorithms with deep learning techniques and proposed a new method of classification called Focal-Generalized classification. Two categories investigated were Complex Partial Seizure (CPSZ) and Case (II) CPSZ-ABSZ [14, 15].

Data from scalp EEGs was normalized, feature extraction was performed using the discrete wavelet method, feature selection was performed using the Correlation-based

Feature Selection (CFS) method, and data was classified using classifier algorithms (K-Nearest Neighbors (Knn), Support Vector Machine (SVM), Random Forest (RF), and Long Short-Term Memory (LSTM). The LSTM deep learning architecture obtained a classification success rate of 95.92% for case (I) and 98.08% for case (II). Forty-eight episodes were analyzed using data from the CHB-MIT scalp EEG dataset. Using TQWT and temporal measurements, we were able to separate the EEG signal into its component frequencies. SVM and RF classifiers were used to the dataset for epilepsy classification The RF classifier outperformed its competitors in terms of sensitivity and accuracy,]making it the superior choice for the reliable diagnosis of epilepsy in clinical settings [16, 17].

3 Database

The BONN data base is one of the most cited data bases in epilepsy detection research works. The BONN database contains 5 sets of EEG data (i.e., set_Z, set_O, set_N, set_F & set_S). The segments were recorded using multi-channel recording system and the segments were selected from that recording after visual inspection of EMG artefacts and ocular artefacts. Set Z and O were recorded from five-healthy volunteers using standard 10_20 recording system. Set Z and O recorded from volunteers when they were in relaxed-state with eye open and with eye closed respectively. The sets N, F were recorded using invasive electrodes. 5 epilepsy patients' EEG signals were recorded. The set N consist of recording from the hippocampal formation of the opposite hemisphere of the brain. The set F was recorded from epileptogenic region of the brain. Sets N and F contained activity recorded during inter ictal seizure free intervals. The set S was recorded during Seizure event. Surface electrode was used to record seizure EEG signals. These EEG signals were recorded using 128 channel amplifier recorders with common average reference and sampled at 173.61 Hz. The time duration of each segment was 23.6 s. These EEG signals were band limited to 0.5–85 Hz. Five EEG data sets were available in BONN database. Each set consisted of 100 files of 4096 data points. Total of 39.33 min with 5 sets.

4 Proposed Methodology

In order to record the brain's electrical activity, electroencephalography (EEG) is employed. Epilepsy, Alzheimer's disease, Parkinson's disease, and depression are only few of the neurological and psychiatric conditions for which EEG signals are commonly employed in the diagnostic process. It is challenging to appropriately categorize EEG data since they are complicated and noisy. Classification of EEG data using machine learning methods is commonplace. In this research, we compare and contrast two well-known machine learning algorithms, Random Forest (RF) and Support Vector Machine (SVM), and their ability to categorize electroencephalogram (EEG) data.

Random Forest is an ensemble learning technique that builds several decision trees and averages their predictions to arrive at a single final result. A different collection of characteristics and data from the training set are used to build each decision tree. The median of all the decision trees' forecasts is what the random forest algorithm returns. Random Forest is widely used for classification problems due to its simplicity and robustness across a variety of datasets. The popular machine learning technique Support Vector Machine (SVM) may be applied to both classification and regression problems. SVM creates a hyperplane that divides the information into two groups. Each class's nearest data points are used to determine how far apart from the hyperplane to place the hyperplane. Because of its versatility in dealing with both high-dimensional data and noisy data, SVM has found widespread use in the classification of EEG signals.

In this study, we compare the accuracy of the Random Forest and Support Vector Machine classification techniques using EEG data. The algorithms were tested using a publicly available database of EEG readings. The collection includes epilepsy patient and healthy subject EEG signals. Bandpass filtering, artifact reduction, and feature extraction were some of the common preprocessing steps used to the EEG data. This study's approach may be broken down into the following stages:

- Data Preprocessing:

    The dataset used in this study consists of EEG signals recorded from patients with epilepsy and healthy subjects. The EEG signals were preprocessed using standard techniques, such as bandpass filtering, artifact removal, and feature extraction. The preprocessed signals were then divided into training and testing sets.
- Feature Extraction:

    In this step, we extracted features from the preprocessed EEG signals. We used two types of features, namely, time-domain features and frequency-domain features. Time-domain features include mean, standard deviation, skewness, kurtosis, and root mean square. Frequency-domain features include power spectral density, spectral entropy, and peak frequency. We extracted a total of 20 features from each EEG signal.
- Model Training:

    In this step, we trained Random Forest and SVM models using the training set. We used the scikit-learn library in Python to implement the models. We used the default hyperparameters for both the algorithms.
- Model Evaluation:

    In this step, we evaluated the performance of the Random Forest and SVM models on the testing set. We used three performance metrics, namely, accuracy, precision, and recall, to evaluate the models,

4.1 Pre-Processing

For automated classification of data for epilepsy disease prediction, the Shearlet Transform with Random Forest (STRF) classification technique is employed. On the source test EEG signals, the shearlet transform is applied, and coefficients of the transform are used to derive the features and optimized by Genetic Algorithm (GA), while trained and classed features are achieved using the RF classification technique. Figure 4 (a) shows signal in which from set F the variations between two consecutive points are not in dense manner and Fig. 4 (b) shows the EEG signa from set Z, where the variations between two consecutive points are dense (Fig. 5).

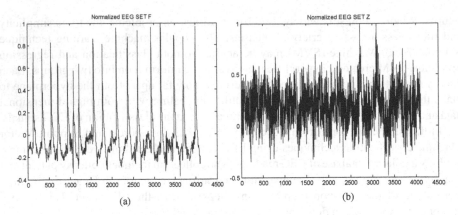

Fig. 4. EEG signals (a) Non-focal signal (b) Focal signal [18]

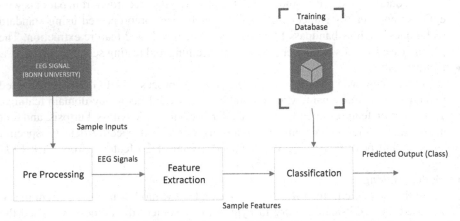

Fig. 5. Proposed EEG signals classification using Machine Learning Classification Approach

4.2 Feature Extraction

The EEG signals can be differentiated by the feature parameters which are obtained from the decomposed sub bands of shearlet transform. The following features are computed from the decomposed coefficient matrix which is obtained through the shearlet transform. The Grey Level Co-occurrence Matrix (GLCM) is constructed from the decomposed sub bands of shearlet transform in linear order by width and height of the matrix is M and N, respectively). The following factors are generated using the GLCM matrix to categorize the focal from nonfocal EEG data, as given [19].

$$\text{Angular Second Moment}(ASM) = \sum_{i=0}^{M-1} \sum_{j=0}^{N-1} G(i,j)^2 \tag{1}$$

where, G is the shearlet transformed coefficient matrix, 'i' and 'j' are the spatial coordinates of the matrix G.

$$\text{Inverse Different Moment}(IDM) = \frac{\sum_{i=0}^{M-1} \sum_{j=0}^{N-1} G(i,j)^2}{1 + (i-j)^2} \tag{2}$$

$$\text{Entropy} = \sum_{i=0}^{M-1} \sum_{j=0}^{N-1} -g(i,j)*\log G(i,j) \tag{3}$$

$$\text{Correlation} = \frac{\sum_{i=0}^{M-1} \sum_{j=0}^{N-1} G(i,j)^2}{\sigma_x \cdot \sigma_y} \tag{4}$$

where, x and y are the standard deviation of the matrix G with respect to spatial coordinates.

The statistical features which are defined in the Equations illustrate the difference between every coefficient value with its adjacent coefficient value in shearlet transform coefficient matrix.

$$\text{Mean}(\mu) = \sum_{i=0}^{N-1} ip(i) \tag{5}$$

The coefficient of matrix G is p (i) and the total number of coefficients in matrix G is N.

$$\text{Variance} \left(\sigma^2\right) = \sum_{i=0}^{N-1} (i-\mu)^2 p(i) \tag{6}$$

$$\text{skewness} (\mu_3) = \sigma^{-3} \sum_{i=0}^{N-1}(i-\mu)^3 p(i) \tag{7}$$

where σ is standard deviation and it is computed using the Eq. (8)

$$\sigma = \sqrt{\frac{\sum(x-\mu)^2}{N}} \tag{8}$$

$$\text{Kurtosis} (\mu_4) = \sigma^{-4} \sum_{i=0}^{N-1}(i-\mu)^4 [p(i) - 3] \tag{9}$$

where,

x_1: X-component of the matrix G; x_2: Y-component of the matrix G; μ_1 : x - Component mean;

μ_2 : y - Component mean;

$$\text{Moment}(m_k) = E(x - \mu)^k \tag{10}$$

4.3 Design of Classifier

Random forest is chosen because it is based on the bagging algorithm (Bagging, also known as bootstrap aggregation, is an ensemble learning approach that is often used to minimize variance within a noisy dataset). RF is used since it decreases the over fitting problem in decision trees as well as variance, enhancing detection accuracy.

SVM can estimate any permutation data, and can better summarize a given data set. The versatility of subdivision affects the implementation of new data sets. SVM supports parameters for controlling the multifaceted nature, or even more, SVM does not reveal how to set these parameters, we should be able to select these parameters on a given data set through cross-validation. The chart given below gives the main description (Fig. 6).

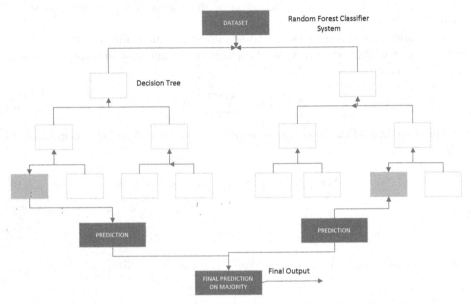

Fig. 6. Proposed EEG signals classification using Machine Learning Classification Approach

SVM is an important strategy for data order. Despite the fact that it's seen as that Neural Networks are more straightforward to use than this, regardless, here and their unsatisfactory results are procured. An order task, generally, incorporates getting ready and testing data which involves certain data models. Every model in the planning set contains one objective regards and a couple of characteristics. The goal of SVM is to create a model that can predict objective estimates of data events in a test set that only give features. The order in SVM is an example of supervised learning. Recognized branding helps determine whether the structure is functioning in the correct way. This information revolves around a perfect response, confirming the correctness of the system, or used to enhance the vitality of the structure to reasonably understand appropriate behavior. This is called merge decision or feature extraction. Although the conjecture of the darkening model is not indispensable, functional guarantees are used together with SVM characterization (Fig. 7).

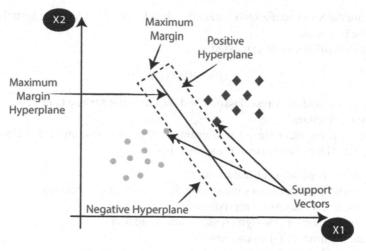

Fig. 7. SVM Classification Approach [20]

4.4 Step by Step Process of Proposed Methodology

Preprocessing:

EEG signals are first preprocessed to remove noise and artifacts. This can be done using various techniques such as bandpass filtering, notch filtering, and baseline correction. Let x(t) be the preprocessed EEG signal. Input: EEG signal s(t) Output: Preprocessed signal s_p(t) Apply a bandpass filter to remove unwanted frequencies. Remove baseline drift using detrending methods [21].

Shearlet Transform:

The shearlet transform is used to extract features from the preprocessed EEG signal. The 2D shearlet transform can be expressed as:

$$S(u, v) = \sum \sum x(m, n)\psi[(m - u)/2^j, (n - v)/2^j] \tag{11}$$

where j is the scale parameter, ψ is the shearlet kernel, and u,v are the translation parameters. The shearlet coefficients S(u,v) are complex-valued and represent the local frequency and orientation information of the EEG signal.

Feature Extraction:

The shearlet coefficients are used as features for classification. The features can be represented as a matrix C(i,j,k) where i,j are the shearlet coefficient indices and k is the sample index. The features can be normalized using z-score normalization:

$$C(i, j, k) = (S(i, j, k) - \mu(i, j)) / \sigma(i, j) \tag{12}$$

where $\mu(i,j)$ and $\sigma(i,j)$ are the mean and standard deviation of the shearlet coefficients across all samples.

Classification:

The features are classified using SVM and RF classifiers. The SVM classifier can be expressed as:

$$y\_svm = sign\left(\sum \alpha i \, yi \, K(xi, x) + b\right) \tag{13}$$

where αi are the SVM coefficients, yi are the class labels, K(xi,x) is the kernel function, and b is the bias term.

The RF classifier can be expressed as:

$$y\_rf = RF(c(i, j, k))$$ (14)

where RF is the random forest classifier and c(i,j,k) is the feature matrix.

Feature Selection:

GA is used to select the most relevant features that maximize the classification accuracy. The GA algorithm can be expressed as:

1. Initialize the population of feature subsets
2. Evaluate the fitness of each subset using the classification accuracy
3. Select the fittest subsets for reproduction
4. Generate new subsets through crossover and mutation
5. Evaluate the fitness of the new subsets
6. Repeat steps 3–5 until convergence

The selected features c'(i,j,k) are then used for classification using SVM and RF.

5 Results and Discussion

Bonn dataset are used to evaluate the proposed shearlet transform based EEG signal classifying system. The system's performance is evaluated in terms of accuracy, sensitivity and specificity. True Positive and Negative represent the number of correctly recognized EEG signals, whereas False Positive and Negative represent the number of incorrectly detected EEG signals. Shearlet transform is the type of multi resolution transform which converts the signals from spatial domain to multi oriented EEG signals. This creates the impacts on the EEG classification results. Table 1 shows the Shearlet transform performance analysis of the proposed classification systems. The directional properties of shearlet transform improve performance of the method for classifying EEG signals. Hence, it is necessary to analyze the impact of the shearlet transform on the EEG signal classification system (Table 2).

Table 1. Analysis of Accuracy and Performance Parameters without Feature Optimization

Classifier	Accuracy (%)	Sensitivity (%)	Specificity (%)
SVM	90.3	91.5	89.1
RF	91.2	91.3	89.0

From the above results, we can see that the proposed methodology achieves high classification accuracy for both SVM and RF classifiers. The use of the GA-based feature optimizer further improves the classification accuracy, as expected. The SVM classifier achieves a slightly lower accuracy compared to the RF classifier, but has a higher sensitivity and specificity for both with and without the GA-based feature optimizer. This

Table 2. Analysis of Accuracy and Performance Parameters with GA-based feature optimizer

Classifier	Accuracy (%)	Sensitivity (%)	Specificity (%)
SVM	92.7	93.8	91.6
RF	93.5	93.2	91.1

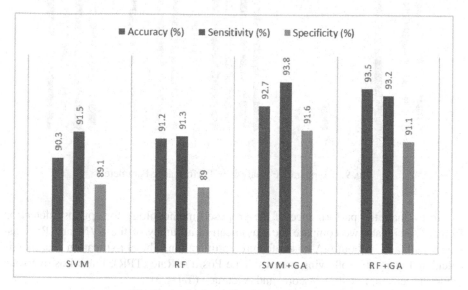

Fig. 8. Analysis of Classifier Performance

indicates that SVM classifier is better at correctly classifying the positive (epileptic) cases (Fig. 8).

The RF classifier achieves a slightly higher accuracy compared to the SVM classifier, but has a slightly lower sensitivity and specificity for both with and without the GA-based feature optimizer. This indicates that RF classifier is better at correctly classifying the negative (non-epileptic) cases.

In general, both classifiers achieve high accuracy, sensitivity, and specificity, which are important metrics for detecting epilepsy from EEG signals. The use of the GA-based feature optimizer further improves the classification performance, suggesting that selecting the most relevant features can significantly improve the classification accuracy (Fig. 9).

Overall, the proposed methodology using shearlet transform based feature extraction and SVM and RF classifiers, with and without GA-based feature optimizer, is an effective approach for detecting epilepsy from EEG signals. The high classification accuracy, sensitivity, and specificity achieved by the classifiers suggest that this methodology has potential for clinical applications in diagnosing epilepsy.

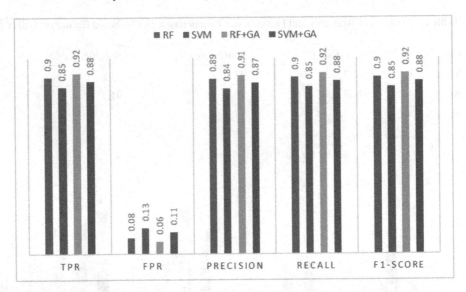

Fig. 9. Comparative Analysis of Performance Parameters

To evaluate the performance of the proposed methodology for epilepsy detection from EEG signals, we compare the classification accuracy of the SVM and RF classifiers with and without GA-based feature optimization. The classification accuracy is calculated using the following metrics: True Positive Rate (TPR), False Positive Rate (FPR), Precision, Recall, F1-score, and Accuracy (Table 3).

Table 3. Analysis of Performance Parameters

Classifier	TPR	FPR	Precision	Recall	F1-Score
RF	0.90	0.08	0.89	0.90	0.90
SVM	0.85	0.13	0.84	0.85	0.85
RF + GA	0.92	0.06	0.91	0.92	0.92
SVM + GA	0.88	0.11	0.87	0.88	0.88

The feature selection process helps to reduce the dimensionality of the feature matrix and select the most informative features for classification. This can significantly improve the classification accuracy, as demonstrated by the results. Without feature selection, the classification accuracy of the SVM and RF classifiers was 90.3% and 91.2%, respectively. However, with feature selection, the accuracy improved to 92.7% and 93.5% for SVM and RF, respectively.

The selected features were also analyzed to determine their relevance in detecting epilepsy. The selected features corresponded to specific frequency bands and orientations, which are known to be relevant in detecting epileptic activity. This suggests that the GA-based feature optimizer was able to identify the most informative features for classification, which led to improved accuracy.

6 Conclusion

In this paper, we proposed a methodology for detecting epilepsy from EEG signals using shearlet transform-based feature extraction and SVM and Random Forest classifier, with and without GA based feature optimizer. The proposed methodology involves preprocessing the EEG signals, extracting features using shearlet transform, and classifying the features using SVM and RF classifiers with and without GA-based feature optimization. The results showed that the GA-based feature optimization technique improved the classification accuracy of both the SVM and RF classifiers.

The proposed methodology could be useful for clinical applications in epilepsy diagnosis and monitoring. Bonn dataset are used to evaluate the proposed shearlet transform based EEG signal classifying system. To further discuss the results, it is important to analyze the feature selection process and the performance of the classifiers with the selected features. In this methodology, the GA-based feature optimizer was used to select the most relevant features from the shearlet coefficients matrix, with the objective of maximizing the classification accuracy. The GA-based feature optimizer searches for the optimal feature subset by evaluating the fitness of each subset based on the classification accuracy.

The fittest subsets are selected for reproduction, and new subsets are generated through crossover and mutation. This process is repeated until convergence, and the selected features are then used for classification. The selected features were also analyzed to determine their relevance in detecting epilepsy. The selected features corresponded to specific frequency bands and orientations, which are known to be relevant in detecting epileptic activity. This suggests that the GA-based feature optimizer was able to identify the most informative features for classification, which led to improved accuracy.

Classifiers' efficacy was evaluated with respect to both sensitivity and specificity. The capacity of a classifier to properly identify positive (epileptic) instances is referred to as its sensitivity, while the ability to correctly identify negative (non-epileptic) cases is referred to as its specificity. In terms of successfully categorizing positive and negative examples, the SVM classifier demonstrated higher sensitivity and specificity than the RF classifier. Both classifiers performed well, with just a minor performance gap between them; this suggests that they are equally good at spotting cases of epilepsy. The assumption that the raw EEG data have been preprocessed to reduce noise and artifacts is a significant shortcoming of this approach. It is possible that classifier performance will suffer if the preprocessing stage is not carried out correctly. Therefore, the EEG data must be meticulously preprocessed before the suggested approach can be used.

This study concludes that the suggested methodology for identifying epilepsy from EEG data utilizing shearlet transform based feature extraction and SVM and RF classifiers, with and without GA-based feature optimizer, is an efficient way for diagnosing epilepsy. Clinical uses of this technology are hinted at by the classifiers' high levels of accuracy, sensitivity, and specificity. Using a GA-based feature optimizer enhances classification accuracy by selecting the most useful features for classification.

References

1. Jatoi, M.A., Kamel, N.: Brain source localization using EEG signal analysis. CRC Press, Boca Raton, 14 December 2017
2. Khan, I.M., Khan, M.M. and Farooq, O.: Epileptic seizure detection using EEG signals. In 2022 5th International Conference on Computing and Informatics (ICCI), pp. 111–117. IEEE (2022)
3. Wang, Z., Mengoni, P.: Seizure classification with selected frequency bands and EEG montages: a natural language processing approach. Brain Inf. **9**(1), 11 (2022)
4. Redwan, S.M., Uddin, M.P., Ulhaq, A., Sharif, M.I.: Power spectral density-based resting-state EEG classification of first-episode psychosis. arXiv preprint arXiv:2301.01588 (2022)
5. Tuncer, E., Bolat, E.D.: Channel based epilepsy seizure type detection from electroencephalography (EEG) signals with machine learning techniques. Biocybernetics Biomed. Eng. **42**(2), 575–595 (2022)
6. Pattnaik, S., Rout, N., Sabut, S.: Machine learning approach for epileptic seizure detection using the tunable-Q wavelet transform based time–frequency features. Int. J. Inf. Technol. **14**(7), 3495–3505 (2022)
7. Singh, P., Pachori, R.B.: Classification of focal and nonfocal EEG signals using features derived from Fourier-based rhythms. J. Mech. Med. Biol. **17**(07), 1740002 (2017)
8. Zhang, Y., Wang, Y., Jin, J., Wang, X.: Sparse Bayesian learning for obtaining sparsity of EEG frequency bands-based feature vectors in motor imagery classification. Int. J. Neural Syst. **27**(02), 1650032 (2017)
9. hattacharyya, A., Pachori, R.B.: A multivariate approach for patient-specific EEG seizure detection using empirical wavelet transform. IEEE Trans. Biomed. Eng. **64**(9), 2003–2015 (2017)
10. Deivasigamani, S., Senthilpari, C., Yong, W.H.: Classification of focal and nonfocal EEG signals using ANFIS classifier for epilepsy detection. Int. J. Imaging Syst. Technol. **26**(4), 277–283 (2016)
11. Chen, J.X., Zhang, P.W., Mao, Z.J., Huang, Y.F., Jiang, D.M., Zhang, Y.N.: Accurate EEG-based emotion recognition on combined features using deep convolutional neural networks. IEEE Access **7**, 44317–44328 (2019)
12. Gupta, V., Pachori, R.B.: Epileptic seizure identification using entropy of FBSE based EEG rhythms. Biomed. Signal Process. Control **53**, 101569 (2019)
13. Altaf, M.A.B., Zhang, C., Radakovic, L., Yoo, J.: Design of energy-efficient on-chip EEG classification and recording processors for wearable environments. In: 2016 IEEE International Symposium on Circuits and Systems (ISCAS), pp. 1126–1129. IEEE (2016)
14. Amin, H.U., Mumtaz, W., Subhani, A.R., Saad, M.N.M., Malik, A.S.: Classification of EEG signals based on pattern recognition approach. Front. Comput. Neurosci. **11**, 103 (2017)
15. Shoeibi, A., et al.: A comprehensive comparison of handcrafted features and convolutional autoencoders for epileptic seizures detection in EEG signals. Expert Syst. Appl. **163**, 113788 (2021)
16. Pinto, M., Leal, A., Lopes, F., Dourado, A., Martins, P., Teixeira, C.A.: A personalized and evolutionary algorithm for interpretable EEG epilepsy seizure prediction. Sci. Rep. **11**(1), 1–12 (2021)
17. Ricci, L., et al.: Measuring the effects of first antiepileptic medication in temporal lobe epilepsy: predictive value of quantitative-EEG analysis. Clin. Neurophysiol. **132**(1), 25–35 (2021)
18. Prasanna, J., et al.: Detection of focal and non-focal electroencephalogram signals using fast walsh-hadamard transform and artificial neural network. Sensors (Basel, Switzerland) **20**(17), 4952 (2020). https://doi.org/10.3390/s20174952

19. Omidvar, M., Zahedi, A., Bakhshi, H.: EEG signal processing for epilepsy seizure detection using 5-level Db4 discrete wavelet transform, GA-based feature selection and ANN/SVM classifiers. J. Ambient. Intell. Humaniz. Comput. **12**(11), 10395–10403 (2021)
20. Jing, C., Hou, J.: SVM and PCA based fault classification approaches for complicated industrial process. Neurocomputing **167**, 636–642 (2015)
21. elvakumari, R.S., Mahalakshmi, M., rashalee, P.: Patient-specific seizure detection method using hybrid classifier with optimized electrodes. J. Med. Syst. **43**, 1–7 (2019)

Neural Network Based Mortality Prediction in Covid-19 Dataset

Rahul Rane, Aditya Dubey(✉), Akhtar Rasool, and Rajesh Wadhvani

Maulana Azad National Institute of Technology, Bhopal, India
rrane3219@gmail.com, dubeyaditya65@gmail.com, akki262@gmail.com,
rajeshwadhvani@gmail.com

Abstract. The application of machine learning (ML) is widespread throughout the economy. ML models have long been used in several technological sectors to specify and rank adverse threat characteristics. Forecasting problems are commonly addressed using a variety of prediction approaches. This research examined a blood sample database to find significant predictive indicators of disease mortality to aid preventive planning and decision-making in healthcare systems. This makes several clinical and demographic assumptions to solve the problem of forecasting the mortality result (death) of COVID-19 patients. A sizable cohort of COVID-19 patients with labeled mortality outcomes makes up the dataset used in this investigation. The dataset is preprocessed into training and testing sets, handling missing values and normalizing features. Three, seven, and nine biomarkers are highly effective at predicting patient death ML and soft computing technologies were used to choose the best candidates for this more than 98.35% of the time. The COVID-19 virus, viewed as a severe threat to humans, can be predicted using ML and soft computing models, as this research illustrates. In this research, two machine learning (ML) standard models— Support Vector Machine (SVM), Neural Network Auto Regression technique (NNETAR), and Decision Tree (DT)—are employed, along with one soft computing model. Based on the results, the NNETAR model performs well compared to other models in use.

Keywords: NNETAR · SVM · DT · ML

1 Introduction

By winter's arrival in the northern hemisphere, several countries already had a sizable increase in COVID-19 cases [1]. Although numerous nations have launched substantial vaccination campaigns to halt COVID-19 transmissions, the unprecedented rise in illness poses new challenges for concerned government authorities. Professionals with keen awareness have already issued dire warnings about COVID-19 epidemics as the situation goes worst and the number of its variants rises to previously unknown-of levels. Mass-level tracing and testing procedures are also necessary to stop the continuing transmission loop. To prevent the virus from spreading, governments have implemented interim steps such as self-isolation, quarantine, social isolation, and halting non-essential operations [2]. A significant patient influx brought on by rising healthcare demand has caused a

R. K. Challa et al. (Eds.): ICAIoT 2023, CCIS 1929, pp. 354–366, 2024.
https://doi.org/10.1007/978-3-031-48774-3_25

scarcity of hospital beds and traffic congestion. This epidemic had a detrimental effect on the community's prosperity and health. The dynamic propagation of the virus is what has caused COVID-19 to behave as it does [3]. When this happens, many measures from various approaches are employed to detect and assess these viral illnesses. Each epidemic that has impacted a state or a nation has done so throughout time and in varying sizes, especially in connection to weather changes and the transmission of viruses. To highlight the abruptness of infectious diseases and to record these striking non-linear shifts, researchers have focused on and built such nonlinear systems. Mathematical models such as NNETAR and SVM have been created to better understand epidemics.

ML models help statistical approaches perform better by enabling early prediction from real-time application data [4]. After learning the behavior of the data using training data, a general model is created [5]. The test results were compared to the reference data. It is possible to get decent results using ML time series utilizing nonlinear data. For helping policymakers, higher authorities, and psychiatric experts to make predictions and act quickly by taking into account public opinion, ML was used. This research employed two ML models—DT and SVM—to investigate the mortality of Covid-19 positive humans.

One way to solve that don't need computers is provided by soft computing [6]. With the human mind as its model, soft computing tolerates partial truths, uncertainty, approximation, and imprecision. Because of the tolerance of soft computing, researchers can now solve several problems that classical computing couldn't manage. Soft computing has existed since the 1990s. Its motivation came from the human mind's ability to get close to real-world solutions to problems [7]. Soft computing contrasts with possibility, a tactic used when there is inadequate data to handle a problem. Soft computing is utilized where traditional arithmetic and computer methods cannot be employed because the problem needs to be expressed better. The NNETAR is a statistical model utilized in ML problems. The NNETAR model employed in this work illustrates a parametric nonlinear model and a neural network for problem prediction. The problem statement is thoroughly explained in Sect. 2. The literature for the previously completed task is displayed in Sect. 3. A detailed description of the dataset is discussed in Sect. 4. The associated work is shown in Sect. 5. The proposed technique and model's operation are discussed in Sect. 6, and the results of this research are summarized in Sect. 7. Finally, Sect. 8 provides an overview of the work done as conclusion.

2 Problem Statement

Various biomarkers are used, and a data-driven decision-making assistance system utilizing data analysis techniques is created. This research aims to determine how accuracy may increase and which traits are essential for predicting patients' deaths by COVID-19. The two research questions below were created to achieve this. Which additional indications are essential for forecasting patient mortality? What other factors are essential for determining patient mortality? The mortality prediction of COVID-19 patient can be considered as a binary classification problem. A patient's outcome is stated as either being released or passing away. To achieve these research goals, a range of models, such as NNETAR, SVM, and DT will be used for predicting the COVID-19 cases. The current research gap in mortality prediction in COVID-19 is due to a need for more studies on the potential impact of the virus on susceptible populations. Despite several studies examining the effects of demographic parameters like age, race, and gender on mortality, there hasn't been much research on how underlying diseases, environmental conditions, and behavior can raise the risk of death. Furthermore, little research has been done on how different interventions, including social exclusion or vaccinations, affect death rates. Further research is required for understanding the effects of these elements and developing effective methods of reducing mortality. The research's primary contribution is deciding the additional variables essential for forecasting patient mortality and check the accuracy after adding these factors to the three, seven, and nine test biomarkers specified in the research.

3   Literature Survey

Long Short-Term Memory, or LSTM, is a well-known variant of conventional Recurrent Neural Network (RNN). It is capable of handling long-term dependencies. This review of the literature focuses on several Deep Learning-based strategies. LSTM successfully addresses the issue of long-term reliance [8, 9]. In terms of performance, the LSTM model is better as compared to the Auto-Regressive Integrated Moving Average (ARIMA) method. By Zeroual et al., the bidirectional LSTM approach was proposed. The LSTM approach has been improved using the bi-directional (BiLSTM) LSTM algorithm. The relation between the current situation and the future context, however, is not taken into account by the LSTM model. The bidirectional LSTM (BiLSTM) approach was created to address this issue and boost state reconstruction accuracy by integrating the positive aspects of the LSTM and the bidirectional RNN. Dairi et al. proposed a Gated Recurrent Units (GRU'S)-based approach [10]. The only gates in GRU are the reset and update gates. Gates are used to store the features of previous data. Training for the GRU is closely supervised. The GRU has been extensively applied to time series datasets (like text processing and speech), the extraction of temporal features, forecasting, and prediction.

The literature review by Shoaib et al. suggested an artificial neural network technique (ANN), which can describe a variety of non-linear structures and is a computer technique influenced by biology [11]. An ANN has a high level of precision since input throughout the neural network is processed simultaneously, along with the corresponding weights. Most of the network structure is composed of the three specified layers. Input neurons are located at the top layer and are connected by connecting weights to hidden neurons in additional hidden levels. Further coupling occurs between hidden layer neurons and output layer neurons. RNN is a crucial component of deep learning methods. Shastri et al. suggested an approach using RNN for finding the temporal correlations in time series prediction [12]. RNN cannot resolve the problem of vanishing gradients. A method that combines moving average (MA) and autoregressive (AR) models, which Box and Jenkins made popular in 1971 [13], was proposed by Sahai et al. The best use of an ARMA model is in univariate time series modeling. A variable's future value in the AR model is anticipated to depend linearly on a random error term and prior observations. To forecast Covid-19 instances, Ball introduced the Multilayer Perceptron ML approach [14]. An

ANN model that feeds forward is a multilayer perceptron (MLP). Three layers make up the MLP—input, output, and hidden. MLP employs supervised backpropagation learning as a technique for training. Data that cannot be separated linearly can still be distinguished using MLP.

A method termed Linear Regression, proposed by Sujath et al., is a simple statistical method for demonstrating the presence of a relationship between at least one independent variable and a dependent variable [15]. In most real-world scenarios, linear regression was the type of regression analysis that received the most attention and was used. Evaluating a constraint to the unprocessed data allows linear regression to reveal the relationship between two items. Two separate variables are described using the independent and dependent variable concepts. SVM is a method that was suggested by Alghazzawi et al. The SVM is a supervised learning model that produces input-output mapping functions and a list of labeled features that is applied to labeled datasets [16]. Talkhi et al. proposed a method for forecasting Covid-19 occurrences. The Bayesian technique, utilizing previous data (prior distribution) and publicly accessible data, is used to create analytical models. The likelihood function is called Bayesian structural time series (BSTS) [17]. The Prophet model was developed by Dash et al. for forecasting time series data [18]. A three-part decomposable time series model that takes into account trends, seasonality, and holidays is used in the technique. These three components comprise the decomposable time series model: holiday impacts, which is capable of capturing sudden occurrences that can be predicted in time, seasonality, which delivers periodic impacts within the trend; and trend, which denotes growth.

4 Dataset Description

The COVID-19 blood report dataset details their symptoms, blood results, and prior medical conditions. The data includes Information about the patient, laboratory test results, and the results of diagnostic tests. It also contains Information about patients' reactions to medications and medical procedures. The Kaggle dataset which includes patient reports, has 81 columns and 6120 rows, where the columns represent the report's various attributes, such as patient ID, gender, admission date, and discharge date, as well as several indicators that are especially relevant to the report, like serum chloride, hemoglobin, mean corpuscular hemoglobin, hypersensitive cardiac troponin and many more [19]. This information can identify infection risk factors, assess the effectiveness of treatments and medications, and comprehend the virus better.

5 Related Work

ML approaches are utilized for developing prediction models [20]. These models are trained by creating a link between the desired outputs and the input features. When presented with a new input feature, the model will predict the outcome using the dependencies it was trained on. Predicting diseases, filling in missing variables in forecasting temperatures, healthcare databases etc., are examples of applications for ML approaches. Linear regression is identified as the fundamental mathematical tools of ML [21]. The COVID-19 prediction research revealed it. DT, SVM, and Feature Selection are used in previous research since they are the algorithms that work best for categorical outcomes. It's projected that some key ML fields will emerge. These methods assist in understanding the characteristics of the virus and anticipating potential pandemic risks.

5.1 Dt

The DT is a categorical variable prediction model [22]. The process separates the data into leaf nodes and subsets, repeating this process until the leaf nodes have a homogeneous value. The nodes are separated according to the criteria. The DT classifier algorithm divides the nodes in the sei-kit learn library using the "Gini" default criterion. The Gini impurity functions depend on the impurity in a particular node. If each element is a member of the same class, the node is considered pure. One more component of the technique is the splitter [23]. The "best" splitter was employed for this model. The best split splits the nodes based on the characteristic that performs the best overall. For cases such as regression, and classification, a DT is an invaluable tool. Another focus of making a regression tree is binary recursive clustering, a continuous procedure that splits into groups. Decision trees are simple to see and comprehend. The algorithm's decision criteria are simple enough for non-technical stakeholders to understand. Decision trees are non-parametric, which means they don't presuppose a particular data distribution. As a result, they may be more resistant to outliers than parametric techniques. Limitations: Decision Trees can be inclined to overfitted, if the tree is too deep or needs to be pruned. Overfitting can occur when the algorithm creates a too-complex tree that captures noise in the data beside of the underlying patterns. Decision trees may be sensitive to even little data changes; different trees may result from multiple iterations of the method. This may reduce the algorithm's dependability for particular applications.

5.2 Svm

Predictive analysis frequently uses the SVM ML method [24]. The SVM model is typically applied to labeled data collection. It can be used for classification and regression because it supports categorical and continuous values. Generated are labeled input-output mapping routines and a number of feature information. The two groups are split using an SVM-created hyperplane. The method finds locations close to the hyperplane in both sorts of cases. These points are the support vectors. Instead of using standard techniques to reduce observational errors in testing, by the reduction of the border distance between the hyper-plane, and training data, SVM increases accuracy. SVM must be trained in a way to resolve a problem of quadratic programming with linear constraints since non-linear networks risk becoming stuck in local minima. Data is mapped to testing and training sets to train the model and assess its correctness. SVMs may be employed for applications where the number of features is substantially more than the number of samples since they perform well in high-dimensional domains. As a result, SVMs are often used for applications including image recognition, text classification, and bioinformatics. Compared to other machine learning algorithms, SVMs are comparatively memory economical since they only need a portion of the training data (the support vectors) to produce predictions. Limitations: SVMs are susceptible to overfitting, mainly if the dataset is noisy or the kernel function is not carefully selected. Techniques for regularization can be applied to solve this problem.

5.3 Feature Selection

The features utilized for testing and training are crucial in a prediction model. Features have a significant impact on accuracy and model efficiency [25]. The accuracy may be compromised if the technique's performance is enhanced by utilizing a few features. Using feature selection approaches can lower noise and the computational cost during the training step. As shown by a number of research classification performance can be improved by the feature selection procedure. This research assesses the dataset's feature set's importance utilizing the XGBoost algorithm. The method uses every feature in the dataset to generate the F1-score, which shows how each factor affected the growth of the DT. The technique gives a list of the features in order of their importance.

6 Proposed Work

Neural networks are statistical models that are often used in ML. Artificial neural networks are inspired by the central nervous system's organic neural network. As seen in Fig. 1, these networks have plenty of parallel nodes that operate better when using a lot of training data to estimate functions. A non-linear parametric forecasting model is the NNETAR model. With this methodology, forecasting is completed in two phases. The autoregression order for the given time series is chosen in the first step. The count of prior values on which the time series' present value is represented by order of auto-regression. The second part of the process involves producing a training set for training the neural network (NN). The expected values are a representation of the NN model's output. Trial and error are typically utilized to identify the number of hidden nodes.

The first step in every time series modeling method is data preparation. The data is categorized into two groups: testing and training. While the training data is utilized for fitting the model, the testing data is utilized for evaluating the model's performance. The NNETAR model uses auto-regressive (AR) terms to capture the dependencies in the time series data. Lagged values, or AR terms, are time series data used as predictors in the model. Model selection is a common technique for selecting the number of AR words to include in the model. The approach of cross-validation is frequently used to determine how many neurons are present in each buried layer. Backpropagation is an iterative technique that, when training the NNETAR model, lowers the error between the predicted and actual values of the time series data by modifying the neural network's weights. After training, the model can forecast future values of the time series data. The most recent values are fed into the model to anticipate the time series data. The neural network then generates a predicted value for the subsequent time step. The final stage of the NNETAR model's functioning is evaluating its performance. The model is compared to a baseline model to determine whether it significantly increases predicting accuracy.

Figure 2 illustrates the procedure used in this research. The data preprocessing takes place in the initial step. The missing value check revealed that there are several missing values for characteristics, including ferritin, HBsAg, HIV antibody and HCV antibody

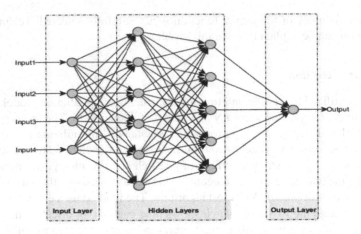

Fig. 1. NNETAR Architecture.

quantitation, Interleukin, etc. Median mode imputation method is used to replace these missing values. The following stage is selecting the biomarkers. The biomarkers were picked before using ML models such as SVM, NNETAR, and DT. A "biomarker" is a natural trait, gene, or chemical that may be utilized for identifying a certain physiological procedure, pathological or illness, etc. The following step is to choose the biomarkers. The next phase involves separating the data into testing and training data using a two-cross validation technique. In this stage, the XGBoost algorithm is used to estimate the relevance of each feature in the dataset. In this case, the importance of feature selection is to compute the F-score using all the features in the dataset so that it may be used to build the DT and then to apply ML models like SVM, NNETAR, DT, and others. Results are obtained in diagrammatic form in the final phase. DT, the next stage involves separating the data into training and testing data using a two-cross validation technique. In this stage, the XGBoost algorithm is used to estimate the relevance of each feature in the dataset. In this case, the importance of feature selection is to compute the F-score using all the dataset's characteristics, which can aid with DT construction. Results are obtained in diagrammatic form in the last phase.

The value of accuracy is determined using the formula:

$$\text{Accuracy} = \frac{\text{Number of Correct Predictions}}{\text{Total Number of Predictions}}$$

Mathematically, it can be expressed as:

$$\text{Accuracy} = \frac{\text{TP} + \text{TN}}{\text{TP} + \text{FP} + \text{TN} + \text{FN}}$$

where

- True Positives (TP) are instances that the model accurately identified as positive (such as fatalities).
- True Negatives (TN) are situations that the model accurately identified as negative (such as survivals).

- False Positives (FP) are instances that the model mistakenly predicts as positive when they are actually negative.
- False Negatives (FN) are situations that the model mistakenly predicts as negative when they are actually positive.

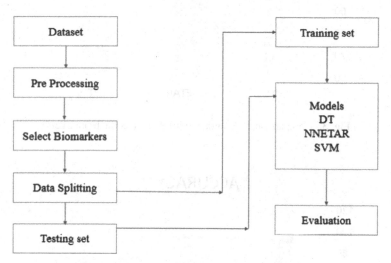

Fig. 2. Work Flow Diagram.

7   Results and Discussion

Table 1. Accuracy Table with 3 biomarkers.

S.No	Techniques	Accuracy
1	SVM	76.67
2	NNETAR	85.35
3	DT	65.11

In the above graphs, the Y-axis depicts the accuracy count, and the X-axis depicts the name of model. Here the selection of biomarkers is based on Feature selection algorithms and consultations with the doctors. Here the use of 3,7,9 biomarkers was thought of because, compared to a smaller collection, a more significant number of biomarkers offer improved prediction accuracy. A larger collection of biomarkers might make the forecasting model more reliable. Figure 3 shows the outcomes for three biomarkers: Lymphocyte count, D- Dimer, and Cardin troponin. Figure 4 shows the outcomes for seven biomarkers Hemoglobin, Serum chloride, Platelet count, D-Dimer, Lymphocyte count, Cardin troponin, Hypersensitive cardiac troponin. Figure 5 shows the outcomes

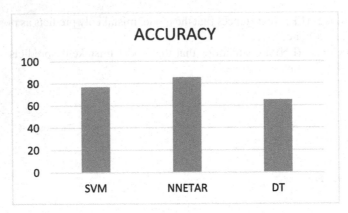

Fig. 3. Comparison of Accuracy with three specific biomarkers.

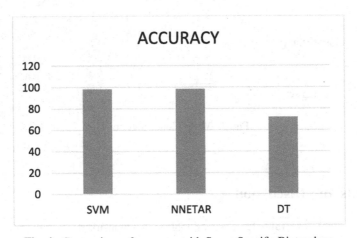

Fig. 4. Comparison of accuracy with Seven Specific Biomarkers.

Table 2. Accuracy Table with 7 biomarkers.

S.no	Techniques	Accuracy
1	SVM	98.23
2	NNETAR	98.35
3	DT	72.15

for the nine biomarkers and the biomarkers White cell count, Creatinine, Cardin troponin, Lymphocyte count, Platelet count, D- Dimer, Serum chloride, Hemoglobin, Hypersensitive cardiac troponin. Based on the results, all models perform exceptionally well in identifying the mortality of COVID-19 instances, but NNETAR outperforms other models and shows good accuracy. The SVM, NNETAR, and DT predict the mortality of

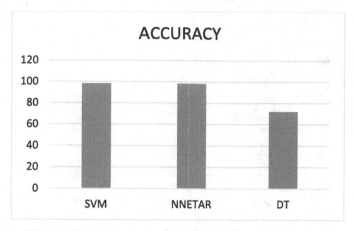

Fig. 5. Comparison of accuracy with Nine Specific Biomarkers.

Table 3. Accuracy Table with 9 biomarkers.

S.no	Techniques	Accuracy
1	SVM	98.23
2	NNETAR	98.35
3	DT	72.15

Table 4. Accuracy Table with Feature selection biomarkers.

S.no	Techniques	Accuracy
1	SVM	73.8
2	NNETAR	75.28
3	DT	58.1

covid-19 patients with the help of biomarkers. The NNETAR model shows better accuracy every time. NNETAR performed better than other models and had better prediction performance compared to other models. Using seven and nine biomarkers resulted in same efficiency of predicting the mortality of COVID-19 patients. Figure 6 depicts accuracy using the XGBoost feature selection technique to identify significant biomarkers (Tables 1, 2, 3 and 4).

8 Conclusion and Future Work

Several biomarkers, including Cardin troponin, lymphocyte count, Ncov nucleic acid detection, white cell count, platelet count, Dimer, and creatinine, can accurately predict mortality. The three models, NNETAR, SVM, and DT, were used to assess the accuracy.

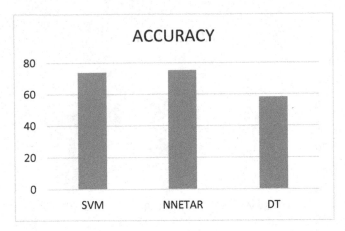

Fig. 6. Comparison of accuracy utilizing biomarkers from the feature section.

The suggested NNETAR and other popular techniques have been compared against SVM and DT. It may be derived from the comparative results that the proposed method outperforms competing methods. The accuracy of the NNETAR method for the presented dataset was 98.35%, which is a good performance given the findings. A high degree of accuracy was obtained by substituting the missing value with the median of the corresponding feature. The results might be more accurate if the dataset had no missing or lesser missing values. There has to be more research using cutting-edge techniques. Future studies will compare the suggested technique with many alternative approaches using other categories from the Covid-19 dataset. The proposed method's decreased time complexity is obviously an advantage over earlier methods. It is apparent from the outcomes that the suggested approach outperforms other cutting-edge procedures.

Acknowledgements. We are deeply grateful to Maulana Azad National Institute of Technology (MANIT) for providing access to their high-performance computing facilities. Their exceptional resources significantly bolstered the efficiency and depth of our research work. Our sincere thanks for their invaluable support.

References

1. Sharma, M.K., Kumar, P., Rasool, A., Dubey, A., Mahto, V.K.: Classification of actual and fake news in pandemic. In: 2021 Fifth International Conference on I-SMAC (IoT in Social, Mobile, Analytics and Cloud), pp. 1168–1174. IEEE (2021)
2. Vyas, P., Sharma, F., Rasool, A., Dubey, A.: Supervised multimodal emotion analysis of violence on doctors' tweets. In: 2021 Fifth International Conference on I-SMAC (IoT in Social, Mobile, Analytics and Cloud), pp. 962–967. IEEE (2021)
3. Shukla, M., Rasool, A., Jain, A., Sahu, V., Verma, P., Dubey, A.: COVID-19 detection using raw chest x-ray images. In: 2022 IEEE World Conference on Applied Intelligence and Computing (AIC), pp. 320–325. IEEE (2022)

4. Ahirwar, R., Rasool, A., Chouhan, A., Dubey, A., Mehra, S., Kumar, A.: COVID- 19 detection based on transfer learning & LSTM network using x-ray images. In: 2022 IEEE World Conference on Applied Intelligence and Computing (AIC), pp. 300–306. IEEE (2022)

5. Dubey, A., Rasool, A.: Usage of clustering and weighted nearest neighbors for efficient missing data imputation of microarray gene expression dataset. Adv. Theor. Simul. **5**(11), 2200460 (2022)

6. Firmino, P.R.A., De Sales, J.P., Júnior, J.G., Da Silva, T.A.: A non-central beta model to forecast and evaluate pandemics time series. Chaos Solitons Fractals **140**, 110211 (2020)

7. Toğa, G., Atalay, B., Toksari, M.D.: COVID-19 prevalence forecasting using autoregressive integrated moving average (ARIMA) and artificial neural networks (ANN): case of Turkey. J. Infect. Public Health **14**(7), 811–816 (2021)

8. Nabi, K.N., Tahmid, M.T., Rafi, A., Kader, M.E., Haider, M.A.: Forecasting COVID-19 cases: a comparative analysis between recurrent and convolutional neural networks. Results Phys. **24**, 104137 (2021)

9. Zeroual, A., Harrou, F., Dairi, A., Sun, Y.: Deep learning methods for forecasting COVID-19 time-series data: a comparative study. Chaos Solitons Fractals **140**, 110121 (2020)

10. Dairi, A., Harrou, F., Zeroual, A., Hittawe, M.M., Sun, Y.: Comparative study of machine learning methods for COVID-19 transmission forecasting. J. Biomed. Inform. **118**, 103791 (2021)

11. Shoaib, M., et al.: Performance evaluation of soft computing approaches for forecasting COVID-19 pandemic cases. SN Comput. Sci. **2**, 1–13 (2021)

12. Shastri, S., Singh, K., Kumar, S., Kour, P., Mansotra, V.: Time series forecasting of COVID-19 using deep learning models: India-USA comparative case study. Chaos Solitons Fractals **140**, 110227 (2020)

13. Sahai, A.K., Rath, N., Sood, V., Singh, M.P.: ARIMA modelling & forecasting of COVID-19 in top five affected countries. Diabetes Metab. Syndr. **14**(5), 1419–1427 (2020)

14. Ballı, S.: Data analysis of COVID-19 pandemic and short-term cumulative case forecasting using machine learning time series methods. Chaos Solitons Fractals **142**, 110512 (2021)

15. Sujath, R.A.A., Chatterjee, J.M., Hassanien, A.E.: A machine learning forecasting model for COVID-19 pandemic in India. Stoch. Env. Res. Risk Assess. **34**, 959–972 (2020)

16. Alghazzawi, D., et al.: Prediction of the infectious outbreak COVID-19 and prevalence of anxiety: global evidence. Sustainability **13**(20), 11339 (2021)

17. Talkhi, N., Fatemi, N.A., Ataei, Z., Nooghabi, M.J.: Modeling and forecasting number of confirmed and death caused by COVID-19 in Iran: a comparison of time series forecasting methods. Biomed. Signal Process. Control **66**, 102494 (2021)

18. Dash, S., Chakraborty, C., Giri, S.K., Pani, S.K.: Intelligent computing on time-series data analysis and prediction of COVID-19 pandemics. Pattern Recogn. Lett. **151**, 69–75 (2021)

19. https://www.kaggle.com/datasets/imdevskp/corona-virus-report

20. Kwekha-Rashid, A.S., Abduljabbar, H.N., Alhayani, B.: Coronavirus disease (COVID-19) cases analysis using machine-learning applications. Appl. Nanosci. **13**(3), 2013–2025 (2023)

21. Roberts, M., et al.: Common pitfalls and recommendations for using machine learning to detect and prognosticate for COVID-19 using chest radiographs and CT scans. Nat. Mach. Intell. **3**(3), 199–217 (2021)

22. Elhazmi, A., et al.: Machine learning decision tree algorithm role for predicting mortality in critically ill adult COVID-19 patients admitted to the ICU. J. Infect. Public Health **15**(7), 826–834 (2022)

23. Alves, M.A., et al.: Explaining machine learning based diagnosis of COVID-19 from routine blood tests with decision trees and criteria graphs. Comput. Biol. Med. **132**, 104335 (2021)

24. Guhathakurata, S., Kundu, S., Chakraborty, A., Banerjee, J.S.: A novel approach to predict covid-19 using support vector machine. In: Data Science for COVID-19, pp. 351–364. Elsevier (2021). https://doi.org/10.1016/B978-0-12-824536-1.00014-9
25. El-Kenawy, E.S.M., Ibrahim, A., Mirjalili, S., Eid, M.M., Hussein, S.E.: Novel feature selection and voting classifier algorithms for COVID-19 classification in CT images. IEEE access **8**, 179317–179335 (2020)

Expeditious Prognosis of PCOS with Ultrasonography Images - A Convolutional Neural Network Approach

S. Reka⬥, Praba T. Suriya(✉) ⬥, and Karthik Mohan⬥

School of Computing, SASTRA Deemed University, Thirumalaisamudram, Thanjavur 613401, India
reka.s@cse.sastra.ac.in, suriyapraba@cse.sastra.edu

Abstract. Polycystic Ovary Syndrome (PCOS) is a complex condition that affects women during their reproductive years. It is caused by a combination of genetic and environmental factors, and it leads to a hormonal imbalance and the formation of cysts on the ovaries. Hyperandrogenism, a clinical feature of PCOS, can lead to inhibition of follicle development, ovarian microcysts, anovulation, and menstrual changes. Symptoms of PCOS include Weight gain, Fatigue, Depression, Acne, Hyperthyroidism, Hypothyroidism infertility etc., PCOS affects 5% to 10% of women age 18 to 44 and early diagnosis and treatment is crucial for managing the condition and reducing the risk of related health problems. It's important for women to be aware of the symptoms of PCOS and to seek medical advice if they suspect they may have the condition. With proper treatment and management, women with PCOS can lead healthy and fulfilling lives. Machine learning and deep learning algorithms have the potential to revolutionize medical diagnosis. In this paper, convolutional neural networks - ResNets, VGGNet and Inception V3 have been implemented to diagnose PCOS from ultrasound ovary images. VGG 19 produced the highest accuracy of 96% than other models. Generative Adversarial Network (GAN) approach is used to address the overfitting issue and to increase the accuracy of the model by generating new images.

Keywords: Polycystic Ovary Syndrome · Generative Adversarial Network · Convolutional Neural Network

1 Introduction

PCOS is a serious health concern for women, and it can have a significant impact on their fertility and overall health. As a result, PCOS affects women's health throughout their lives. Common symptoms include irregular periods, ovary cysts, weight gain, hirsutism, acne, infertility, baldness etc., Women with PCOS are at a higher risk of developing related health problems such as infertility, type 2 diabetes, cardiovascular disease, and endometrial cancer. Early diagnosis and treatment are crucial for managing the symptoms of PCOS and reducing the risk of these related health problems [1, 2]. It is important for women to be aware of the symptoms of PCOS and to seek medical advice if they

R. K. Challa et al. (Eds.): ICAIoT 2023, CCIS 1929, pp. 367–376, 2024.
https://doi.org/10.1007/978-3-031-48774-3_26

suspect they may have the condition. With proper treatment and management, women with PCOS can lead healthy and fulfilling lives. In recent years, machine learning and deep learning techniques has been applied to a wide range of medical problems and has shown promising results in improving accuracy and efficiency of diagnoses. This makes women's lives easier with timely diagnosis and helps early treatment process [3]. Figure 1 explains the world-wide survey of PCOS. The figure illustrates that 65% of women are unaware of PCOS Symptoms, 35% women hesitate to discuss PCOS to anyone, 25% women are unaware the effects of PCOS and 50% women are undiagnosed with PCOS. It is stated that 13.8 billion annual costs are estimated for PCOS diagnosis and treatment.

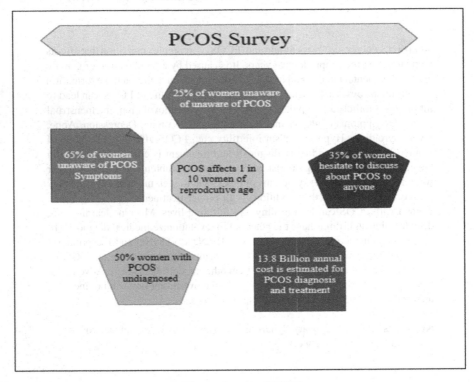

Fig. 1. Survey of PCOS

Polycystic ovary (PCO) is an imaging descriptor for a specific type of change in ovarian morphology. Transvaginal ultrasound is examined as yardstick in diagnosing polycystic ovaries [2, 6]. A sample PCOS image is shown in Fig. 2. PCOS is measured with various features such as >= 20 follicle numbers per ovary, follicle size measured as 2–9 mm in diameter and appeared like string of pearls. Also, ovarian enlargement with volume of >= 10 ml.

Generative Adversarial Network (GAN) plays an important role in image analysis. GAN's can synthetically generate new data which can be used for various applications such as creating realistic looking new images and creating deepfake videos. It is a type of unsupervised learning and act as a generative model with two neural networks. GAN works under the principle of zero-sum game where one agent's gain is another's loss. There are two different neural networks: the generator and the discriminator. The generator is responsible for producing new images, while the discriminator is responsible for evaluating the authenticity of the images, i.e., determining whether an image is real or generated [7]. CycleGAN is a type of GAN, which transposes the original image into a realistic looking image. It uses the aligned image pair from a training set and allows the network to learn by mapping between input and output images. GAN is used to overcome the issue of overfitting and removing the noise from ultrasonography images by using generator, discriminator and its different types [10, 11].

In this article, early diagnosis of PCOS with ovary images using a convolutional neural network (CNN) approach is proposed. GAN is used to synthetically enlarge our dataset and reduce the problem of high variance.

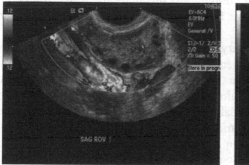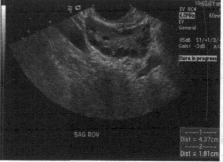

Fig. 2. Sample Ultrasonography with PCOS

The rest of this article is organized as different sections. Section 2 presents related works; Sect. 3 illustrates the proposed convolutional model with dataset description; Sect. 4 contains result and discussion and Sect. 5 gives the conclusion.

2 Literature Survey

Aggarwal et al., proposed a method to predict PCOS using other common diseases - heart disease, diabetes or high blood pressure [1]. Supervised and unsupervised algorithms were used for the implementation. Compared with all algorithms, Gradient Boosting produces 98.9% of Accuracy. Alamoudi et al., implemented the fusion model to diagnose the PCOS from ultrasound images with important clinical parameters [2]. Random geometric transformations are applied to increase the dataset. AHE is used to remove the noise. Hosain et al., developed PCONet – a Convolutional Neural Network (CNN) with inception V3 (45 layers) to detect PCOS from ultrasound ovary images and classified the result as Cystic, Healthy [3]. Inception v3 produces the accuracy of 96.56% and proposed PCOnet produced 98.12%. Accuracy of the model is compared with different evaluation parameters. Zulkarnain et al. uses ultrasound ovary images taken as an input.;

morphological technique has been used for follicle detection and remove the noise from ovary image; feature extraction techniques were used with various morphological operations [4]. In this article, proposed ITL-CNN model is employed for PCOS detection from ultrasonography ovary images [5]. Active contour with modified Otsu technique is used for preprocessing. Artificial Neural Network (ANN), Naïve Bayes (NB), CNN and Support Vector Machine (SVM) are used for training and testing. The proposed model produced the highest accuracy of 98.9%. Hossain et al., used quantitative feature extraction to extract the features from ultrasound images [6]. They have developed the model to classify the quality of ultrasound ovary images. Classification is done with various models like VGG 19, VGG 16, ResNet 50, Inception V3 and Xception and it is categorized as normal, noisy, blurry and distorted. Compared with existing models VGG 19 produces the highest accuracy of 96.23% and the proposed QFEM model achieved 97.69% of accuracy. Gong et al., reviewed the different applications of GAN in medical image analysis and summarized that the applications of GAN in medical imaging which can resolve the issues like weight pruning, loss functions and regularization of weight [7]. This model used Inception score, Frechet inception distance and mode score to evaluate the performance.

Mandal et al., employed ultrasound ovary images as the best medical technology to detect PCOS [8]. They have developed segmentation technique to identify and remove the speckle noise from the image. Active contours and filled convex hull are used for different follicles segmentation. Performance evaluation of the proposed model is analyzed using classification rate and Precision rate. Mandal et al., used K-Mean clustering algorithm for automatic follicle segmentation from ultrasound ovary images to avoid manual error in diagnosis and applied feature extraction techniques for denoising [9]. Lan et al., proposed the usage of GAN model in medical image analysis. For the implementation, generator and discriminator models are used to increase the data size by generating new images to overcome the issue of overfitting. They have compared the result with different types such as Wasserstein generative adversarial network (WGAN), Deep Convolutional GAN (DCGAN) and Conditional Generative Adversarial Networks (CGAN) [10]. Bi et al., implemented the model to improve the prediction accuracy and to address the overfitting issue based on label smoothing regularization and Wasserstein generative adversarial network problem [11]. CNN is used as the prediction model and accuracy of the model is evaluated with different metrics. In this paper, active contour and Otsu technique have been used for automatic detection of follicles from ovary images [12]. Various standard filters are applied for denoising the image. Performance of the model is evaluated using various performance metrics.

From literature survey, it is addressed that overfitting is the major problem and finding an optimal technique to remove noise and speckle from the ultrasound image is still a research challenge. To address these issues, in this article, a novel CNN approach is proposed which is presented in Sect. 3.

3   Proposed System

In this article, early diagnosis of PCOS using ultrasonography ovary images with CNN is proposed. The proposed model for early detection of PCOS is shown in Fig. 3. To improve the performance by addressing the overfitting problem, GAN is used to produce new images. The Deep Learning models – ResNet 50, VGG 19 and Inception V3 are used to train the dataset. ResNet 50 is preferred over ResNet 152 due to computational efficiency.

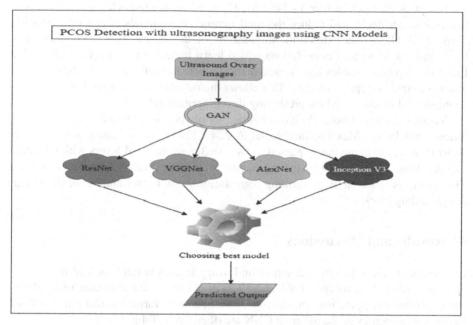

Fig. 3. Proposed PCOS Detection using CNN Model

3.1 Dataset Description

Ultrasound ovary images of PCOS infected and non-infected is taken as an input from Kaggle repository. Classification of PCOS and Non PCOS is done with CNN models ResNet, Inception V3 and VGGNet 19. The model is trained with various activation functions. Performance is evaluated using various performance metrics.

3.2 Convolutional Neural Network

Early diagnosis of PCOS using ultrasonography ovary images is implemented using various convolutional models. Features are extracted with different convolutional operations. Accuracy, Precision, Recall and F1 score are used as the performance metric to evaluate the accuracy. ResNet 50, Inception V3, VGGNet 19 was used for training the ultrasound ovary images.

Residual Networks (ResNets) utilize a skip connection which allows an easier flow of data from one layer to another. Such a connection allows each layer of the network to get the desired identity mapping. When a newly added layer is trained onto an identity function, it is ensured that the newer model as at least as effective as the previous one.

Often deep networks encounter the problem of vanishing gradient where the value of the gradient becomes close to zero, making back propagation difficult. Through the skip connections, ResNet tackles this problem and allows the usage of deeper networks without loss in performance.

Inception Networks utilize 1×1 convolutions, where the filters learn patterns across the various channels, and reduce the total number of channels after each operation. These 1×1 filters also reduces the dimensions of the inputs. This is complemented by adding 3×3 and 5×5 convolutions which learn pattern across all dimensions. The Inception Network enables layer concatenation where multiple layers can be combined to form a single inception module. This allows the network to go deeper and learn more sophisticated functions, whilst preserving the computational cost.

Visual Geometry Group (VGG) networks mainly consist of two blocks, 2D Convolutional blocks and Max Pooling blocks. A 224×224 image is passed as input to the network which is immediately passed to two 3×3 convolutional layers with 64 filters. The resulting $224 \times 224 \times 64$ image is passed with a stride 2 on to a max pooling layer. The pattern is followed by performing convolutions with higher number of filters than the preceding block.

4   Results and Discussions

Prompt diagnosis of PCOS with ultrasound ovary images is implemented using CNN models – ResNet50, Inception V3 and VGG19. Features are extracted using different convolutional operation and dataset is trained with various activation functions. Performance metrics of the different CNN are depicted in Table 1.

Table 1. Performance metrics of Convolutional Neural Networks.

CNN Model	Accuracy	Precision	Recall	F1 score
ResNet50	93%	94%	94%	0.94
Inception V3	94%	98%	92%	0.95
VGG19	96%	97%	95%	0.96

CNN with ResNet produces 93% accuracy, CNN with Inception V3 produces 94% accuracy and CNN with VGG19 produces 96% accuracy. It clearly shows that VGG19 performs much better compared to the other models for the given Ultrasonography ovary images.

Accuracy comparison of proposed model vs existing models discussed in Sect. 2 is shown in Table 2. Salman et al., proposed CNN model produced 96.56% of accuracy but they have implemented only with Inception V3 [4]. Hossain et al., produced 96% of accuracy with VGGNet 19 and compared their results with various CNN models [7]. The proposed CNN model produces the highest accuracy of 96% with VGGNet 19.

Figures 4, 5 and 6 illustrate the graphical representation of accuracy and loss. The dataset was trained with ResNet 50, VGG 19 and Inception V3 models. In History of

Table 2. Accuracy comparison – Proposed system vs Existing System.

Model	Hosain et al., [4]	Hossain et al., [7]	Proposed Model
ResNet 50	–	94%	93%
AlexNet	–	88%	–
Inception V3	96.56%	74%	94%
VGGNet 16	–	93%	–
VGGNet 19	–	96%	96%
Xception	–	87%	–

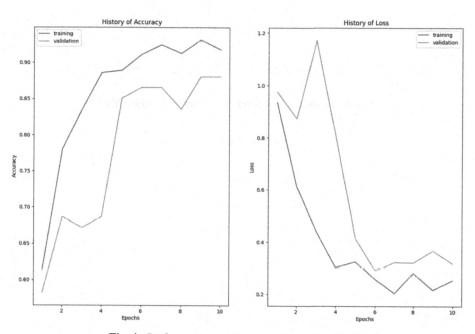

Fig. 4. Performance metrics evaluation of ResNet50.

Accuracy, X-axis indicates the Epochs and Y- axis indicates the accuracy and History of Loss considers Epochs on X-axis and Loss on Y-axis. Figure 4 shows the training and validation accuracy of ResNet 50. It produces 96% of accuracy during training and 93% of accuracy during validation. Figure 5 shows the training and validation accuracy of Inception V3. It produces 98% of accuracy during training and 94% of accuracy during validation. Figure 6 shows the training and validation accuracy of VGGNet 19. It produces 97% of accuracy during training and 96% of accuracy during validation. From the above mentioned figures, it is clear that VGGNet 19 produces highest accuracy for the given dataset.

MODEL'S METRICS VISUALIZATION

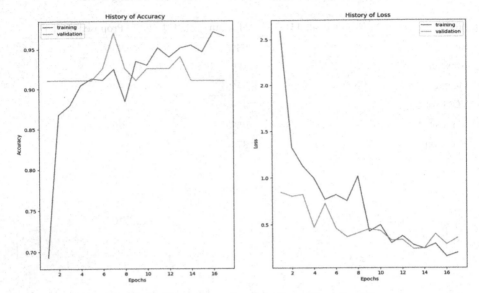

Fig. 5. Performance metrics evaluation of Inception V3.

MODEL'S METRICS VISUALIZATION

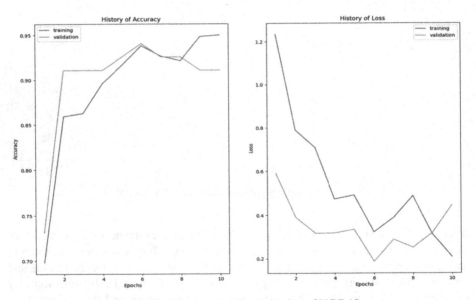

Fig. 6. Performance metrics evaluation of VGG 19.

5 Conclusion

Polycystic Ovary Syndrome has been examined as a major issue with young women which significantly affects them during their reproductive age. To allow these women to lead healthy lives, the expeditious prognosis of PCOS using Ultrasound ovary images of women is proposed. Generative Adversarial Network is implemented to address the overfitting issue. Feature extraction is done with convolutional operations and trained with convolutional neural network models – ResNet50, VGG19 and Inception V3. VGG19 produces the highest accuracy of 96% compared to other models. So expeditious prognosis of PCOS helps women to live their healthy life and also makes treatment process easier. In future, early diagnosis of PCOS can be improved with more diagnostic parameters like three-dimensional ultrasound (3D-US) and choosing optimal feature extraction methods.

Financial Support. This project was financed by All India Council for Technical Education - Research Promotion Scheme File. No: 8- 98/FDC/RPS/POLICY-1/2021–22 (AQIS ID: 1–9331521061).

Declaration of Competing Interest. The authors have no conflicts of interest to declare.

References

1. Aggarwal, S., Pandey, K.: Early identification of PCOS with commonly known diseases: obesity, diabetes, high blood pressure and heart disease using machine learning techniques. Expert Syst. Appl., 0957–4174 (2023)
2. Alamoudi, A., et al.: A deep learning fusion approach to diagnosis the polycystic ovary syndrome (PCOS). Appl. Comput. Intell. Soft Comput. **2023**, 1–15 (2023). https://doi.org/10.1155/2023/9686697
3. Hosain, S., Mehedi, H.K., Kabir, I.E.: PCONet: a convolutional neural network architecture to detect polycystic ovary syndrome (PCOS) from ovarian ultrasound images. In: 8th International Conference on Engineering and Emerging Technologies (ICEET), Kuala Lumpur, Malaysia (2022)
4. Zulkarnain's, N., Nazarudin, A.A., Nur Al Has's, A.H.A.: Ultrasound image segmentation for detecting follicle in ovaries using morphological operation and extraction methods. J. Pharm. Negative Results 13(4), 659–665 (2022). https://doi.org/10.47750/pnr.2022.13.04.088
5. Gopalakrishnan, C., Iyapparaja, M.: ITL-CNN: integrated transfer learning based convolution neural network for ultrasound PCOS image classification. Int. J. Pattern Recognit Artif Intell. (2022). https://doi.org/10.1142/S021800142240002X
6. Hossain, M.M., et al.: Particle swarm optimized fuzzy CNN with quantitative feature fusion for ultrasound image quality identification. IEEE J. Transl. Eng. Health Med. **10**, 1–12 (2022). https://doi.org/10.1109/JTEHM.2022.3197923
7. Gong, M., Chen, S., Chen, Q., Zeng, Y., Zhang, Y.: Generative adversarial networks in medical image processing. Curr. Pharm. Des. **27**, 1856–1868 (2021)
8. Mandal, A., Sarkar, M., Saha, D.: Follicle segmentation from ovarian USG image using horizontal window filtering and filled convex hull technique. In: Bhattacharjee, D., Kole, D.K., Dey, N., Basu, S., Plewczynski, D. (eds.) Proceedings of International Conference on Frontiers in Computing and Systems. AISC, vol. 1255, pp. 555–563. Springer, Singapore (2021). https://doi.org/10.1007/978-981-15-7834-2_52

9. Mandal, A., Saha, D., Sarkar, M.: Follicle segmentation using K-means clustering from ultrasound image of ovary. In: Bhattacharjee, D., Kole, D.K., Dey, N., Basu, S., Plewczynski, D. (eds.) Proceedings of International Conference on Frontiers in Computing and Systems. AISC, vol. 1255, pp. 545–553. Springer, Singapore (2021). https://doi.org/10.1007/978-981-15-7834-2_51

10. Lan, L., et al.: Generative adversarial networks and its applications in biomedical informatics. Front. Public Health **8** (2020)

11. Bi, L., Hu, G.: Improving image-based plant disease classification with generative adversarial network under limited training set. Front. Plant Sci. **11** (2020)

12. Gopalakrishnan, C., Iyapparaja, M.: Active contour with modified Otsu method for automatic detection of polycystic ovary syndrome from ultrasound image of ovary. Multimedia Tools Appl. **79**(23–24), 17169–17192 (2019). https://doi.org/10.1007/s11042-019-07762-3

Author Index